Unless Recalled Earlier

DEMCO, INC. 38-2931

Bioethics for Scientists

Bioethics for Scientists

Edited by
Professor JOHN BRYANT
School of Biological Sciences, University of Exeter, Exeter, UK

Dr LINDA BAGGOTT LA VELLE
Graduate School of Education, University of Bristol, Bristol, UK

Revd Dr JOHN SEARLE
Exeter & District Hospice, Exeter, UK

JOHN WILEY & SONS, LTD

British Library Cataloguing in Publication Data

A catalogue record for this book is available from the British Library

ISBN 0 471 49532 8

Typeset in 10/12 pt Times by Vision Typesetting, Manchester
Printed and bound in Great Britain by Biddles Ltd., Guildford and King's Lynn
This book is printed on acid-free paper responsibly manufactured from sustainable forestry in which at least two trees are planted for each one used for paper production.

Contents

List of Contributors

Mr Alex Aylward, 64 Waterside, Haven Road, Exeter EX2 8DP, UK
aylward@btinternet.com   Phone 01392 496392

Dr Linda Baggott la Velle, Graduate School of Education, University of
Bristol, 35 Berkeley Square, Clifton, Bristol BS8 1JA, UK
Linda.Lavelle@bristol.ac.uk   Phone 0117 928 7012   Fax 0117 928 7110

Professor Barry Barnes, Department of Sociology, University of Exeter, Exeter
EX4 4QJ, UK
S.B.Barnes@exeter.ac.uk   Phone 01392 263279   Fax 01392 263285

Dr Geeta Bharathan, Department of Ecology and Evolution, State University
of New York, Stony Brook, NY 11794-5245, USA
geeta@life.bio.sunysb.edu   Phone 00-1-631-632-9508
Fax 00-1-631-632-7626

Professor John Bryant, School of Biological Sciences, University of Exeter,
Exeter EX4 4QG, UK
J.A.Bryant@exeter.ac.uk   Phone 01392 264608   Fax 01392 264668

Dr Shanti Chandrashekaran, Division of Genetics, Indian Agricultural
Research Institute, New Delhi 110012, India
shantic@vsnl.net

Revd Dr Audrey R. Chapman, American Association for the Advancement of
Science, 1200 New York Avenue NW, Washington, DC 20005, USA
achapman@aaas.org   Phone 00-1-202-326-6795
Fax 00-1-202-289-4950

Professor R.G. Frey, Department of Philosophy, Bowling Green State
University, Bowling Green, OH 43403, USA
rfrey@bgnet.bgsu.edu   Phone 00-1-419-372-2117
Fax 00-1-419-372-8191

Dr Harry Griffin, Roslin Institute, Roslin BioCentre, Midlothian, EH25 9PS,
UK
harry.griffin@bbsrc.ac.uk   Phone 0131 527 4478   Fax 0131 527 4309

Professor Stephen Hughes, School of Biological Sciences, University of Exeter,
Exeter EX4 4QG, UK
S.G.Hughes@exeter.ac.uk   Phone 01392 263776
Fax 01392 264668

Professor Bartha Maria Knoppers, Centre de Recherche en Droit Public,
Université de Montréal, Montréal, Québec, H3C 3J7, Canada
Phone 00-1-514-343-6714   Fax 00-1-514-343-2122

Mr Tony May, 100 Rhodrons Avenue, Chessington, Surrey KT9 1AZ, UK
tonymay@suto.demon.co.uk   Phone 0181 544 3430

Dr Sue Mayer, GeneWatch UK, The Mill House, Manchester Road,
Tideswell, Derbyshire SK17 8LN, UK
mail@genewatch.org   Phone 01298 871898   Fax 01298 872531

Revd Dr David de Pomerai, School of Life Sciences, University of Nottingham,
University Park, Nottingham NG7 7RD, UK
David.Depomerai@nottingham.ac.uk

Revd Professor Michael J. Reiss, University of London Institute of Education,
20 Bedford Way, London WC1H 0AL, UK
m.reiss@ioe.ac.uk

Revd Dr John Searle, Exeter and District Hospice, Dryden Road, Exeter
EX2 5JJ, UK
johnlizex@aol.com   Phone 01392 432153

Dr Christopher Southgate, Department of Theology, University of Exeter,
Exeter EX4 4QJ, UK
C.C.B.Southgate@exeter.ac.uk

Dr Peter Turnpenny, Department of Clinical Genetics, Royal Devon and
Exeter Hospital, Exeter, EX2 5DW, UK
turnpenn@eurobell.co.uk

Preface

We can trace the origins of this book back to two sources. The first of these is the place of science and technology within wider society. Science is not value free and we take issue with those who claim that it is so. Almost every new development in biomedical science has social and/or ethical implications. Furthermore, professionals in all fields, including science, are being reminded increasingly frequently of their responsibilities, not just within their own profession but to the wider community. Those responsibilities certainly include professional codes of practice but should also embody an appropriate concern for the way that 'society' makes use of, for example, scientific discoveries and inventions.

Scientists need to be able to enter the ethical debate: too much of the debate, especially in the media, is conducted with little scientific understanding. Scientists who recognise and understand the ethical dimension can make a major contribution. Increasing numbers of science (and especially, but not exclusively, biomedical science) students are recognising this and are thus eager to participate in courses that enable them to apply ethical principles to problems and situations arising within their academic disciplines.

This introduces the second source of the book. It has grown out of our work over the last 30 years: teaching undergraduates and postgraduates, involvement in adult education at different levels and engaging with high school students and other young people. Further, our particular areas of activity – molecular biology (JAB), human embryology and fertilisation (LBL), intensive care and palliative medicine (JFS) – have led to our grappling with the implications of the rapid developments in science and medicine for individuals and for society. Complex questions have arisen about how these developments should be used; about whether or not particular courses of action are right or wrong; what is the balance between benefit and harm of developments and treatments? Our thinking has often been clarified as we have debated these issues with colleagues in Exeter from a wide range of disciplines, some of whom have contributed to this book.

Not only have those colleagues contributed to this book but they also contribute to courses at Exeter for BSc, BSc(Ed) and MSc students. Indeed, it was the establishment of our teaching programme that actually led to the confluence of the two streams in the book's history. Courses were started partly because of a growing interest amongst students and partly because of our professional concerns and interests. Our teaching activities led us to realise that there was no single text that would help, for example, an undergraduate biologist or a student teacher who wished to become informed about a wide range of issues in what we call Bioethics. In the USA, the term *Bioethics* has, until very recently, meant *Medical Ethics*. In the UK the word has always had a wider meaning but the subject has been the preserve of ethicists and philosophers. Worthwhile and relevant as those disciplines are, their pre-eminence in the bioethical literature has not been helpful to science undergraduates (nor indeed to some professional scientists!). And thus a book was born.

In assembling this book, as scientists we have endeavoured to engage with the modes of thinking employed by ethicists and philosophers because they play such an important part in analysing the issues and formulating ethical responses. However, we believe that ethical decision making in all the areas discussed in this book must be informed by a clear understanding of the science involved. Hence considerable space is given to presentation of the background science. Further, we have tried to discuss the thinking of ethicists and philosophers in a way that is helpful for and understandable by scientists who may not be (indeed, probably are not) familiar with the patterns of thinking in the humanities and social sciences. And then comes the hard part, the application of ethics usually evolved or derived in the context of human society, to issues in environmental science, biology, biotechnology and medicine. In order to do this we have assembled a team of authors who have responded eagerly to the challenge. We hope that the resulting text fulfils our wishes to be helpful, informative and thought provoking. Mary Warnock[1], in the context of a discussion of public and private morality in bioethical issues, said '. . . we must try to get it right, otherwise we shall be dominated for ever by the ethics of the pub bore, who will say, "I think it's disgusting. There ought to be a law against it". . .'. For us, getting it right means providing the tools to inform our readers *how* to think about the issues without trying to tell them *what* to think. We hope, therefore that this book is a contribution to the ongoing debate in bioethics.

In producing this book we have been helped and encouraged by many other people. Our co-writers, the authors who, with us, have provided the text, have shared in our vision and we thank them for their excellent contributions. We are especially grateful to Steve Hughes for his ready willingness to help and advise. Early on in our thinking about this venture we had much help and support from Suzi Leather (now Deputy Chair of the UK's Food Standards Agency), Iain Markham (now Professor of Ethics at Hope University College, Liverpool) and Christopher Southgate, who also contributed extensively to the final product. We want to thank too, those who, although not contributing to the book itself, nevertheless helped us in discussion of particular topics, especially Donald Bruce (of the Church of Scotland *Science, Religion and Technology* project), Avice Hall (University of Hertfordshire) and Gordon Dunstan (formerly of Kings College, University of London). We are also grateful to several cohorts of undergraduate and postgraduate students who have participated enthusiastically in our

---

[1] Warnock, M. (1998) *An Intelligent Person's Guide to Ethics* Duckworth, London, UK.

courses, made helpful suggestions for improvement and constantly requested that we recommend a useful text that brings all the issues together between its two covers. Doubtless, succeeding cohorts of students will tell us whether this is the book that meets that demand! And of course, we need to say that if the book does have its shortcomings, the responsibility for those is ours and ours alone. Finally, we must thank our colleagues at John Wiley & Sons, especially Sally Betteridge, who shared our enthusiasm for this book, strongly supporting the proposal through the company's decision-making process, and Suzanne Kriston, who looked after us from the signing of the contract through to production, showing admirable qualities of tact, patience and persistence.

John Bryant
Linda Baggott la Velle
John Searle
October 2001

I   Setting the Scene

1 Introduction to Ethics and Bioethics

Michael J. Reiss

1.1 THE SCOPE OF ETHICS

Ethics is the branch of philosophy concerned with how we should decide what is morally wrong and what is morally right. Sometimes the words 'ethics' and 'morals' are used interchangeably. They can, perhaps, be usefully distinguished (Reiss and Straughan, 1996), though many languages do not allow for distinctions to be made. We all have to make moral decisions daily on matters great or (more often) small about what is the right thing to do: Should I continue to talk to someone for their benefit or make my excuse and leave to do something else? Should I give money to a particular charity appeal? Should I stick absolutely to the speed limit or drive 10% above it if I am sure that it is safe to do so? We may give much thought, little thought or practically no thought at all to such questions. Ethics, though, is a specific discipline which tries to probe the reasoning behind our moral life, particularly by critically examining and analysing the thinking which is or could be used to justify our moral choices and actions in particular situations.

1.2 THE WAY ETHICS IS DONE

Ethics is a branch of knowledge just as other intellectual disciplines, such as science, mathematics and history, are. Ethical thinking is not wholly distinct from thinking in other disciplines but it cannot simply be reduced to them. In particular, ethical conclusions cannot be unambiguously proved in the way that mathematical theorems can. However, this does not mean that all ethical conclusions are equally valid. After all most philosophers of science would hold that scientific conclusions cannot be unambiguously proved, indeed that they all remain as provisional truths, but this does not mean that my thoughts about the nature of gravity are as valid as Einstein's were. Some conclusions – whether in ethics, science or any other discipline – are more likely to be valid than others.

Bioethics for Scientists. Edited by John Bryant, Linda Baggott la Velle and John Searle.
© 2002 by John Wiley & Sons Ltd.

One can be most confident about the validity and worth of an ethical conclusion if three criteria are met (Reiss, 1999): first, if the arguments that lead to the particular conclusion are convincingly supported by reason; second, if the arguments are conducted within a well established ethical framework; third, if a reasonable degree of consensus exists about the validity of the conclusions, arising from a process of genuine debate.

It might be supposed that reason alone is sufficient for one to be confident about an ethical conclusion. However, there are problems in relying on reason alone when thinking ethically. In particular, there still does not exist a single universally accepted framework within which ethical questions can be decided by reason (O'Neill, 1996). Indeed, it is unlikely that such a single universally accepted framework will exist in the foreseeable future, if ever. This is not to say that reason is unnecessary but to acknowledge that reason alone is insufficient. For instance, reason cannot decide between an ethical system which looks only at the consequences of actions and one which considers whether certain actions are right or wrong in themselves (i.e. *intrinsically* right or wrong), whatever their consequences. Then feminists and others have cautioned against too great an emphasis upon reason. Much of ethics still boils down to views about right and wrong informed more by what seems 'reasonable' than what follows from reasoning.

The insufficiency of reason is a strong argument for conducting debates within well established ethical frameworks, when this is possible. Traditionally, the ethical frameworks most widely accepted in most cultures arose within systems of religious belief. Consider, for example, the questions 'Is it wrong to lie? If so, why?'. There was a time when the great majority of people in many countries would have accepted the answer 'Yes. Because scripture forbids it'. Nowadays, though, not everyone accepts scripture(s) as a source of authority. Another problem, of particular relevance when considering the ethics of biotechnology, is that while the various scriptures of the world's religions have a great deal to say about such issues as theft, killing people and sexual behaviour, they say rather less that can directly be applied to the debates that surround many of today's ethical issues, particularly those involving modern biotechnology. A further issue is that we live in an increasingly plural society. Within Europe there is no longer a single shared set of moral values. Even the various religions disagree on ethical matters and many people no longer accept any religious teaching.

Nevertheless, there is still great value in taking seriously the various traditions – religious and otherwise – that have given rise to ethical conclusions. People do not live their lives in isolation: they grow up within particular moral traditions. Even if we end up departing somewhat from the values we received from our families and those around us as we grew up, none of us derives our moral beliefs from first principles, *ex nihilo*, as it were. In the particular case of moral questions concerning biotechnology, a tradition of ethical reasoning is already beginning to accumulate. For example, most member states of the European Union and many other industrialised countries have official committees or other bodies looking into the ethical issues that surround at least some aspects of biotechnology. The tradition of ethical reasoning in this field is nothing like as long established as, for example, the traditions surrounding such questions as abortion, euthanasia, war and trade protectionism. Nevertheless, there is the beginning of such a tradition and similar questions are being debated in many countries across the globe.

Given, then, the difficulties in relying solely on either reason or any one particular ethical tradition, we are forced to consider the approach of consensus (Moreno, 1995). It is true that consensus does not solve everything. After all, what does one do when consensus cannot be arrived at? Nor can one be certain that consensus always arrives at the right answer: for example, a consensus once existed that women should not have the vote.

Nonetheless, there are good reasons both in principle and in practice in searching for consensus. Such a consensus should be based on reason and genuine debate and take into account long established practices of ethical reasoning. At the same time, it should be open to criticism, refutation and the possibility of change. Finally, consensus should not be equated with majority voting. Consideration needs to be given to the interests of minorities, particularly if they are especially affected by the outcomes, and to those – such as young children, the mentally infirm and non-humans – who are unable to participate in the decision-making process. At the same time it needs to be borne in mind that while a consensus may eventually emerge there is an interim period when what is more important is simply to engage in valid debate in which the participants respect one another and seek for truth through dialogue (cf. Habermas, 1983; Martin, 1999).

1.3 IS IT ENOUGH TO LOOK AT CONSEQUENCES?

The simplest approach to deciding whether an action would be right or wrong is to look at what its consequences would be. No-one supposes that we can ignore the consequences of an action before deciding whether or not it is right. This is obvious when we try to consider, for example, whether imprisonment is the appropriate punishment for certain offences – e.g. robbery. We would need to look at the consequences of imprisonment, as opposed to alternative courses of action such as imposing a fine or requiring community service. Even when complete agreement exists about a moral question, consequences may still have been taken into account.

The deeper question is not whether we need to take consequences into account when making ethical decisions but whether that is all that we need to do. Are there certain actions that are morally required – such as telling the truth – whatever their consequences? Are there other actions – such as betraying confidences – that are wrong whatever their consequences? This is about the most basic question that can be asked in ethics and it might be expected by anyone who is not an ethicist that agreement as to the answer would have arisen. Unfortunately this is not the case. There still exists genuine academic disagreement amongst moral philosophers as to whether or not one needs only to know about the consequences of an action to decide whether it is morally right or wrong.

Those who believe that consequences alone are sufficient to let one decide the rightness or otherwise of a course of action are called *consequentialists*. The most widespread form of consequentialism is known as utilitarianism. Utilitarianism begins with the assumption that most actions lead to pleasure (typically understood, at least for humans, as happiness) and/or displeasure. In a situation in which there are alternative courses of action, the desirable (i.e. right) action is the one which leads to the greatest net increase in pleasure (i.e. excess of pleasure over displeasure, where displeasure means the opposite of pleasure, that is, harm).

Utilitarianism as a significant movement arose in Britain at the end of the 18th century with the work of Jeremy Bentham and J.S. Mill. However, its roots are much earlier. In the fifth century BCE[1], Mo Tzu in China argued that all actions should be evaluated by their fruitfulness and that love should be all embracing. In Greece, Epicurus (341–271 BCE) combined a consequentialist account of right action with a hedonistic (pleasure-seeking) theory of value (Scarre, 1998).

Utilitarianism now exists in various forms. For example, some utilitarians – preference utilitarians – argue for a subjective understanding of pleasure in terms of an individual's own perception of his or her well-being. What all utilitarians hold in common is the rejection of the view that certain things are intrinsically right or wrong, irrespective of their consequences.

Consider the question as to whether or not we should tell the truth. A utilitarian would hesitate to provide an unqualified 'yes' as a universal answer. Utilitarians have no moral absolutes beyond the maximisation of pleasure principle. Instead, it might be necessary for a utilitarian to look in some detail at particular cases and see in each of them whether telling the truth would indeed lead to the greatest net increase in pleasure.

There are two great strengths of utilitarianism. First, it provides a single ethical framework in which, in principle, any moral question may be answered. It does not matter whether we are talking about the legalisation of cannabis, the age of consent or the patenting of DNA; a utilitarian perspective exists. Secondly, utilitarianism takes pleasure and happiness seriously. The general public may sometimes suspect that ethics is all about telling people what not to do. Utilitarians proclaim the positive message that people should simply do what maximises the total amount of pleasure in the world.

However, there are difficulties with utilitarianism as the sole arbiter in ethical decision making. For one thing, an extreme form of utilitarianism in which every possible course of action would have consciously to be analysed in terms of its

---

[1] CE means common era and thus BCE is a culturally wider term than BC although dates are numbered in the same way, i.e. fifth century BCE = fifth century BC.

countless consequences would quickly bring practically all human activity to a stop. Then there is the question as to how pleasure can be measured. For a start, is pleasure to be equated with well-being, happiness or the fulfilment of choice? And, anyway, what are its units? How can we compare different types of pleasure, for example sexual and aesthetic? Then, is it always the case that two units of pleasure should outweigh one unit of displeasure? Suppose two people each need a single kidney. Should one person (with two kidneys) be killed so that two may live (each with one kidney)?

Utilitarians claim to provide answers to all such objections (e.g. Singer, 1993). For example, rule-based utilitarianism accepts that the best course of action is often served by following certain rules – such as 'Tell the truth', for example. Then, a deeper analysis of the kidney example suggests that if society really did allow one person to be killed so that two others could live, many of us might spend so much of our time going around fearful that the sum total of human happiness would be less than if we outlawed such practices.

The major alternative to utilitarianism is a form of ethical thinking in which certain actions are considered right and others wrong in themselves, i.e. intrinsically, regardless of the consequences. Consider, for example, the question as to whether a society should introduce capital punishment. A utilitarian would decide whether or not capital punishment was morally right by attempting to quantify the effects it would have on the society. Large amounts of empirical data would probably need to be collected, comparing societies with capital punishment and those without it with regard to such things as crime rates, the level of fear experienced by people worried about crime and the use to which money saved by the introduction of capital punishment might be put. On the other hand, someone could argue that regardless of the consequences of introducing capital punishment, it is simply wrong to take a person's life, whatever the circumstances. Equally, someone could argue that certain crimes, for example first degree murder, should result in the death penalty – that this simply is the right way to punish such a crime.

There are a number of possible intrinsic ethical principles and because these are normally concerned with rights and obligations of various kinds, this approach to ethics is often labelled *deontological* (i.e. 'rights discourse'). Perhaps the most important such principles are thought to be those of autonomy and justice. People act autonomously if they are able to make their own informed decisions and then put them into practice. At a common sense level, the principle of autonomy is why people need to have access to relevant information, for example before consenting to a medical procedure. Autonomy is concerned with an individual's rights. Justice is construed more broadly: essentially, justice is about fair treatment and the fair distribution of resources or opportunities. Considerable disagreement exists about what precisely counts as fair treatment and a fair distribution of resources. For example, some people accept that an unequal distribution of certain resources (e.g. educational opportunities) may be fair provided certain other criteria are satisfied (e.g. the educational opportunities are purchased with money earned or inherited). At the other extreme, it can be argued that we should all be completely altruistic. However, as Nietzsche pointed out, it is surely impossible to argue that people should (let alone believe that they will) treat absolute strangers as they treat their children or spouses. Perhaps it is rational for us all to be egoists, at least to some extent.

If it is the case that arguments about ethics should be conducted solely within a consequentialist framework, then the issues are considerably simplified. Deciding whether anything is right or wrong now reduces to a series of detailed, in depth studies of particular cases. As far as modern medicine and biotechnology are concerned, ethicists still have a role to play but of perhaps greater importance are scientists and others who know about risks and safety, while sociologists, psychologists, policy makers and politicians who know about people's reactions and public opinions also have a significant role.

Much energy can be wasted when utilitarians and deontologists argue. There is little if any common ground on which the argument can take place, though some philosophers argue that there can be no theory of rights and obligations without responsibility for consequences, and no evaluation of consequences without reference to rights and obligations. The safest conclusion is that it is best to look both at the consequences of any proposed course of action and at any relevant intrinsic considerations before reaching an ethical conclusion.

1.4   WIDENING THE MORAL COMMUNITY

Traditionally, ethics has concentrated mainly upon actions that take place between people at one point in time. In recent decades, however, moral philosophy has widened its scope in two important ways. First, intergenerational issues are recognised as being of importance (see e.g. Cooper and Palmer, 1995). Second, interspecific issues are now increasingly taken into account (see e.g. Rachels, 1991). These issues go to the heart of 'Who is my neighbour?'. The term *bioethics* is often used when such questions are being considered, though in the USA and some other countries 'bioethics' often simply means 'medical ethics'.

Interspecific issues are of obvious importance when considering biotechnology and ecological questions. Put at its starkest, is it sufficient only to consider humans or do other species need also to be taken into account? Consider, for example, the use of new practices (such as the use of growth promoters or embryo transfer) to increase the productivity of farm animals. An increasing number of people feel that the effects of such new practices on the farm animals need to be considered as at least part of the ethical equation before reaching a conclusion. This is not, of course, necessarily to accept that the interests of non-humans are equal to those of humans. While some people do argue that this is the case, others accept that while non-humans have interests these are generally less morally significant than those of humans.

Accepting that interspecific issues need to be considered leads one to ask

'How?'. Need we only consider animal suffering? For example, would it be right to produce, whether by conventional breeding or modern biotechnology, a pig unable to detect pain and unresponsive to other pigs? Such a pig would not be able to suffer and its use might well lead to significant productivity gains: it might, for example, be possible to keep it at very high stocking densities. Someone arguing that such a course of action would be wrong would not be able to argue thus on the grounds of animal suffering. Other criteria would have to be invoked. It might be argued that such a course of action would be disrespectful to pigs or that it would involve treating them only as means to human ends and not, even to a limited extent, as ends in themselves. This example again illustrates the distinction between utilitarian and deontological forms of ethical reasoning, as the issue of pain can be separated from that of rights and obligations in this case.

Intergenerational as well as interspecific considerations may need to be taken into account. Nowadays we are more aware of the possibility that our actions may affect not only those a long way away from us in space (e.g. acid rain produced in one country falling in another) but also those a long way away from us in time (e.g. increasing atmospheric carbon dioxide levels may alter the climate for generations to come). Human nature being what it is, it is all too easy to forget the interests of those a long way away from ourselves. Accordingly, a conscious effort needs to be made so that we think about the consequences of our actions not only for those alive today and living near us, about whom it is easiest to be most concerned.

1.5 THE LEVEL AT WHICH ETHICAL DECISIONS ARE MADE

1.5.1 INTRODUCTION

Ethical decisions are taken at a number of levels. Failure to distinguish between these can lead to sterile debates in which one side argues that X is good while the other side argues that X is bad. Consider, for example, the case of genetically modified food crops (Reiss, 2001) (see also Chapters 8 and 9, this volume). Controversies over such crops show no sign of abating; indeed, recent years have seen both an intensification of the debate about them and, if anything, a polarisation of attitudes towards them. At the one pole are those who see genetically modified crops as being dangerous or intrinsically unacceptable. Genetically modified pollen and seeds are said to be 'contaminating' conventional crops and wild plants and to pose a threat to human health (see e.g. Friends of the Earth, 1998). At the other pole are those who see these crops as satisfying consumer choice in the West and as providing a lifeline to the prevention of famine and the alleviation of suffering in many rural economies (see e.g. Dale, 1999).

The ethical issues surrounding genetically modified crops are revealed by looking at the various stages or levels in their development, production and consumption. To a certain extent we can envisage the production and use of genetically modified crops as requiring the following steps:

- research by scientists
- commercial development
- regulatory approval
- planting by farmers
- the stocking of genetically modified food by retailers
- the purchase and consumption of genetically modified food by consumers.

While this list simplifies the situation in various ways (for example, scientists are, of course, involved in both 'pure' research and commercial development, while individuals at each of these levels also act as consumers) categorising genetically modified food production into these six levels is convenient for analysis.

1.5.2   RESEARCH BY SCIENTISTS

Various ethical reasons can be advanced for allowing a particular piece of scientific research to go ahead (Reiss, 2000). One argument is the standard one that one needs strong arguments before one bans things. One of the lessons of history is that in earlier times practices were banned which now most of us consider appropriate. For example, many countries now allow women and non-property owning men to vote at elections. In those cases where countries have decided to ban practices permitted in previous time, e.g. slavery and torture, this is usually because the practices are now widely considered intrinsically unacceptable for reasons to do with respect of persons.

It is noteworthy just how much we allow people to do even when we know it would, in at least some senses, be better for them if they did not. For example, it would be better for most people, from a narrow, physical health perspective, if they did not smoke cigarettes. We allow adults to smoke precisely because we think it is better to live in a society where, roughly speaking, the granting of autonomy is considered a higher good than the imposition of beneficence – i.e. making people do what is 'good' for them.

A particular form of the 'arguments against banning things' reason is that scientists should have autonomy with respect to their work. This is not to assume naïvely that scientists choose without any constraints their subjects for study. Indeed, it may be that one of the strongest arguments for encouraging scientists to believe that they are acting autonomously is simply the consequentialist one that many people, including, I suspect, academics such as research scientists and moral philosophers, produce their best work when they believe that they are doing what they want to do.

For most people, including members of the general public, by far the most significant reason for it being right for a scientist to carry out a piece of research is that there is a reasonable chance that the information gained or ideas generated will be of worth, helping to increase the sum total of human happiness or produce some other desirable benefits. This is most clearly the case when we are talking about research in the applied sciences, such as medicine, electronics and the development of genetically modified crops.

In addition to ethical reasons for allowing a particular piece of research to go ahead, there are arguments against doing a particular piece of research. For a start, we can imagine pieces of research where most people would consider it wrong for someone even to want to do the research. Then we can imagine cases where carrying out the research programme itself would be wrong because of the direct consequences for those involved in the research. An example is provided by Stanley Milgram's classic work on obedience in which experimental subjects thought they were administering large electric shocks to people when in fact they were not (Milgram, 1963). Despite the undoubted worth of the study, from which a considerable amount of valuable information about human psychology has been learned and which fulfils all of the criteria listed above in favour of scientists undertaking research, the codes of many psychological societies would nowadays preclude such research from being undertaken on the grounds that the deception it requires is excessive.

Another reason for holding that it would be wrong to carry out a particular research programme is if the programme would be expected to lead to undesired consequences. Some would hold that such an example is provided by research into genetically modified crops. However, it is difficult to defend the argument that the likely consequences of laboratory and greenhouse research into genetically modified crops are negative, in part precisely because such research ought to be able to identify certain harms (principally ecological and food safety ones) before they occur. Of course, once one considers field trials for research into genetically modified crops, it becomes easier for it to be argued that the net consequences of these will be harmful as field trials obviously entail a significant degree of non-containment of genetically modified organisms.

A final and very significant argument against a proposed piece of research is simply that, whatever its merits, given that research funds are limited, the money could be spent better elsewhere. This argument has an undoubted validity and academics in general, not just research scientists, are all too familiar with it (see also Chapter 15, in relation to the Human Genome Project). However, this argument is difficult to apply to genetically modified crops since the claims made on their behalf are considerable and difficult to reject out of hand. There would, therefore, seem to be a strong case for testing such claims. Further, if it might be the case that genetically modified crops will not alleviate world hunger but instead put small farmers out of work and lead to certain wild plant populations being damaged through genetic introgression or local extinctions, then there is a strong argument for carrying out rigorous research to establish whether these would indeed be the results of genetically modified crop technology, without having first or simultaneously allowed commercial development to proceed to the point where it is much more difficult to prevent the technology from proceeding any further.

1.5.3   COMMERCIAL DEVELOPMENT

There is an enormous literature on business ethics (e.g. Jackson, 1995). At one extreme it can be argued that 'the business of business is business'. In other words, companies only have a duty to those who own them – shareholders, banks, private individuals and so on. And this duty is a financial one, i.e. maximisation of profit – or some more refined version of this taking into account discounted cash flows and economic value added (Grant, 1998).

At the other extreme, for example from certain utilitarian perspectives, it can be argued that companies (more formally, those who control the actions of companies) have a duty to ensure that they operate so as to maximise global happiness or some other utility (e.g. preferences). Pharmaceutical companies are particularly prone to this 'analysis'. After all, it does seem wrong that people should be dying of AIDS simply because they and their governments cannot afford to buy drugs that would extend their lives considerably. It can be argued, though, that it is unfair of people to expect the shareholders or other owners of a particular company to forgo income for this reason. Rather, the onus is on the governments (and the electorates) in wealthy countries to provide such medica- tion for those who cannot afford it. The most straightforward way for this would be for the governments of wealthy countries to use their powers (financial and regulatory) to ensure that drugs are purchased from the pharmaceutical com- panies and sent to those parts of the world where they are most needed. Such purchases should be made at prices which still allow the companies to make a small profit. Taking into account economies of scale and the distinctions be- tween the variable costs involved in making more of the drugs and the fixed costs (principally the earlier massive investments in the development of the drugs), such prices should be markedly lower than the current market ones, reducing the cost to the donor governments very considerably.

A rather different approach is to adopt a customer or stakeholder focus. This somewhat pragmatic approach is based on the belief that a company only remains in existence – and whatever else a successful company is, it is certainly a company that survives – if it satisfies its customers to a reasonable degree and maintains sufficiently good relationships with its various stakeholders (cus- tomers, employees, regulators, suppliers, etc.) for it to continue to operate reasonably smoothly. Under this reasoning, companies need to consider not just the 'bottom line' of financial performance. Indeed, certain companies are now talking about the 'triple bottom line', taking into account their social/ethical responsibilities.

Companies genuinely seeking to fulfil what they feel are their social/ethical responsibilities – e.g. fair treatment of their employees, reduction of pollution – still strive to be financially successful but take on board a wider range of aims and objectives. In addition, their relationships with their various stakeholders are likely to be more open and more transparent and they are likely to place more weight on frequent dialogue with them. The nature of this dialogue is likely to be very different from a traditional company, which may simply rely on market research to identify customer needs, irregular formal negotiations with any unions and networking or the occasional piece of espionage to identify competitors' intentions and capabilities.

1.5.4 REGULATORY APPROVAL

Why do we need regulation rather than simply allowing those harmed by actions to take those responsible for the harm to court? One answer is that people often want regulation to prevent such harms from happening. (I may be able to sue you for harming me because you have failed to show a duty of care, for example by driving too fast, but I may prefer that the harm had never happened, for example by my government having passed laws about speed limits.) Another answer is that legal recourse is a very imperfect way of attempting to redress harms. In particular, those with little money, power or persistence stand only a small chance of successfully taking anyone, certainly a large company, to court.

These considerations suggest that the particular duties of a regulatory system are (i) to prevent certain harms; (ii) to provide especial protection for those unable to take legal actions against those responsible for harms. In the case of genetically modified crops, one would therefore expect regulators to pay particular attention firstly to those especially likely to be harmed and secondly to those who lack agency. Some actors, for example non-human organisms and young children, potentially fall into both categories. Other actors, for example adults with certain food allergies, fall into just one. At the same time, it needs to be remembered that there are arguments which suggest that certain genetically modified crops may lead to a lessening of such potential harms or at least to them becoming rarer. If these crops lead to fewer pesticides being used, there may be benefits both to wildlife and to human health. Regulators need to keep such possibilities in mind and be especially resistant to listening only to partial points of view – whether from companies or from pressure groups.

A further duty of a regulator may be to allow people to act autonomously. The persistent cry for legislation for labelling of genetically modified foods (see Chapter 9) validly has this behind it. However, a distinction can usefully be made as to whether a regulator has a duty to insist that labelling actually occurs or a duty to insist that labelling is permitted. Usually it is the former that is argued for, so in the EU there are now regulations about the labelling of genetically modified foods. I think it can be argued that it would have been better simply to have allowed labelling. (At one point companies in the USA were forbidden even from stating that their products were GM-free.) This would mean that food companies and retailers would have been free (i) to choose to label GM foods; (ii) to choose to label non-GM foods; (iii) not to bother. This certainly would have saved quite a bit of money and would have suited those consumers who are not too bothered (at least when it comes actually to making food purchase decisions rather than being interviewed for the Eurobarometer or other surveys) whether their food contains genetically modified ingredients or not, simply wanting it to be affordable, safe, tasty and to look good.

1.5.5   PLANTING BY FARMERS

Are there particular ethical considerations that apply to farmers? A farm can be seen simply as a small company, in which case there is little to add to what I wrote earlier about commercial duties. At the same time, farms are unusual partly in that they persist much longer than many companies and partly in that they are rooted in one place. Farmers have a particular and enduring relationship with land. This line of argument might suggest that farmers have a particular duty to preserve soil and (endangered) wildlife. Preserving soil, fortunately, is also, pretty much, a duty of self-interest. No farmer seeks soil erosion. Indeed, certain of the farming practices introduced in recent years, for example, direct drilling, reduce erosion while herbicide-resistant crops – whether genetically modified or not – hold out the prospect of avoiding pre-emergence spraying, reducing erosion further (and, of course, also reducing the level of chemical input into the soil).

Perhaps farmers also have a duty to reduce fluctuations in yields on the grounds that it is fluctuations – e.g. two very bad harvests after 20 years of good harvests – that lead to malnutrition, suffering and death. It might be that fluctuations would be reduced simply by maximising yields but it might be that they would better be reduced by farmers not, as it were, putting all their eggs into one basket. Such a principle might need to be backed up by some sort of state subsidy (because individual farmers would not be able to maximise their incomes) but one could imagine a situation in which farmers (collectively rather than necessarily individually) were encouraged to have a diversity of types of agricultural production (e.g. conventional, genetically modified and organic), a diversity of crop species and a diversity of varieties within each crop species (so as to reduce the chances of a pest or disease suddenly wiping out a high proportion of a nation's agricultural yield).

1.5.6   THE STOCKING OF GENETICALLY MODIFIED FOOD
         BY RETAILERS

Retailers are companies so the earlier consideration of commercial interests again applies. Is there anything distinctive about retailers and genetically modified food? One line of argument would be that perhaps retailers ought to provide choice to consumers. Certainly, major supermarket chains frequently claim that they provide choice to their customers, though this may be more rhetoric than the result of careful ethical reflection. Indeed, the late 1990s in the UK saw a move towards the near elimination of choice for the consumer in terms of genetically modified foods. Whereas there was a time when some small supermarket chains (notably Iceland) sought to provide only GM-free foods while the major chains labelled the few genetically modified foods they knowingly sold (notably tomatoes), now all supermarket chains are doing what they can to eliminate genetically modified foods from their shelves, thereby removing choice from those who would buy these products.

In fairness to retailers, they can legitimately argue that in doing this they are simply responding to the wishes of the majority of their customers. Do retailers have a duty to stock certain products for a minority of consumers even when it would be commercially better for them not to do so? It is difficult to argue that retailers do have such a duty – along much the same lines as I argued earlier that pharmaceutical companies do not have a duty of charity. But just as I argued that governments in wealthy countries may have a duty to help poorer countries, perhaps governments have a duty to protect minority food interests. Perhaps governments should subsidise food products desired only by minorities – e.g. organic foods, welfare-friendly meats and genetically modified foods – so that they can be purchased at non-premium prices. I think this argument only has any force where there are genuinely convincing ethical arguments in favour of subsidising minority products (cf. Rippe, 2000). I doubt that I have a right to expect governments to subsidise my wish to buy wild rice as cheaply as cultivated rice. On this reasoning the arguments for subsidising genetically modified foods are weaker than those for subsidising, for example, welfare-friendly meats.

1.5.7 THE PURCHASE AND CONSUMPTION OF GENETICALLY MODIFIED FOOD BY CONSUMERS

Finally we come to the duties of individual consumers. Suppose that I buy food not only for myself but also for my children or aged parents. If I believe, providing that my belief is not entirely irrational, that there is a chance, even if only a small one, that a food – whether genetically modified corn, shellfish or beef – is dangerous to their health, then I surely have a duty not to purchase that food unless the probability of the danger is exceptionally low and there are extremely pressing reasons why I should buy the foods (e.g. buying the food significantly aids the farmer/fisherperson of that food more than the alternative I would purchase). Given the suggestion by some scientists, albeit a very small minority of them, that genetically modified foods might be dangerous for your health, this line of argument would mean that it is the duty of parents and others with people under their care not to buy genetically modified food for their consumption. Despite my own belief that the few genetically modified foods permitted to be sold in the EU are at least as safe as conventional foods, I therefore support, for example, the decision by a number of Local Education

Authorities in England not to purchase genetically modified foods for school meals.

Suppose, though, that I buy food only for myself (or for myself and a mentally competent partner – and that the two of us genuinely discuss whether we ought to buy genetically modified foods). It seems obvious that I have a right to buy either genetically modified foods or GM-free foods but is one of these courses of action better than the other? Given the present empirical uncertainty as to the environmental, health and socio-economic consequences of genetically modified crops I do not think we can say so. However, perhaps I do have a duty to keep myself informed so that if the evidence begins to weigh one side of the scale downwards I can, if needs be, amend my food purchasing decisions. The same line of reasoning makes it incumbent upon me to consider (and act upon the considerations) whether, for example, I should buy welfare-friendly food. I note in passing that as I have a large income and no children there is a greater burden of duty on me to bear such considerations in mind and act upon them than if I had a small income with several dependants.

1.6 CONCLUSIONS

There is no single way in which ethical debates about biotechnology or almost any other matter can unambiguously be resolved. However, that does not mean that all ethical arguments are equally valid. Ethical conclusions need to be based on reason, take into account well established ethical principles and be based, so far as possible, on consensus. Education and debate play an important role, helping to enable people to clarify their own thinking, express their views and participate in the democratic process. As far as biotechnology and modern medicine are concerned, both intrinsic and consequentialist arguments for and against their deployment can be advanced. Deciding whether or not particular instances of modern biotechnology and biomedical science are acceptable means looking in detail at individual cases. To a large extent, this is what the rest of this book does.

REFERENCES

Cooper, D.E. and Palmer, J.A. (eds) (1995) *Just Environments: Intergenerational, International and Interspecies Issues*. Routledge, London, UK.

Dale, P. (1999) Public reactions and scientific responses to transgenic crops. *Current Opinion in Biotechnology*, **10**, 203–208.

Friends of the Earth (1998) *Genetically Modified Food: Briefing*. Friends of the Earth, London, UK.

Grant, R.M. (1998) *Contemporary Strategy Analysis: Concepts, Techniques, Applications* (3rd edn). Blackwell, Oxford, UK.

Habermas, J. (1983) *Moralbewusstsein und Kommunikatives Handeln*, Suhrkamp, Frankfurt am Main, Germany.

Jackson, J. (1995) Reconciling business imperatives and moral virtues. In *Introducing Applied Ethics*. Almond, B. (ed), Blackwell, Oxford, UK, pp. 104–117.

Martin, P.A. (1999) Bioethics and the whole: pluralism, consensus, and the transmutation of bioethical methods into gold. *Journal of Law, Medicine & Ethics*, **27**, 316–327.

Milgram, S. (1963) Behavioural study of obedience. *Journal of Abnormal Psychology*, **67**, 371–378.

Moreno, J.D. (1995) *Deciding Together: Bioethics and Moral Consensus*. Oxford University Press, Oxford, UK.

O'Neill, O. (1996) *Towards Justice and Virtue: a Constructive Account of Practical Reasoning*, Cambridge University Press, Cambridge, UK.

Rachels, J. (1991) *Created from Animals: the Moral Implications of Darwinism*. Oxford University Press, Oxford, UK.

Reiss, M. (1999) Bioethics. *Journal of Commercial Biotechnology*, **5**, 287–293.

Reiss, M.J. (2000) The ethics of genetic research on intelligence. *Bioethics*, **14**, 1–15.

Reiss, M.J. (2001) Ethical considerations at the various stages in the development, production and consumption of GM crops. *Journal of Agricultural and Environmental Ethics*, **14**, 179–190.

Reiss, M.J. and Straughan, R. (1996) *Improving Nature? The Science and Ethics of Genetic Engineering*. Cambridge University Press, Cambridge, UK.

Rippe, K.P. (2000) Novel foods and consumer rights: concerning food policy in a liberal state. *Journal of Agricultural and Environmental Ethics*, **12**, 71–80.

Scarre, G. (1998) Utilitarianism. In *Encyclopedia of Applied Ethics volume 4*. Chadwick, R. (ed), Academic, San Diego, USA, pp. 439–449.

Singer, P. (1993) *Practical Ethics* (2nd edn). Cambridge University Press, Cambridge, UK.

2 The Public Evaluation of Science and Technology

Barry Barnes

2.1 INTRODUCTION

An understanding of the formal theories of ethical and moral philosophy may be necessary for anyone who would grapple with substantive bioethical issues but it is not sufficient. Also essential is an awareness of context, and a sense of how ethical debates, involving both experts and ordinary people, are structured by context and by the way that the parties to the debates perceive each other therein. In brief, a broad empirical understanding of the form and setting of ethical debate is needed, and this chapter is intended to provide the first steps toward one[1].

2.2 WHAT IS BEING SPOKEN OF?

Debate in bioethics is part of the larger debate through which we seek to establish, to understand, and to set in better order, our relationship with science and technology, a relationship that is at once a source of hope and anxiety. What though are the science and technology that figure in this debate? The dictionary tells us that science is the knowledge of nature, and technology the technical skills, possessed by a specific culture. On these definitions, the science and technology of a society are a part of its cultural tradition. They are embedded in the culture that all share. And this indeed is how anthropologists encounter the technical lore and skill of simple unspecialised societies. Definitions of this kind, however, seem somehow inappropriate in the societies in which we ourselves live. We do not, in the normal way of things, reckon our cookery books and cake-baking skills to be parts of our science and technology. For us today, science and technology are *other*; we speak of the relationship between us and

---

[1] The discussion will be biased toward Britain, and may reflect some of the unfortunate idiosyncratic features of that setting over the last couple of decades. For a comprehensive survey of what the social sciences can provide by way of background empirical understanding here, see Jasanoff *et al.* (1995).

Bioethics for Scientists. Edited by John Bryant, Linda Baggott la Velle and John Searle.
© 2002 by John Wiley & Sons Ltd.

them, and ask how much in the way of benefit, or threat, they offer us.

For the most part, this sense of otherness is the product of division of labour. As this proceeds, great amounts of knowledge and arrays of technical skills are alienated from ordinary life, and come to reside in esoteric sub-cultures, wherein experts sustain them for us, and make use of them on our behalf. The dictionary commemorates the change by offering alternative definitions, wherein science becomes the systematically acquired knowledge of a field or discipline, and technology the practice of the applied sciences or the mechanical arts. Now science and technology may sensibly be regarded as other, and it may be meaningful to speak of their effects on our everyday culture, even if they are never totally separated from it. Above all, they now exist as a set of occupations, the specialised activities of which are recognised as beyond our ordinary under-standing yet of profound significance to us.

It might be objected that whilst science has indeed become something other and alien, this is not true of technology. Is not technology all around us today? Do we not swim in it? There is indeed a sense in which we do, but in another sense we are alienated from technology more profoundly even than we are from science[2]. If people are asked what technology is today, they tend to speak of it, not as skills, but as tools, machinery and gadgetry. Of course, skilled technical activity has always involved tools and artefacts, and any account of technology must do justice to their role. But in the context of everyday life, technology is now close to being perceived as nothing but artefacts; and the associated skills, and the people who carry them, are rarely mentioned. Technology has been reified in our imagination. Nonetheless, we do remain uneasily aware of a technology behind technology as it were, of a hidden realm of skills and processes, and hence also of people, at work somewhere or other. And if we address biotechnology with this in mind, then we may become aware of what older concepts of technology would have placed in the forefront of our attention from the start: a vast and rapidly growing body of esoteric activities devoted to the manipulation of materials. Biotechnology is a materials technology of quite formidable potency and daunting promise.

2.3   THE TRADITIONAL COLLEGIAL ARRANGEMENT

There are significant differences between our perceptions of science and of technology, and indeed between science and technology themselves, but they can remain in the background here. Most of what I want to say is relevant to both, and bioethical issues tend in any case to arise out of work in fields where the

---

[2] This is well understood by producers of 'consumer technology', who try hard to overcome the problem by flattering and reassuring its purchasers. Individuals are encouraged to regard themselves as empowered by technology. Our experience of this 'technology' is often of consoles with buttons to press and switches to turn and screens to survey – on sealed cases that aspire to invisibility.

boundary between the two is blurred. The discussion here will centre on the occupational activities of possessors of empirical knowledge and expertise, and their relationship with audiences and institutions in the wider society.

A traditional method of effecting a division of intellectual and technical labour in society has been to make specific domains of esoteric knowledge and skill the province of particular occupations organised along collegial lines as professions. How these professions have been organised, how they have oriented themselves to external audiences, and how they have been perceived and evaluated by those audiences are all closely intertwined questions, and so it is important to say a little about the nature of the collegial organisation involved. It is indeed one of the simplest of all forms of organisation. A clear distinction is made between those within and those without the professional group. Those within the group communicate freely with each other on professional matters; all have the same *pro forma* standing as competent peers and the same entitlement to exercise professional judgement. Those outside the group are denied standing on the relevant technical matters and any entitlement to evaluate them; indeed they are expected to recognise their own incompetence on such matters and acquiesce in their exclusion.

Collegially organised disciplines have traditionally accepted the tasks of preserving, transmitting, enlarging, and evaluating the knowledge recognised as in their proper domain. Furthermore, external powers have permitted them to monopolise these tasks and recognised their autonomy in the execution of them. The individual members of such disciplines are not mere employees, but participants in semi-independent communities wherein specific moral and ethical obligations are recognised. Sometimes these may be codified obligations relevant to the specific powers and practices of a particular field. Medical professionals, for example, possess extensive powers of intervention into the operation of human bodies, even against the resistance of the inhabitants of those bodies, and a strong professional ethic constrains their use. But, partly due to the increasing distrust of expert professionals to be discussed later (Sections 2.4–2.6), they are tending to lose powers of this kind, and to be defined simply as carriers and suppliers of advice and skill, who inform and implement the decisions of others (Barnes, 1999). And the discussion will concentrate accordingly on how a collegial framework of ethical and moral obligations characteristic of practically all fields and disciplines informs the performance of this basic role.

A crucial obligation here is that which enjoins individuals to make their findings and judgements freely available to the collective, and to cede any personal rights in them. The findings are then evaluated by the collective in the light of its shared inheritance of esoteric knowledge and understanding, and, if accepted as valid and significant, made available to all. Every member of the collective is thus able rapidly to acquire and utilise results which, rather than merely being idiosyncratic individual reports, have been subject to a collective quality control process and granted the imprimatur of the entire field. But there are important prohibitions to be observed as well as positive injunctions. In particular, individuals may not accept direct financial reward for their work, or for their opinions of the work of others. The recognition and regard of peers must suffice as immediate reward for these, or else professional judgement would be vulnerable to external influence and control, and liable to be less than properly disinterested. The crucial role of peer recognition as the proper currency of reward, and of the taboo on direct monetary incentives, is indeed precisely to maintain the autonomy of the specialised field and especially the independence of the technical judgement of its members. For further discussions of this the reader is referred to the work of Barnes (1985) and Ziman (1968) and particularly to the seminal analysis of academic science by Robert Merton (1973).

The existence of this internal ethical order has important implications as far as external audiences are concerned. A consensus of expertise is presented. The entire collective, and its inheritance of knowledge and competence, is seen to lie behind specific expert pronouncements. Every individual member is able to speak at least to some extent as the authoritative representative of the collective. And the unbiased and disinterested character of such pronouncements is proclaimed in the prohibition upon monetary reward. Thus, the discipline is encountered as a source of knowledge and advice deserving of credibility; and trust and deference on the part of its external audience is confirmed as the correct attitude. Expertise is properly dispensed by such a discipline, we might say, in 'hierarchy' mode, from an authoritative transmitter to a subordinate passive receiver. And a final ethical obligation upon the members of such a discipline is precisely to dispense knowledge honestly, disinterestedly, and only as their competence permits, to such receivers.[3]

It needs to be reiterated that expert authority in this scheme of things extends only over a delimited domain. Viewed in a larger frame, the 'passive receiver' of advice may be one of the politicians or bureaucrats who are 'on top', and the expert 'authoritative transmitter' may be just one of the many minions that they have 'on tap'. Such politicians or bureaucrats are expected to see the advantages of following a self-denying ordinance, against transgressing upon the expert's legitimate domain of autonomy or seeking to influence her judgements therein. This ordinance, along with the reciprocal obligations of experts themselves, must be seen as a constitutive part of an overall ethical frame, within which expertise is dispensed and from which much of its credibility derives. (Whereupon it becomes unsurprising that as politicians and bureaucrats have come to scorn it, as increasingly they have in Britain in recent years, so they have corroded their own already tenuous credibility.)

---

[3] The description does not just apply to academic contexts. Any audience seeking advice from experts requires disinterested and competently scrutinised testimony. If the products of a company are likely to kill people the company tends to prefer its in house experts to say so. The crucial difference between academic experts and their employee equivalents is not any lack of immediate disinterest in the latter, but rather their subsequent respect for norms of confidentiality.

2.4   THE 'MARKET' ALTERNATIVE

Technical expertise requires trust: its acceptance cannot rest upon the visible merits of its advice, nor even upon the manifest efficacy of its practices and artefacts[4]. Even so, our relations with experts are by no means invariably conducted in 'hierarchy' mode, and there are even those who regard it as no more than the residue of an old aristocratic order, wherein scientific professionals borrowed the organisational arrangements of priests and clerics. More common today is a 'market' mode of orientation to expertise, wherein the validity of what we are told is something for us to decide, and expert practitioners routinely recognise our right, even as ignorant outsiders, to evaluate their expertise.

In 'market' mode, experts face fewer formal ethical demands than when they operate collegially in 'hierarchy' mode, although it is arguable that this merely exposes them as individuals to informal ethical demands that are in practice just as great. In 'market' conditions, experts must compete with each other, and it is this that structures and underpins their credibility. An external audience faced with conflicting sources of advice or technical assistance may decide for itself which to accept, comparing one with another, and taking account, if it wishes, of their track records and reputations for veracity and/or efficacy. It is no longer necessary now for different fields of knowledge to be made the exclusive preserves of disciplinary monopolies, and expert judgements need no longer be uncoupled from the biases represented by direct financial rewards. Partiality is attenuated, not by the insulation of experts, but by their exposure as a body to a range of different biases. Experts may seek to remain, or to appear, independent of any and all such biases if they wish, but this is no longer essential. Indeed, experts may make themselves available for hire, by interested parties free to select that one from a range of opinions that they best like the sound of; for the superior reliability and disinterest of expert judgement is no longer presumed in market mode, and lay judgement has institutional precedence over it. Consider how, in a criminal trial, the conflicting 'expert' submissions of defence and prosecution are heard by a 'lay' jury, and the jury's judgement, not the experts' judgements, of the relevant matters of fact is authoritative. So it is whenever expertise is addressed in 'market' mode.

The evaluation of technical expertise in modern societies proceeds in both 'hierarchy' and in 'market' mode but there are fundamental deficiencies in both. Whilst an exclusively hierarchical system offers experts the opportunity to make the very most of their special knowledge and skill, the complete absence of any external accountability or lay evaluation permits abuse of privilege: a purely hierarchical approach will in theory support charlatans as readily as genuine experts. On the other hand, a pure 'market' system expects too much of lay evaluation. If lay persons acquire the knowledge they need properly to evaluate

---

[4] There is frequent controversy over how efficacious technological artefacts actually are: see Mackenzie (1990), Mackenzie and Wacjman (1985), Collins and Pinch (1998).

technical experts, then the efficiency benefits of division of labour are lost and indeed there no longer is expertise; if they do not do so, then lay evaluation will necessarily be incompetent. Indeed for lay persons to presume to evaluate a consensus of technical expertise comes close to their acknowledging the special standing of experts and then promptly denying it[5].

Given the current trend to 'market', it is worth citing some instances that highlight the deficiencies of that mode. Currently (writing at the end of the year 2000), the British media, ever on the lookout for a 'food scare', are making much of policy issues raised by genetically modified foods. Debate on these issues is, of course, much to be desired. But what must it be like for a competent expert to read the relevant 'science', courtesy of the newspapers; or to hear the foods denounced as dangerous in passer-by interviews; or to come to realise how widespread is the belief that there are no genes in food until scientists put them there (but see also Chapter 9)? Indeed, it must be dispiriting to experts of all kinds to find that entities they regard as intrinsic to human beings as natural organisms, genes for example, or 'chemicals', are widely regarded by outsiders as extrinsic intrusions, 'unnatural' and dangerous. And it is an intriguing ethical question whether there is a duty on a society, or its citizenry, to acquaint itself with the rudimentary elements of the 'cosmology' of the scientists on whose knowledge it chooses to rely.

It is not only the misapprehensions of 'the public', however, that can undermine authentic expertise. The claims of formal organisations and organised pressure groups, possessed of their own in-house specialists, may also do so. Consider the campaign famously waged by Greenpeace to prevent the Shell Company from disposing of its Brent-Spar oil container in the deep ocean (Rose, 1998)[6]. As its activists were seen rushing to the rescue of 'the environment', it was (falsely, as came to light later) alleged that vast quantities of toxic chemicals and pollutants were present in the structure. And the mixture of televisual spectaculars and breathtaking misinformation cynically deployed was indeed strikingly successful. How must it have felt to someone with a genuine concern for the technical issues to watch this campaign deliver victory to an organisation that apparently has even less care for truthfulness than it has for legality?

Not all threats to the integrity of expertise, however, are external ones. Experts may become their own worst enemies, in systems that operate in 'market' mode. The BSE ('mad cow disease') outbreak and its eruption into the public domain placed immense ethical demands on scientists and experts. Some of them will now remember episodes wherein they stood silent, as their controllers intoned expedient accounts of what 'the science' they claimed to be guided

---

[5] A fascinating complication here is the existence of 'lay expertise', in the possession of ordinary members, to which professional experts do well to defer; see Epstein (1996).

[6] The example of a pressure group is cited since instances of scandalous misrepresentation by companies, pharmaceuticals manufacturers for example, are all too familiar. For documentation of what is claimed, even the Greenpeace 'authorised version', in Rose (1998), will serve.

by 'really implied'. No doubt they, like those of their colleagues who risked their careers by putting their heads above the parapet, now have a special insight into the value of traditionally constituted, genuinely independent bodies of expertise[7].

Anyone who knows him- or herself to possess genuine expertise, yet is treated nonetheless as just another 'interested' contributor to a plethora of competing claims, is bound to be struck by the limitations of the 'market' mode. Indeed there can be few more tragic experiences than that of such an expert watching lives being lost and resources wasted because his or her skills have been disdained and distrusted. Yet the trend is all to 'market'. Always dominant in some contexts, it is extending ever further at the expense of 'hierarchy'; and is now apparent in our orientation to expertise even in the most esoteric and mathematical areas of the natural sciences. It is important to consider why this is so.

2.5 THE MOVE TO 'MARKET'

The shift to 'market' is part of a larger development associated with systematic secular changes in the nature of our society. At the macro-level, the key changes involve differentiation, division of labour, a shift toward democratically ordered institutions, and the spread of powers and resources to a larger and larger proportion of people. At the level of informal experience, they are manifest as a reduced willingness to accord deference and honour to others, and to acknowledge their claims to status and authority without compelling reasons. We appear to have lost the knack of deferring to those of superior standing, in whatever respect, whilst remaining fully secure in our own self-esteem. In our empowered society, the existence of independent expertise is sometimes perceived, less as an efficient institutional arrangement, and more as the cause of a 'democratic deficit' in our decision making. Deference is equated with an unreciprocated dependence, and such dependence is felt as inferiority and indignity (a feeling that some scientists are amazingly skilled at intensifying!). Use of hired expertise is a widely favoured response. We remain ready to reward experts richly for their services, but we increasingly insist on offering only direct monetary rewards, that deny, or even invert, status relations, and erode expert autonomy rather than reinforcing it as deference does. The result is a shrinking supply of genuinely independent expertise and a situation wherein most major technical decisions are fought over by the 'experts' of conflicting parties. And such experts find themselves addressing lay audiences that, far from deferring to them, are increasingly inclined to question their advice, and even to evaluate its technical basis.

Moreover, as the taboo on direct monetary reward is eroded, expertise becomes ever more closely coupled to sources of finance, and ever more liable to be

---

[7] Britain has been intensely affected in recent years, not just by BSE, but also by a rabid aversion to genuinely independent expertise and indeed independent professional activity of any kind.

perceived as in their pocket. And because those sources are plural and divided, expertise itself becomes plural and divided, often even at the collective and institutional level, since different fields may serve different client-audiences. Consequently, experts cease to speak with one voice, and come to be found on both sides of any tendentious issue, acting not just as sources of technical advice but as advocates as well. Every interest group able to purchase it secures the advice it wants to hear, from experts of some sort or other. And advisors routinely express firmly held convictions, reflecting, they tell us, their own expert knowledge, in the face of the equally firm convictions of expert opponents. All this is beautifully displayed and analysed, in study after study, in the work of Dorothy Nelkin (1975; 1992).

Needless to say, where experts disagree it is not possible to relate to them in 'hierarchy' mode. Indeed, their disagreement will encourage a generalised distrust of expertise, and rightly so, since one side must be incorrect (at least within the limits of available data) in such cases, yet both represent the authority of expertise as such. Similarly, experts associated with powerful interests, even to the extent of being in their pay, cannot but engender suspicion in lay audiences and an inclination to question their authority[8]. And distrust and suspicion of this kind is indeed ever more apparent. One highly visible way in which it expresses itself today is in a readiness to associate risks and uncertainties with scientific and technological advances, and to question the risk assessments offered by technical professionals themselves. In particular, there is an intense public interest in the risks and dangers science and technology pose to the environment.

In his book, *Risk Society* (1988), Ulrich Beck sets out in an uncompromising fashion what is now the ubiquitous method of argument used by individuals, social movements and bureaucratic organisations alike, when they campaign as environmentalists against technological innovation. Any evident benefit of a programme of innovation must be weighed, it is said, against unintended harms, whether demonstrable pollutions, or purely hypothetical future risks and dangers. Dominant powers and institutions will seek to conceal these risks and dangers, and environmentalists should help to expose them, and facilitate a more enlightened politics that takes account of them. This has indeed become a very important form of utilitarian moral argument, and many studies now exist of disputes conducted in terms of it. And on the face of it they offer support to Beck's vision. Environmental movements have created extensive anxiety about a whole range of previously unacknowledged risks (or alleged risks), enough in

---

[8] Scientists and technologists are monopoly suppliers of instrumental knowledge to political and state power, and this is bound to impact adversely on their credibility. And indeed how far ought outsiders to trust scientists who are content to be gagged, or to have their reports discarded, when their findings are found politically inconvenient? What is to be made of a scientific profession that could lay bare the most recondite physical properties of the radioactive elements with magnificent objectivity, yet long had great difficulty in making known even the most elementary pharmacological properties of the cannabinols? On this last topic, unfortunately, vast amounts of nonsense have been put forth, both by scientists and politicians, and a huge vested interest in untruth has been created.

some cases to bring about substantial revisions, or even reversals, in major technological projects that initially had impressive political support. To cite a topical example in the area of biotechnology, the introduction of genetically modified crops into Britain was effectively halted in its tracks within a few months of its first being subjected to criticism of this kind, notably by Greenpeace (see also Mayer, Chapter 9 in this volume).

On closer examination, however, the lesson to be learned from these controversies is not quite what Beck suggests. For all that a powerful mobilisation of opinion is sometimes achieved, the technical knowledge actually deployed in the public sphere is typically scant and simplistic. Moreover, not only is it selectively deployed as the different versions of things offered by the experts of opposed sides, it is selectively believed as well, with each version securing credibility with its own distinct audience. If Mary Douglas (1992) is right[9], risky side-effects are imputed only to the science and technology of disliked powers and institutions, not to that of friends and allies. Consequently, maps of technological risks and dangers in a society become congruent with its members' maps of the social and moral order. And controversies over science and technology become extensions of political battles ongoing elsewhere. Moral–utilitarian arguments express and rationalise positions in these controversies, but institutional relationships account in a causal sense for their existence.

It is worth adding that controversies over methodology and styles of inference may be structured in this way, as well as substantive expert claims. For example, the need for scientists to recognise the fallibility of their knowledge is everywhere acknowledged, but the in-house experts of environmentalist organisations often give this special emphasis. By combining it with the 'precautionary principle' (Chapters 3 and 9) that nothing should ever be done until we are certain that it is free of risk, they can attack the plans of the 'high-tech' organisations that are their enemies. Needless to say, the experts of these enemies are notably less inclined to dwell on the implications of fallibilism, and more likely to stress the formal inadequacy of a precautionary approach. And on this last point they are of course correct. The ethical issue of what level of protection from risks and uncertainties citizens have a right to expect is indeed increasingly addressed in terms of the precautionary principle. But this profoundly unfortunate development is a triumph of 'the principle' only as rhetoric; for there is no way in which it can be applied authentically in what must always be an incompletely known environment replete with uncertainties.

As the move to market has proceeded, so the encompassing ethical scheme in which expertise was traditionally dispensed has become harder to apply, and no alternative has emerged. This has, of course, been recognised and lamented, and various efforts have been made to fill the 'gap'. For example, the way that the legal system orders adversarial encounters within a shared frame has inspired some largely unsuccessful experiments with 'science courts'. But it is surely

---

[9] See also Douglas and Wildavsky (1982) and Wildavsky (1995), which last serves as an admirable foil for Beck and his 'environmentalist' approach.

utopian to expect that opposed experts, serving diverse and conflicting interests, some of which are actually furthered by the continuation of conflict, are going to evolve an agreed ethical framework within which their differences might be resolved.

2.6   RESPONSES TO BIOTECHNOLOGY

The merits of a traditional deferential orientation to expertise have been emphasised in the previous discussion, but it is now often impossible for individuals to orient to experts in this way. A political and institutional consensus in unconditional support of independent expertise is necessary to sustain the 'hierarchy' mode, and increasingly it is lacking. Social change has produced conditions favourable to the market mode, and for better or worse that is what is likely to flourish, in conditions marked by fragmented expertise and fragmented audiences for them. Indeed, there is now so much fragmentation that it is unwise to speak of the credibility of experts among an undifferentiated public. References to 'the public', including those to 'the public understanding of science', are now rarely more than the rhetorical flourishes of some specific faction or interest group. The expert must expect to attract, not just trust and support, but suspicion and hostility, from audiences moved by the principle that 'my enemy's expert is my enemy'. And experts with enemies must always expect those enemies to take any chance to undermine their credibility and faith in the extent of their powers.

Ironically, it is often the awesome extent of these powers that engenders hostility to experts in the first place. The link is certainly apparent with nuclear physics and engineering, but it seems also to be emerging in the context of biotechnology. Power arises from knowledge and competence, and wherever it emerges it engenders some hostility[10]. Moreover, in the guise of research, scientific and technological expertise is a power that casts a shadow over not just the present but the future. Through its continuing reconstitution of knowledge and competence it forces us to accept a continuing reconstitution of our social life, whether we would or not. Many intellectuals and social scientists have drawn attention to this. Ulrich Beck (1988), for example, has spoken of societies wherein politics chases along vainly in the slipstream of an invisible science and technology, ever seeking *ex post facto* to regulate the powers that they constantly yet unpredictably conjure into being. A nice example here is the difficulties that legislators in Britain have faced in attempting to keep pace with developments in new reproductive technologies and with the pressure for ever more extensive use of human embryos in the research involved (Warnock, 1985; Mulkay, 1997).

Beck (1988) gives marvellous expression to the indignation of the political

---

[10] There may be many reasons for this. Thus Beck (1988) suggests that dependence on the very science they attack is capable of greatly intensifying the bitterness and even irrationality of the attitude to science of some environmentalists.

intellectual confronted by the untouchable power incarnate in specialised technical knowledge. How is it, he asks, specifically referring to this new human
biotechnology, that specialists are able cheerfully to proceed with research on
techniques with truly radical implications for the institution of the family, and
hence for society as a whole, without any kind of democratic control being
brought to bear upon them until it is too late, and they are a *fait accompli?* Why
should there be a constant flow of uncontrolled intrusions into the core of our
social life, leaving political institutions and regulatory agencies to patch some
sort of tolerable order out of their aftermath?

These are important questions. And of course they bear with particular force
upon the new human biotechnology, which has profound implications for
familial and sexual relationships and the associated customs and practices. Nor
should such questions be side-stepped by formulaic references to 'progress', or
to the 'inevitability' of technological advances and the need to make the best of
them. The temptations of such positions to the specialist are obvious of course,
and it is not to be wondered at, for example, that some biotechnologists are
inclined to regard a forthcoming genetically profiled and managed society, even
embodying a moderate eugenics, as laudable and benign, despite the warnings of
recent history (see Chapter 15). But it is important to remain aware of the
complexity of the problems here, and in particular to keep in mind that the social
and institutional implications of technological changes are not simply aggre-
gates of the benefits and advantages they offer to individuals.

A standard feature of many individually beneficial biotechnological innovations is that they create serious problems at the collective level. Consider how
research has made it possible for parents to choose the sex of their children–or
their gender, as is often now erroneously said. No doubt this is a benefit for the
individual 'consumer'. But if it were to lead to a 70/30 ratio of the sexes, or even
of the genders, that would clearly be a collective harm. Indeed this outcome of all
the individual choices involved would amount to a net harm, not just to the
collective as a whole, but to each individual member of it, even though every
individual choice was itself a true expression of rational self-interest. The perverse outcome here is an example of the problem of collective action, also known
as *the free rider problem*, a problem long familiar to economists and social
scientists, and scarcely less so to biologists and biotechnologists.[11]

In the instance above, the ratio of the sexes in an entire society is the product
of very many decisions and actions, and no one such makes a noticeable
difference to it. It may be that a 50/50 ratio is 'the best' for the collective, but no
individual benefits from acting 'morally' to help to bring it about and individuals
do better to follow self-interest, say by choosing a male child for economic
reasons, and to leave others to create the desirable 50/50 ratio. But when
everyone seeks to take a free ride on the 'moral' actions of everyone else in this
way, no 'moral' action at all ensues, and the overall state of affairs that all desire

[11] The *locus classicus* of discussion of this in medicine and biology is vaccination programmes.

does not ensue. And this may happen even when the lack of this desired state is far more harmful to each individual than the loss of the individual advantage for which they acted would have been. (The relevance of the free-rider problem to bioethical issues was famously demonstrated by Hardin (1968), in an account that both clearly set out its basic form and discussed memorable exemplary instances wherein it arose. And indeed one of the best ways of overcoming any difficulty in grasping the general form of the problem is still to turn to Hardin's text and the examples given therein.)

Biotechnological research, in impinging upon sex and kin relations and the domain of the family, intrudes into areas of life of immense significance that are particularly vulnerable to collective action problems as new knowledge is produced. These are areas that so far have been subject to little systematic control and external regulation – areas wherein conduct has only fallen into tolerably coordinated patterns because of the lack of knowledge and power of those engendering it. As knowledge and power become available, this basis of coordination may be eliminated and a profound problem of social order thereby created, precisely the kind of problem that generates anxiety and a sense of moral and ethical disorientation. There is any number of possible illustrations here, besides that given above. Carriers of deleterious genes may pose a collective action problem if they acquire knowledge of their state. For example, someone who is a knowing carrier of a deleterious recessive allele, say for one of the heritable anaemias, may consciously weigh individual good against collective good in deciding whether to procreate and with whom. And others in the know, the partner or the doctor of the carrier, for example, or even the biotechnologist who made the knowledge available, may be similarly empowered (see Chapters 13 and 14).

Societies often move quickly to solve collective action problems by subjecting individuals to social influence, regulation or control. But such moves alter the existing social order, and cumulatively have the potential radically to transform it, a point not always appreciated by technical experts themselves, who, unlike many, may expect to be as powerfully placed in any new order as in the old. Even as new human biotechnology offers individuals new powers, they are likely to encounter efforts to control their use of them, whether from peers, proximate authorities, experts, or bureaucratic creatures of the state. And particular studies have documented both beneficial and tragic consequences associated with all four kinds of intervention (Wilkie, 1993). Probably the greatest cause for anxiety, however, arises at the institutional level. It could well be that the emergence of new powers and possibilities from biotechnology will for the most part be experienced as a greatly intensified medical and bureaucratic intrusion into areas of life that so far have been spared it (Nelkin and Tancredi, 1994).

2.7 PROBLEMS OF MEANS AND ENDS

Clearly, there are impeccable utilitarian arguments with which to rationalise opposition to the benefits of research and innovation, and calls for it to be externally restricted and controlled, even where both the competence of the research and its benefits are recognised by its opponents. But it is not sufficient to address these issues entirely in terms of utilities and the kinds of difficulty represented by the collective action problem. For it may be that the entire utilitarian frame of reference becomes a bone of contention when the standing of specialised expertise is at issue. This is again a point well illustrated by reference to biotechnology (Barnes, 1999).

Biotechnological research projects are commonly justified to external audiences in utilitarian terms. They will provide means to specific ends. But as a vast institutionalised system, human biotechnology must also be justified as a whole, as a means to very general human ends. Commonly, it is justified as conducive to our health and wellbeing. But human health and wellbeing can only be characterised in general terms if we have some conception of the fundamental character of human beings and hence a sense of their normal, 'healthy' condition. Inferences about this are often made from our knowledge of human biology, which is perfectly reasonable. But to rely on such inferences is to create enormous difficulties for the evaluation of a human biotechnology that flaunts its capacity to change anything and everything that constitutes human biology. A technology liable to transform the basic nature of human beings cannot be justified simply by reference to the health and wellbeing of human beings as they currently are. Indeed if humans were to apply such a technology to themselves without limit or restraint, the very distinction between technological skills and artefacts and their human possessors would be eroded, and the whole basis of means/end utilitarian thinking would disappear. A utilitarian frame of justification is viable here only if a taboo is placed on any change in whatever we agree is constitutive of our basic nature (see Chapters 14 and 16 for discussion of specific examples).

It would be far fetched to suggest that formal problems of this kind are sources of anxiety at the level of everyday life. Nonetheless, people are well aware intuitively that ordinary means/end thinking entails a distinction between what are means and what are ends, and that the two have to be marked as separate and made the subjects of different attitudes. And reflection on biotechnology disturbs us at this same intuitive level because, as instrumental action, biotechnology reacts back upon us, the ends of the action, and treats us as means are normally treated. Or, at least, it disturbs because it may be that 'we' are being treated as means, depending on what 'we', as ends, are taken fundamentally to consist in.

This question of what we fundamentally are bothers people. Human beings have the status of independent units in their social relations – each individual is treated therein as a unitary essence – yet they are unsure just where to locate that essence, and how to think of it[12]. It is quite common empirically for individuals to take their entire bodies as constitutive of who or what they fundamentally are. Others, having been told about them, take their genes very seriously. The members of the British House of Lords, for example, insisted on exempting themselves, as nobility, from the legislation that everywhere else gave the husband of the mother full legal standing as the father of her children by donor insemination. For their Lordships inheritance of noble nature had to be by 'blood', as they said (cf. Jones, 1996), and genetic connection. Nor should academics be too quick to criticise them. Moral philosophers are prone to make a similar fetish of mind or reason, and many biologists and social scientists have put forward weird and wonderful ideas of their own on these matters.

There is enormous variability in the responses of individuals to this problem, but many such responses impact inconveniently upon biotechnology as an abhorrence of the use and manipulation of human cellular materials, the genome and germ cells above all, embryos and foetal tissues sometimes, even somatic cells occasionally as with those opposed to transplants or blood transfusions. Among medical researchers in Britain, this is widely referred to in private as 'the yuk factor', and seen as a perverse and irrational source of difficulties. If we can alleviate suffering, eliminate pathologies, or extend life by manipulating these materials, it is asked, then why should we be held back by taboos? But whilst the question is a good one, the presumption of irrationality underlying it is false. Those who observe taboos and avoidances cannot be dismissed out of hand as irrational[13]. Indeed, some of their critics are more deserving of the epithet for deploying a facile utilitarianism that completely fails to reflect on the 'utilities' it invokes, or even to notice that they merit scrutiny. (The debate on 'therapeutic cloning' in the United Kingdom in the closing weeks of 2000 offered several examples of this.)

As we have seen, it is actually necessary to separate off a class of things meriting a special respect, if we are to think rigorously and consistently even in a narrowly utilitarian frame. If this class of things, the ends or goods by reference to which utilities are definable, is allowed to remain implicit, and utilities are simply taken for granted, then an impoverishment of moral and ethical discourse can result. It may, for example, become vulnerable to fetishism, that is, to the mistaken identification of means as ends; for expert practitioners are very easily (and understandably) seduced by the 'technical sweetness' of their procedural innovations into equating what they can do with what ought to be done. The danger of fetishism is particularly acute when evaluating the utility of virtuoso interventions, such as are increasingly involved in, for example, neonatal care and radical surgery.

There are cultures wherein a shared sense of sacredness is routinely sustained,

---

[12] For an extended discussion of the relation of the state and the status of the individual human being with specific relevance to biological accounts of human action, see Barnes (2000).

[13] 'Irrational' has now become a largely meaningless term in these contexts, employed in the indiscriminate abuse of those held to be opposed to scientific or medical 'progress'.

encompassing even a great part of the natural order. But it is a standard theme in sociology how in differentiated societies this shared sense weakens and narrows, and continues to receive explicit linguistic recognition only in references to embodied individuals. One result is that a residual sense of the sacred may sometimes remain in cultures lacking the linguistic resources with which to legitimate and rationalise it. And it may be that intuitions of sacredness are then rationalised and reflected on within the now ubiquitous utilitarian frame, using the vocabulary of motives it provides. Thus, where we find evaluations of technological change rationalised by the standard method of weighing benefits against risks, we need to remain open to the thought that the evaluations may nonetheless actually express an aversion to desecration by technology, and a concern with bodily integrity.

Consider again the recent strong reaction in Britain against genetically modified foods. What it was that moved the mysterious bureaucracy of Greenpeace to attack them can only be conjectured (although it is clear that many members have a genuine concern for 'the environment' whilst some have intrinsic moral objections to moving genes between organisms: see Chapter 9). However, the attack itself was conducted along standard environmentalist lines. There was, allegedly, the unacceptable risk of gene transfer into the environment. But the consequence of the attack was that people who had had no previous engagement with the topic became anxious about their *diet*. Hearing of yet more risks and dangers from the usual sources, they reacted with aversion to the indicted products, often products that they had previously been happily consuming. And as the supermarkets removed the offending boxes from their shelves, so a devastating 'environmentalist' victory was secured and proudly trumpeted as such by Greenpeace.

Whatever is made of episodes of this kind, however, and of the strange eating habits of the modern consumer, it needs to be reiterated that neither the basic tendency to separate and sacralise, nor the associated practices of taboo and avoidance, can sensibly be dismissed as irrational. On the contrary, being practices essential to the ordering of our thought and the structuring of our lives, they are indispensable even to the most uncompromising utilitarianism as we have seen, and to the most rigorous rationalism. Indeed, it is interesting to reflect here on the work of rationalist popularisers of science like Richard Dawkins, wherein an uncompromising picture of human beings as natural objects is aggressively asserted. In truth, the human being is present in this work in two guises, explicitly as profane object and implicitly as sacred object, and this is actually essential to create the dualist structure required by both its moral and its epistemological arguments. Morally, a person is at once a physical system and an end of physical actions. Epistemologically, he or she is at once a physical system, and a rational soul capable of carrying the scientific knowledge that renders him or her as a physical system, and of knowing it to be correct. But how these two versions of the human being are to be related is never considered, and how that part of the physical world that is a human being maps the entire physical world onto itself, including that part of the world that is itself, is not addressed.

2.8 CONCLUSIONS

The task of this chapter has been to examine the institutional and cultural settings in which bioethical issues arise and the form that debates about these issues take therein. It has concentrated on relations between technical specialists and their audiences, and identified three general trends in these relations. First of all, audiences are becoming less deferential to expertise, and increasingly inclined to involve themselves, not only in judging issues, but also in evaluating the specialised knowledge relevant to them. Secondly, the audience for expertise is becoming many audiences, often with strongly opposed interests, and an analogous fragmentation is evident in expertise itself. Thirdly, whilst the utilitarian frame that currently rationalises 'technical' decisions is widely accepted, there are grounds for believing that it fails to give proper expression to all the concerns of audiences, and notably of those audiences seeking to confront the implications of a burgeoning human biotechnology.

All these trends create difficulties for experts seeking to secure external credibility, and all seem set to continue. Some commentators regard a dialogue of equals between experts and a knowledgeable and empowered lay public, and an acceptance by the former of full 'democratic accountability', as the probable and desirable long-term outcome. But in truth there is little evidence to justify such a prediction. Whilst experts face criticism from the public, it is the comfortable criticism of a lay audience that, however active and empowered, gives no sign of being willing to share their responsibilities, or face demands for accountability of the kind that currently fall upon them.[14] Increasingly, in representative democracies, the rule that structures debate, in what increasingly is a single-issue style of politics, is that ordinary individuals are never in the wrong. Large organisations and their specialist advisors are used as sumps for blame. The result is the prominence of pressure groups like Greenpeace, which deploy a radical innovation in accountancy: single-entry bookkeeping.[15] And there is a corresponding failure by the media to provide anything like a comprehensive

---

[14] I am writing these words in the aftermath of a rare fatal accident on the British railways. Directors of the companies involved are facing the synthetic indignation of television interviewers and being called upon to resign. The expert advice they acted on is being instantly 'analysed', and casually held up to ridicule. But in the same week 50 or 60 fatalities have probably occurred on the roads, where daily fatal accidents are generally passed over as the unremarkable accomplishments of ordinary individuals.

[15] Thus, incinerators arouse the ire of Greenpeace and the criticism of their in-house experts, but not questionable 'private' means of waste disposal; and organisationally financed genetic modification is assailed but not other kinds. It is through many such one-sided discourses, each in itself little different from a stream of deceit, that truth on such issues is now pursued in our highly differentiated societies.

picture of the consequences of technical decisions, good and ill, immediate and remote.

Of course, the awkward position in which experts are currently placed is well understood by their professional bodies, which have responded by encouraging adaptation to changing circumstances. They have urged the importance of image and public relations; and of a more sympathetic engagement with outside audiences, in fora readily accessible to them, through agendas that reflect practical concerns as well as narrowly technical ones. And these are perfectly understandable temporising responses to changes that may cause nuisance to individuals but do not deny the standing of expertise and the authority of science at the institutional level.

Unfortunately, this crucial contrast between institutions and individuals is not everywhere so well understood. Science is now more widely respected and trusted than ever before, even if scientists are not; it has no rivals as a source of cognitive authority, even if scientists do; its institutional position remains un-challenged in a context where the scientist no longer stands as a microcosm of the institution. But these are truths that some individual scientists are badly placed to see. Indeed, some scientists have incorrectly interpreted their apparent-ly diminished standing as a sign of a widespread irrationalism and a hostility to science as such, and have lashed out wildly against these things, for example in the highly polemical literature (e.g. Gross and Levitt, 1994; Bricmont and Sokal, 1997; Koertge, 1998) of the current 'science wars'.[16] These scientists should reflect a little more on the fact that scientists and technologists are currently provided with greater resources than ever before, and allowed to monopolise practically all the occupational positions where empirical knowledge is the basis of power. And whilst they are entitled to believe that society would benefit from being still more deferential to the authority of 'science', they should ask whether the increased regulation of judgement and opinion necessary to bring this about would be a price worth paying for it.

This is just part of a larger ethical question raised by our proliferating science and technology. As they inexorably engender new powers, so they engender vastly increased amounts of regulation and bureaucracy. And 'public opposi-tion' to science and technology only serves to encourage this development; since its major effect, in practice, is to encourage pressured politicians to extend bureaucratic control ever further, even into the inner realms of science and technology themselves. There is the danger of death by administration which-ever road we go down in the future, and the moral and ethical issues lying latent here need to be recognised as amongst the most profound of all those raised by

---

[16] This literature represents a wholesale lapse from ethical proprieties. But withdrawal of deference can elicit intense responses, far more than withdrawal of cash and resources. And here it has led legitimately annoyed scientists into careless abuse of their relatively powerless 'enemies' and even into treating them as legitimate targets of hatred. Indeed, the behaviour of some of the science warriors has been redolent of the custom, in earlier times, of burning alive the frail old woman living next door, on the pretext that she was powerful, dangerous, and ill disposed toward her neighbours (i.e. a witch!).

current developments, both in biotechnology and in science and technology more generally.

REFERENCES

Barnes, B. (1985) *About Science.* Blackwell, Oxford, UK.
Barnes, B. (1999) Biotechnology as expertise. In *Nature, Risk and Responsibility.* O'Mahoney, P. (ed), MacMillan, London, UK.
Barnes, B. (2000) *Understanding Agency: Social Theory and Responsible Action.* Sage, London, UK.
Beck, U. (1988) *Risk Society: Towards a New Modernity.* Sage, London, UK.
Bricmont, J. and Sokal, A. (1997) *Impostures Intellectuelles.* Odile Jacob, Paris, France.
Collins, H. and Pinch, T. (1998) *The Golem at Large: What You Should Know About Technology.* Cambridge University Press, Cambridge, UK.
Douglas, M. (1992) *Risk and Blame.* Routledge, London, UK.
Douglas, M. and Wildavsky, A. (1982) *Risk and Culture.* University of California Press, Berkeley, CA, USA.
Epstein, S. (1996) *Impure Science.* University of California Press, Berkeley, CA, USA.
Gross, P. and Levitt, N. (1994) *Higher Superstition.* Johns Hopkins University Press, Baltimore, MD, USA.
Hardin, G. (1968) The tragedy of the commons. *Science,* **162,** 1243–1248.
Jasanoff, S., Markle, G.E., Petersen, J.C. and Pinch T. (eds) (1995) *Handbook of Science and Technology Studies.* Sage, Beverly Hills, CA, USA.
Jones, S. (1996) *In the Blood: God, Genes and Destiny.* Harper Collins, London, UK.
Koertge, N. (1998) *A House Built on Sand.* Clarendon, Oxford, UK.
MacKenzie, D. (1990) *Inventing Accuracy: a Historical Sociology of Nuclear Missile Guidance.* MIT Press, Cambridge, MA, USA.
MacKenzie, D. and Wacjman, J. (eds) (1985) *The Social Shaping of Technology.* Open University Press, Milton Keynes, UK.
Merton, R.K. (1973) *The Sociology of Science.* University of Chicago Press, Chicago, IL, USA.
Mulkay, M. (1997) *The Embryo Research Debate: Science and the Politics of Reproduction.* Cambridge University Press, Cambridge, UK.
Nelkin, D. (1975) The political impact of technical expertise. *Social Studies of Science,* **5,** 35–54.
Nelkin, D. (ed) (1992) *Controversy: Politics of Technical Decisions.* Sage, Beverly Hills, CA, USA.
Nelkin, D. and Tancredi, L. (1994) *Dangerous Diagnostics: the Social Power of Biological Information* (2nd edn). Chicago University Press, Chicago, IL, USA.
Rose, C. (1998) *The Turning of the Spar.* Greenpeace, London, USA.
Warnock, M. (1985) *A Question of Life: the Warnock Report on Human Fertilisation and Embryology.* Blackwell, Oxford, UK.
Wildavsky, A. (1995) *But is it True?* Harvard University Press, Cambridge, MA, USA.
Wilkie, T. (1993) *Perilous Knowledge.* Faber, London, UK.
Ziman, J. (1968) *Public Knowledge.* Cambridge University Press, Cambridge, UK.

II Ethics and the Natural World

3 Introduction to Environmental Ethics

Christopher Southgate

3.1 INTRODUCTION

What value does the non-human world have? There are straightforward commercial answers in respect of farmland, planted forests, and fishing-streams, but the answers are much more complex in respect of public parks, uncultivated land, waterways, and indeed the air itself. What value does any given bird have? Any given beetle? The last two questions may seem to some artificial, but the failure to answer them in a consistent way, and to match human aspirations to the continuing health of the non-human world, has given rise to a global ecological crisis of enormous proportions. There are many accounts of the extent to which humans are damaging the environment. A recent and telling one is Reg Morrison's in *The Spirit in the Gene* (Morrison, 1999). His analysis is important because he shows that humans have been over-using environments for at least 3600 years.[1] What is different now is

- that our capacity to stress ecosystems is now enormously greater, because of the extent of human population and the power of our technology, and
- that the effects of that technology are global in their extent. Classic examples are CFCs, largely produced in the higher-tech countries of the Northern Hemisphere, affecting the ozone hole over Antarctica, radioactive fall-out from the nuclear accident at Cernobyl in the Ukraine affecting lambs on UK hill-farms, and most dramatically the whole issue of global warming.

However, we also need to be aware of the presuppositions behind the questions that are asked: as soon as we ask 'What value has a certain element of the non-human world?' we are in effect separating the biosphere into (a) the non-human, which may be valued or non-valued, and (b) humans, who are detached beings with the right and privilege of assessing value. Again, what we regard as valuable will depend on our own situations as individuals or societies, according

---

[1] Morrison (1999), pp. 98–99. Morrison also presses the important question – what is it about human beings and their genetic and cultural inheritance that makes us fail to confront this situation appropriately?

Bioethics for Scientists. Edited by John Bryant, Linda Baggott la Velle and John Searle.

to whether we are, for example, trying to survive a night in wild country, to start a farm, or to locate raw materials for computer technologies.

3.2 SCHEMES BY WHICH TO VALUE THE NON-HUMAN WORLD

The essential categories of value to note in this discussion are the following.

- *Instrumental value* – the non-human world valued in terms of its usefulness to human beings, as an instrument for their use. In practice the great majority of decisions about the non-human world are made in terms of instrumental value. That is not however to say that this is the most appropriate ethical scheme.
- *Intrinsic value* – the non-human world regarded as valuable in and of itself, whether it is used by or useful to human beings or not. Schemes based on intrinsic value might accord value to all non-human entities, including rocks, or only to living organisms. Again, living organisms might be regarded as fundamentally equal in value, or their value might be graded on a sliding scale, depending on their complexity, or their capacities, so that, for example, the capacity that higher animals have to feel pain might give them special status.

    Within non-instrumentalist or intrinsic value approaches to the non-human world, an effort is sometimes made to establish a category of *inherent value*. This recognises that non-human entities and systems can have value which does not derive from human use, but stresses that the value arises from human valuing. However, for most purposes the central distinction is between instrumental and intrinsic value.

A deeper question lying behind these categories is: on what basis is the value assigned? If human use is deemed to be the prime or sole cause of value, that must stem from a conviction as to the specialness of human beings. That in turn could rest on a naturalistic conclusion that humans have capacities such as self-consciousness, free rational choice, not possessed by other species, or from a religious conviction of human distinctiveness. Historically, religion has been the principal source of underlying ethical convictions of this type, and it still exerts more of an effect than might be supposed (see Chapter 1).

The conviction that human needs and benefits are in a completely different category of importance to those of the non-human is usually termed *anthropocentrism*[2]. By contrast, the conviction that humans are merely one species among many, not special in any way, and that all species are of value, is called *biocentrism*. An important third category is *theocentrism*, the conviction that

---

[2] Michael Northcott in his valuable book *The Environment and Christian Ethics* (Northcott, 1996) uses the term 'humanocentrism' to mean the same thing. This is more correct in terms of derivation from the Latin, but 'anthropocentrism' remains the standard term.

entities are of value because God created them and remains in relationship to them.

It would seem that the doctrine of creation underlying Judaism, and by derivation also Christianity and Islam, must give rise to a theocentric ethic, since the opening chapter of the Book of Genesis unfolds around the conviction that everything was made by the Lord God, and that God found it good. That in itself would seem to argue strongly for the intrinsic value of all non-human entities. However, in a profoundly influential article written in 1967 Lynn White Jr. called Christianity 'the most anthropocentric religion the world has seen' (White, 1967).

White was one of the first to link environmental degradation to religious convictions, and he did so citing Genesis Chapter 1, verse 26, the divine command to the first humans to 'dominate the earth and subdue it'. White linked this to Christianity's role in fostering the development of modern science[3], and the technology that stemmed from it. It is Western science, derived from Christian culture, which has succeeded – in Francis Bacon's notorious phrase – in 'putting Nature to the test'. Though White's analysis has been heavily criticised, for example by Nash (1991), it serves to point to the ambiguous role that talk of God has had in the development of humans' relation to their environment. To make a religious statement like 'the earth is the Lord's and all that therein is' (Psalm 24.1) does not of itself solve the questions of environmental ethics.

This question of values in the non-human world is discussed further at the end of the chapter.

3.3 SYSTEMS FOR APPLYING ASSIGNMENTS OF VALUE

The last section addressed the question of what value a non-human entity might have, but a further question in ethics is: how is the value-system to be *applied*? Chapter 1 introduced the broad categories of ethical application. In deontological schemes individuals or groups have duties towards what is regarded as being of value. At the other end of the spectrum is the Nietzschean rational egoist position in which a person, or even an organisation, simply calculates the outcome most favourable for itself. It was also noted that many schemes of ethical application are consequentialist, in which the outcomes for a *range* of beings are considered (in Chapter 1, the main focus was on human beings) and an effort is made to calculate the outcome that is most advantageous *overall*. These systems are illustrated in Figure 3.1.

In connection with the figure it should be noted that an ethics based on rational egoism is not in a position to value non-human systems for themselves: by definition, value is only represented by the good of the egoist him or herself. On the other hand, an ethics that accords intrinsic value to the non-human world

[3] For a careful assessment of this relationship see John H. Brooke's *Science and Religion: a Historical Perspective* (1991).

SYSTEM OF VALUING THE NON-HUMAN WORLD

INSTRUMENTAL   INTRINSIC
VALUE ONLY   ◄────────►   (as well as
                               instrumental)
                               VALUE

ETHICAL CALCULUS

RATIONAL EGOISM

CONSEQUENTIALISM

DEONTOLOGICAL ETHICS

Figure 3.1. Ethical approaches to assessing situations involving possible consequences for the non-human world

will often impose duties on humans to respect, care for, or indeed reverence that world. However, environmental decisions taken by politicians are much more likely to be consequentialist. A decision such as that of the UK Government in the 1990s to build a controversial by-pass road round the town of Newbury (completed in 1998) involved a complex risk–benefit analysis – a classic tool of consequentialism. The Government came under pressure to include in this the rare species affected. The final analysis might seem to have accorded intrinsic value to non-human species – the survival even of Desmoulin's whorl snail (*Vertigo moulinsiana*) was considered. However, the attitude – 'they won't be around for our children's children to know about if we destroy them now' – could be made on the basis of instrumental value; it is not of itself an indicator that intrinsic value was being assigned in the present.

3.4 POSITIONS IN THE ENVIRONMENTAL DEBATE

A whole range of positions are adopted in the contemporary debate on the environment. Two extremes will help to illustrate the character of the debate more clearly.

• The first might be termed *technocratic cornucopianism*. '*Cornucopia*' is the Latin term for a 'horn of plenty'. In this view there is no shortage of resources on Earth; it is merely a matter of humans devising more and more ingenious technologies to harvest them. The planet Earth, in effect, is regarded as being an infinite resource for humans, and able to take anything humans can throw at it. That is clearly an extreme view, but it lies behind quite a lot of anti-environmentalist thinking on the right wing of US politics, and in-fluenced policy particularly during the Reagan presidency (1981–89).

Figure 3.2. Campaigners protesting about the building of the by-pass road around Newbury in the UK. Photographs reproduced by permission of Mark Lynas

- The second is the movement known as '*deep ecology*', which derives from a seminal article by the Norwegian philosopher Arne Naess first published in 1972 and reprinted in 1995. The argument of this paper illustrates a number of key themes in environmental ethics.[4]

[4] The following analysis of deep ecology is adapted from Southgate (1999).

Naess contrasted 'environmental fixes' within the existing attitudes of modern society ('shallow ecology') with a real change of mind-set in relation to the planet. Characteristics of a deep-ecological mind-set would include the following (Naess, 1995).

1. Rejection of what Naess called 'the man-in-environment image' (in which humans are distinct from the world and can choose what value to place upon it) in favour of a 'relational, total-field image' *(therefore rejecting the form of our original question on value, and insisting on seeing every system as a whole, of which humans are only parts)*.
2. Emphasising biospherical egalitarianism *(i.e. a radically biocentric approach, insisting on the real equality of all species)*.
3. Stressing principles of diversity and symbiosis.

The profound respect for all creatures inherent in 1 and especially in 2 – the conviction of the need for a biocentric scheme – has its roots in Buddhism. Naess however acknowledges that it is necessary to kill some fellow-creatures to live – this expedient should be kept to a minimum (see also Chapter 12). Very central to deep ecology is the concept of wilderness – those areas of the planet which are as they would be if no humans ever went there. One can sense here the influence of such early American environmentalists as John Muir, and of Aldo Leopold's *A Sand County Almanac* (1949) (see below). If there is a deity for deep ecologists it is wilderness.

For some writers the natural corollary of such thinking has been that there are far too many humans, and that reduction of the human population to some notional 'carrying capacity' – a number of humans which the Earth could sustainably support – should be welcomed. (A figure sometimes used is 0.5 billion, as against an actual population for the year 2000 of over 6 billion.) There is a strain of 'environmental fascism' about some of these comments, very far from the stark but visionary tone of Naess's original article. Deep ecologists' suggestions as to how humans might live in the future are often very radical. In the standard work on deep ecology by Devall and Sessions (1985) they quote Paul Shepard's view that the Earth's population should be stabilised at 8 billion by 2020 and that this population should be positioned in cities strung around the edge of the continents, with the interior allowed to return to the wild. The contact with this wilderness, so essential to our ecological health, would all be by journeys made on foot. Incidentally, this would mean feeding the world on biotechnologically engineered microbial food (Devall and Sessions, 1985).

Deep ecology (DE) and technocratic cornucopianism (TC), then, show how polarised the debate is – between a sense that no alteration in our ways with the Earth is necessary, and Naess's sense that no amount of 'shallow' fixes will do. In TC nature is reduced to a mere commodity – a commodity, remarkably, which is believed to be in endless supply – and is to be exploited at will. In DE wild nature is given a status akin to holiness, and humans are to abandon, in effect, the whole programme of interacting with it that began with the dawn of agriculture.

> *The term 'nature'*: Precisely because 'nature' has a range of very different connotations, e.g. that which is essential to a being; the sum total of physical reality; the sum total of physical reality apart from humans; and the created world apart from God and divine grace (see Ruether, 1992) – we avoid using the term in the rest of this chapter, preferring 'non-human entities/systems/world' as being clearer and less prejudicial.

3.5   PRINCIPLES USED IN FORMULATING ENVIRONMENTAL POLICY

We have seen how environmental ethics embraces some extreme solutions. We move now to look at how these would actually operate in ethical situations where different proposed actions are being considered, and different value-schemes are in tension. We consider five options, which we list here in order of ascending care for the non-human world:

  (i) free-market economics;
 (ii) the polluter-pays principle;
(iii) environmental economics;
 (iv) the precautionary principle; and
  (v) radical biocentrism.

At the top left of the grid in Figure 3.1 are the schemes on which *pure free-market economics* (i) would operate, i.e. no intrinsic value is attached to the non-human world, since 'values' are the financial values of the marketplace, and the system assumes that its participants will pursue their own self-interest. However, it is very widely conceded within economics that the environment constitutes a 'free-market failure' because many of the resources concerned are not owned, nor do they have financial values attached to them. The classic examples are the air itself and the oceans. Since the usefulness of these is not owned or priced, yet one person's activity can deprive others of their benefit, the free-market principle that the market itself will establish appropriate patterns of exchange is not likely to succeed. When insecticide residues from the tropics arrive in the breast milk of Inuit mothers (see Chapter 5) there is no market relationship between user and sufferer to counteract this. Likewise, nuclear power has not failed economically, in free-market terms, but has undergone radical re-evaluation because of its possible effects on the environment (see Chapter 5). Daly and Cobb (1994) sum up the problem when they say 'the market sees only efficiency – it has no organs for hearing, feeling or smelling either justice or sustainability'.

Let us look now at option (v) – *radical biocentrism*. Can we, as Naess (1995) urged, undergo a total shift of attitude and see ourselves as just one species among many? Can we pursue a policy that responds to his observation that 'the flourishing of non-human life *requires* a smaller human population' and his injunction that 'if (human) vital needs come into conflict with the vital needs of nonhumans, then humans should defer to the latter.' It is extremely difficult to see how this system can operate throughout the world, particularly given the desperate straits in which so many humans find themselves. For example, a tree looks very different if it is the only fuel source within ten miles of your home from if it is the central feature of a historic park.

However, as a spiritual exhortation to individuals and societies whose impact on the planet is greatest (that of the *average* U.S. citizen approaches 100 times that of the average citizen of India, and many Westerners' consumption is far higher yet) radically biocentric thinking will remain important. As the ecological theologian Sallie McFague (1993) comments, deep ecological thinking is to be commended 'for its poetic power more than its conceptual adequacy' (McFague, 1993). If its 'poetry' served only to persuade any significant number of humans to become vegetarians, this would, over time, lead to a greatly reduced load on the biosphere and the supplies of fresh water.[5]

So, if we accept that the free market cannot manage by itself, but that radical biocentrism cannot do justice to the way very many humans have to live, what about option (ii), *the polluter-pays principle*? On this basis, an organisation can damage the environment as much as is necessary to fulfil its purposes (of e.g. manufacturing a chemical, producing electric power etc.) but must be responsible for the full costs of clean-up. The principle is consequentialist in essence, but it insists that environmental damage is a consequence implying costs for the damager. Thus, for example, after the sinking of the oil tanker *Exxon-Valdez* disgorged great quantities of oil into Prince Edward Sound in 1989, the only concern of the international community, on this principle, should have been to ensure that the Exxon Corporation paid the whole costs of removing the oil. Likewise after the destruction of the nuclear plant at Cernobyl in the Ukraine in 1986, the concern should have been that the Soviet Government fully compensate all concerned (including the sheep farmers in Scotland who found that their lambs were unsaleable because of radioactive fall-out)[6]. The value-system implicit in the principle is entirely instrumentalist; the polluter returns the system to its previous state of human usefulness.

The first problem with this principle is analogous to that noted above in respect of the free-market approach. Some elements of the environment are not owned, or convertible into cash value, but they can still be damaged by environmental abuse. The soil of part of the Ukraine was damaged on a very long-term basis by the Cernobyl accident, in a way that defies effective clean-up or easy

---

[5] Meat production is a very inefficient way to use the solar energy trapped in cereal crops, and it has recently been calculated that it takes almost $16 \, m^3$ of fresh water to produce a quarter-pound hamburger (Morrison, 1999) – again, vastly more than would be required to generate the same food value in a vegetarian diet.

[6] See Chapter 5 regarding the debate on nuclear power.

costing. Thousands of seabirds died in Prince Edward Sound, but they were not owned, and could not therefore be subject to straightforward compensation of the owner by the polluter.

The examples chosen should make clear the other major objection to the polluter-pays principle as the sole ingredient in an environmental policy. Neither accident should have happened – those responsible had no right to risk the sort of ecological damage that was caused (let alone the loss of human life at Cernobyl). The principle seems to imply that anything can be permitted which can be paid for, whereas it has been a truism of environmental thinking since Rachel Carson's chilling book *Silent Spring* (1962)[7] that some courses of action cannot be permitted, and that much environmental damage is invisible and may be irremediable.

A natural foil for the polluter-pays principle is option (iv) – *the precautionary principle.* This states that no course of action should be undertaken if there is a risk that it could give rise to unacceptable damage. If there is a question as to whether damage might occur from a particular action, the principle says that the action should not be pursued until it can be shown that it will be safe. In effect it introduces a deontological element into the prevailing consequentialism by which governments and organisations frame environmental decisions. Certain courses of action are excluded from consideration – there is a duty not to do them, because their consequences might be unacceptable[8]. A classic case in which the precautionary principle is much invoked at present is that of the proposed commercial growth in the UK of genetically modified (GM) crops (discussed in detail in Chapters 8 and 9; see also Chapter 1).

At first sight the precautionary principle seems like common sense, but an argument often advanced against it is that no human development of any sort would have taken place had the principle been rigidly observed. Every new initiative carries elements of risk, and one of the major ingredients in the discussion of GM crops is about risk – not only what level is acceptable, but also who should calculate the risk. This in itself represents a realisation that total avoidance of risk is unfeasible. In the light of this, the principle operates as a reminder of the need to gather as much information about the potential risks of an action as possible, and then to investigate whether it is possible to minimise those risks, or to eliminate the most harmful.

We have now looked at four of our options for trying to formulate environmental policy. It is interesting that the Rio Declaration, stemming from the United Nations Conference on Environment and Development, affirms the importance of free trade *and* of the polluter-pays principle *and* of the precautionary principle[9]. This shows the tensions governments experience in formulating environmental policy within an economic context which sets a very high pre-

---

[7] See Chapter 5 on *Silent Spring* and the debate on DDT.
[8] As often where a deontological element is introduced, the principle finds itself expressed in legislation, both in the European Union and the US.
[9] See Section 4.2.

mium on unfettered operation of global markets, and has very limited mechanisms for costing ecological damage.

This naturally poses the question: can option (iii), *environmental economics*, a modification of classical economics that takes into account ecological costs, provide a reliable guide to appropriate policy-making? An example of an equation in environmental economics would be

$$\text{NPV} = \sum_{t=1}^{n} [(B_t - C_t - aR_t)/(1 + i)^t$$

where NPV is the net present value of continuing from time t_1 to t_n a particular process which earns an income. B is the income-generating benefit, C the cost of controlling the pollution, R the risk to population health, the coefficient a the risk that converts that risk into financial terms, t each unit of time under consideration, and i the discount rate relating present risks to future risks (from McMichael, 1995).

Immediately one can see how many assumptions lie behind such an equation.

* That the effect of the environmental damage is such that it can be cleared up, and the damage can be summed up totally in the cost of that clear-up. We have already noted the weakness of this assumption.
* That the risk to health is likewise representable in pure financial terms – immediately one has to note that in terms of contribution to GDP an American life can be 'worth' over 80 times the life of a person of the same age and sex from a less developed country. Should this sort of economics correct that inequality to render all human lives of equal value, or recognise the economic realities of the enormous difference in income-generation and resource-use between the two types of society?
* That there is a discount rate to be applied to the future. In conventional economics future situations are always valued lower than present ones, to recognise both the realities of inflation and the different priorities that future generations might have, but it might be argued that we should actually value the future more highly than the present, because of the vital need to preserve the ecological health of the planet as was commented on in Chapter 1 in respect of inter-generational responsibility. (A vital, if controversial, concept here is that of *sustainability* – we discuss this further in Chapter 4.)

As Holmes Rolston III (1988) warns, 'Numbers are usually thought to make values explicit, but numbers can disguise rather than expose value judgments. . . The data seldom change anyone's mind, but they are gathered and selected to justify positions already held, ignored or reinterpreted if in conflict with favored positions. We should decide about the latent ideology, only secondarily about the number analysis.'

So quantitative approaches are not as helpful as might have been hoped. They form an important part of the cost–benefit analyses that precede such decisions as to whether to build the Newbury Bypass (see Section 3.3), but it is vital to note Rolston's point that numbers are theory laden. Daly and Cobb (1994) illustrate this by recalculating gross national product for the USA from 1950 to 1990, taking into account pollution and the depletion of natural resources. Gross national product *per capita* rose by 120.8% in that time. The revised index, the 'Index of Sustainable Economic Welfare' rose only by 31.7%, and reached its maximum in 1976. Between 1976 and 1990, including the great Reagan years of economic boom, the ISEW, a measure of America's real wealth, fell by 9%[10].

Economics, then, is a human science, the science of the *oikos*, the household, but it looks very different if parts of the household other than short-term human interest receive a higher weighting. For a recent contrasting of neo-classical and ecological economics see McFague (2001).

3.6   THE CONTRIBUTION OF ENVIRONMENTAL ETHICS

We have seen just how difficult it is to develop an ethical system that actually allows us to make real choices in the conflicts of interest that are found wherever humans are utilising their environment. We illustrate these conflicts of interest in the case studies in Chapters 4 and 5, but first we ask whether the young field of environmental ethics can enable us, in Rolston's words, to 'decide the latent ideology' and develop systems which avoid unworkable extremes and hold in creative tension the needs of humans, present and future, and the needs and values of the non-human world.

Earlier we mentioned two seminal writings in this field: the papers by Lynn White Jr. (1967) and Arne Naess (1972/1995). White's paper yearns for a revival of the thought of Francis of Assisi. Deep ecology, springing from Naess's paper, looks back to the love affair with the American wild in such writers as Thoreau and Muir, and more recently in Aldo Leopold and Annie Dillard. In this section we look first at Leopold's proposal for a 'land ethic', then at two more recent environmental philosophers, John Passmore and Holmes Rolston III.

Aldo Leopold's *A Sand County Almanac*, his book of essays derived from his time in rural Wisconsin (1949; reprinted 1966), has had a great influence on recent American attitudes to being with the land: partly because the observations of nature are very beautifully and tenderly written[11]; partly because Leopold moved beyond mere observation to propose a very different way of understanding human being from the one which had dominated Western thought. Leopold's key phrase is 'plain member and citizen' – this for him is what a human is in the biotic community. He saw human ethics as having

---

[10]   For a quantitative estimate of the 'natural capital' of the earth see Costanza et al. (1977).
[11]   Interestingly, unlike many who have espoused his ethic, Leopold's quality of observation was fed by his practice of hunting and fishing – he does not develop an ethic of 'non-harming' as do many biocentric writers influenced by Buddhism (such as for example Jay McDaniel in *Of God and Pelicans* (1989)).

already evolved from the ancient state of regarding women and slaves as property, and now needing to evolve beyond mere economic exploitation of the non-human world. He saw the reasons for our failure in this regard as twofold:

(i) he realised what we have already noted in this chapter, that most of the inhabitants of the 'land community' have no economic worth, yet they make a contribution to the integrity and stability of the biosphere;
(ii) he maintained that we simply do not know enough about the biosphere to manage its affairs. He called therefore for a radical change both in the attitudes of landowners and in the overall way we educate ourselves about the land – such that we learn how to *perceive* rather than to *acquire*.

A great lover of wilderness, Leopold nevertheless has a much more matter-of-fact attitude to it than many ecological writers – wilderness is, among other things, where we see the greatest diversity of evolutionary strategies, and therefore the place which most broadens our understanding of the things that go wrong when humans stress biotic systems.

Leopold's ethical conclusion (1966) was summed up in his so-called 'land ethic' – 'a thing is right when it tends to preserve the integrity, stability and beauty of the biotic community. It is wrong when it tends otherwise'. Note the strong emphasis on community, on the overall system as the carrier and determinant of value. While Leopold's point about our ignorance of ecosystems, and therefore our questionable competence to be their managers, is well taken, we noted above how difficult it would be to pursue the radical biocentrism that the land ethic, taken at its most literal, would imply. Humans find it very difficult to live in any place in any numbers without profoundly changing the non-human 'community'. A further difficulty is one that afflicts all ethical systems framed in terms of communities rather than individuals, the question being *who is to decide* what is in the best interests of the community. There is the danger of some humans losing all their status in the implementation of the land ethic by others.

John Passmore's *Man's Responsibility for Nature* (1974/1980) was one of the first major philosophical essays on the challenge of the ecological crisis. Passmore affirmed the achievements of human civilisation, particularly those of the West, and the distinctiveness of humanity. He rejected Leopold's notion of 'the land community' on the grounds that there is true community only between those who share rights and mutual responsibilities – this cannot include the non-human world.

Nevertheless Passmore recognised both the extent of the developing crisis and the political difficulty of changing human societies so as to minimise it. He saw 'seeds' of a solution both in the notion that humans could think of themselves as stewards of nature, and in the perception that it is better to co-operate with the ways of biological systems than to make changes which radically transform them. He ends by concluding that humans need not so much a radically new ethic but a sensibly extended version of the traditional Western humanistic ethic that it is wrong to injure one's neighbour – that must include injuring the ecology of which the neighbour is a part and on which the neighbour depends. He looked for the sorts of developments in environmental economics on which we touched above (the present author finds it disheartening, over a quarter of a century after Passmore first wrote, to see how little influence such developments have yet had).

What of future generations? For Passmore it would be wrong to reduce those generations' freedom of thought or action *either* by destroying the natural world *or* by destroying the social traditions (philosophical, scientific, artistic, or political) that permit and encourage that freedom. This again emphasises the inter-generational element that is a necessary part of thinking in bioethics and in environmental ethics (Chapter 1).

3.7 VALUES RECONSIDERED

The contribution of the philosopher Holmes Rolston III, in particular in his book *Environmental Ethics* (1988), places the question of value once again at the centre of the debate, and deploys most helpfully the categories of instrumental and intrinsic value at which we began to look in Section 3.2. Rolston shows in how many different ways the non-human world can be important to us – he lists 14 types, including life-support, biodiversity, and economic, recreational, scientific, medical, aesthetic, historical, and symbolic value. Such a long list could be used to try and show that *all* our valuing is instrumental in character – it is just that we are able to extract satisfaction from the world in a great many different ways. However, Rolston makes the important point that just because a value could be thought of in instrumental terms does not mean that it is how it actually functions – that would be to assume that all human valuing is self-interested (cf. Nietzsche's rational egoism). In fact altruism is one of the most marked – and controversial – aspects of human behaviour. Altruism in an ecological context means precisely the recognition that other entities may have value whether or not humans make use of them.

Rolston is not afraid to affirm the distinctive character of human life. He considers that the complex of characteristics we call human personality (containing, for example, self-awareness, hope, belief, altruistic relationship, and philosophical exploration) has no parallel in the non-human world and confers a richness of experience and moral value – though also of responsibility – found nowhere else in the biosphere. But he also affirms the (intrinsic) value of all life and of all systems of organisms. He calls his view '*bio-systemic* and *anthropo-apical*' – this last word meaning that he concedes that humans are at the apex of the range of moral standing.

Rolston's scheme allows him to embrace human ethical distinctiveness, and at the same time to locate value in the following.

(a) Sentient life (organisms which have nervous systems sufficiently advanced that they may be said to feel, in particular to feel pain). Sentient organisms have experience and can suffer. Rolston infers that humans therefore have a duty not to *increase* their suffering.

(b) Non-sentient life, which does not suffer, but which is still processing its environment, 'evaluating' it against the ideal for that species, with the (unconscious) goal of reproducing itself. Such organisms are 'evaluators', in however a crude a sense, and value therefore inheres in them.

(c) Non-living elements, insofar as they testify to the universe's capacity to generate order and beauty and to act as the matrix for the development of living systems. We are aware that we could vandalise or trivialise a non-living entity of beautiful or unusual character – a cliff-face, or a newly arisen volcanic island. Therefore we discern a value in that entity.

Importantly, Rolston also sees how intrinsic and instrumental value combine. Non-living entities have the least intrinsic value, but the greatest instrumental value to other entities. Humans must be viewed as having the greatest individual intrinsic value, and should not be used merely instrumentally, but at the same time they have a responsibility not to intensify suffering, or to degrade the biotic community, or to trivialise wild places. Where life forms are rare and might be irreversibly damaged by human action, humans have a particular duty to protect their value.

Rolston also finds a place for the value of systems, not as something over against the value of individual entities, but in addition to that value. So the system for him has value as the medium in which biotic values are held and exchanged, and where new kinds of value can be generated. The more diverse the network of exchange, the more valuable the system. This introduces into our original scheme of instrumental and intrinsic value an *additional* category of *systemic value*, which should find a place in any complete ethical analysis.[12] However, as Marietta (1994) discusses, ethical problems can arise where the system is over-emphasised at the expense of the interests of its components (including humans). Overstress on holism in environmental ethics lay behind the extremism we noted above in respect of some deep-ecological proposals (see 3.4). As Marietta comments, consideration of the value of systems as a whole should supplement, rather than supplant, concern for the system's components.

The ultimate invocation of the importance of biological systems is the so-called Gaia hypothesis, the development by James Lovelock of the notion that the whole surface of the Earth with its associated biosphere behaves in some sense as a single organism. This raises the question as to whether environmental ethics could be 'Gaian' – whether all proposed human actions could be referred to the good of a quasi-personal locus of value called Gaia. Although the Gaia hypothesis as such is scientific in nature, Lovelock's chosen name, derived from

[12] Thus Rolston, who like Passmore affirms the specialness of human beings, nevertheless must contradict Passmore's limiting of the understanding of community to communities of rational agents. An ecosystem *is* a community in that entities of value are in interdependent relation.

the Greek name for the goddess of the Earth, has naturally fed personalising and religious understandings of his work.

Gaian rhetoric can be quite helpful in reinforcing awareness of the immense age and complexity of the Earth's systems, and the possible vulnerability of their present 'settings' to human disruption such as global warming. The climatic turbulence associated with *El Niño* reinforces our sense of the complexity of these systems, and of humans' vulnerability to changes within them. But Gaia 'herself' is not a helpful 'mistress' of value, simply because her health and functioning is too cut off from the health and functioning of humans. As Lovelock tellingly put it some years ago (1988),

> Gaia, as I see her, is no doting mother tolerant of misdemeanors, nor is she some fragile and delicate damsel in danger from brutal mankind. She is stern and tough, always keeping the world warm and comfortable for those who obey the rules, but ruthless in her destruction of those who transgress. Her unconscious goal is a planet fit for life.

Global warming might lead to a resetting of the systems that keep the surface of the environment stable – such a resetting might be perfectly suitable for many organisms, and therefore for the 'health' of Gaia, but quite incompatible with the flourishing of all but a very few humans.

3.8   CONCLUSION

Environmental ethics is one of the most essential and underdeveloped aspects of humans' philosophical understanding of their situation. We have seen that central to our ability to understand our relation to the non-human world ethically is the analysis of value. This may be carried out in terms of the non-human world having instrumental or intrinsic value (Sections 3.2, 3.4), and/or also in terms of systemic value (Section 3.6). We saw in considering environmental economics that any analysis in environmental ethics also has a temporal dimension – decisions have to be made about the relative values of beings currently in existence and those which might exist in the future. An important term here is *sustainability*, a concept we consider more fully in Chapter 4. Ethical systems may be deontological or consequentialist (Section 3.3). Principles often invoked include the (consequentialist and instrumentalist) polluter-pays principle and the (deontological) precautionary principle (Section 3.5). Efforts to quantify environmental values, while necessary, and sometimes giving rise to alarming conclusions (Section 3.5), are fraught with all sorts of difficulties, which again relate back to calculations of value and its temporal dimension.

APPENDIX

Readers of this book who are teaching environmental ethics may find the following exercises useful.

1. The class should look out of the classroom window. What is in the field of view which is part of the non-human world?

Comment: The atmosphere is something that may be missed. Even if the view is entirely of buildings it is still unlikely that no plant can be seen, but if the view is very urban the class should extend the view in their imagination to the nearest open country they know of.[13] This will lead into our opening questions, on the theme: what value does the non-human world have?

2 (based on Figure 3.1). Students should be asked to work in pairs to negotiate in what position they would place themselves in Figure 3.1. The exercise can then be extended by inviting each pair to think of a particular context in which the environment is under threat, and to report on what approach to this problem their ethical approach would imply. Three important examples follow in Chapters 4 and 5.

REFERENCES AND RELATED READING

Those marked with an asterisk are particularly recommended for further reading.

Brooke, J.H. (1991) *Science and Religion: a Historical Perspective.* Cambridge University Press, Cambridge, UK.

Carson, R. (1962) *Silent Spring.* Houghton Mifflin, New York, USA.

Costanza, R., d'Arge, R., de Groot, R., Farber, S., Grasso, M., Hannon, B., Limburg, K., Nacem, S., O'Neill, R.V. Parvelo, J., Rashin, R.G., Sutton, P. and van der Belt, M. (1997). The value of the world's ecosystem services and natural capital. *Nature*, **387**, 253–260.

Cromartie, M. (ed.) (1995)* *Creation at Risk: Religion, Science and Environmentalism.* Eerdmans, Grand Rapids, MI, USA.

Daly, H.E. and Cobb, J.B. Jr. (1994)* *For the Common Good.* Beacon, Boston, MA, USA (first published 1989).

Deane-Drummond, C. (2001) *Biology and Theology Today.* SCM Press, London, UK.

Devall, B. and Sessions, G. (1985) *Deep Ecology.* Peregrine Smith, Salt Lake City, UT, USA.

Gottlieb, R.S. (ed.) (1996)* *This Sacred Earth: Religion, Nature, Environment.* Routledge, New York, USA.

Leopold, A. (1966)* *A Sand County Almanac.* Ballantine, New York, USA (first published by Oxford University Press, Oxford, UK, 1949).

Lovelock, J. (1988) *The Ages of Gaia.* Oxford University Press, Oxford, UK.

Marietta, D.E. Jr. (1994) *For People and the Planet: Holism and Humanism in Environmental Ethics.* Temple University Press, Philadelphia, PA, USA.

McDaniel, J. (1989) *Of God and Pelicans.* Westminster, Louisville, KY, USA.

---

[13] An extension to this exercise, which might be returned to later in the course, is provided by Holmes Rolston's questionnaire in his *Environmental Ethics*, (Rolston, 1988, pp. 347–349). The questions there provide a demanding test of individuals' 'bioregional awareness'.

McFague, S. (1993) *The Body of God: an Ecological Theology.* SCM Press, London, UK.

McFague, S. (2001) *Life Abundant.* Fortress Press, Minneapolis, MN, USA.

McMichael, A.J. (1995) *Planetary Overload: Global Environmental Change and the Health of the Human Species.* Cambridge University Press, Cambridge, UK.

Merchant, C. (1992)* *Radical Ecology.* Routledge, London, UK.

Morrison, R. (1999) *The Spirit in the Gene: Humanity's Proud Illusion and the Laws of Nature.* Cornell University Press, Ithaca, NY, USA.

Naess, A. (1995 (1972))* The shallow and the deep, long-range ecology movement. In *Deep Ecology for the Twenty-First Century.* Sessions, G. (ed), Shambhala, Boston, MA, USA.

Nash, J.A. (1991) *Loving Nature: Ecological Integrity and Christian Responsibility.* Abingdon, Nashville, TN, USA.

Northcott, M. (1996) *The Environment and Christian Ethics.* Cambridge University Press, Cambridge, UK.

Passmore, J. (1980)* *Man's Responsibility for Nature.* Duckworth, London (first published 1974).

Rolston, H. III (1988)* *Environmental Ethics: Duties to and Values in the Natural World.* Temple University Press, Philadelphia, PA, USA.

Rolston, H. III (1994)* *Conserving Natural Value.* Columbia University Press, New York, USA.

Ruether, R.R. (1992) *Gaia and God.* SCM Press, London, UK.

Sessions, G. (ed.) (1995)* *Deep Ecology for the Twenty-First Century.* Shambhala, Boston, MA, USA.

Southgate, C. (ed.) (1999) *God, Humanity and the Cosmos: a Textbook in Science and Religion.* T&T Clark, Edinburgh, UK and Trinity Press International, Harrisburg, PA, USA.

White, L. Jr. (1967) *Science,* **155,** 1203–1207 [reprinted in (2000) *The Care of Creation.* Berry, R.J. (ed), Inter-Varsity Press, Leicester, UK].

4 The Use of the Rainforest as a Test Case in Environmental Ethics

Christopher Southgate

4.1 INTRODUCTION

As will have been clear in the last chapter, ethical discussion derives its force, and receives its challenge, from the consideration of real examples. This chapter uses a specific example, namely the question of the proper future use of a piece of tropical rainforest, to illustrate the ethical frameworks introduced in the last chapter. A very complex political, sociological and ecological situation has had to be expressed in shorthand, even in caricature, in order to make the case-study work. We have found that the use of role-play can make ethical tensions much more immediate and more 'real' than most theoretical discussion, thus in Section 4.2 different views about use of a piece of rainforest are presented through the words of hypothetical participants in the debate.

4.1.1 SETTING THE CONTEXT – AN OVERVIEW OF TROPICAL MOIST FOREST DEGRADATION

In writing this summary, the author has drawn on those of Westoby (1989), Adams (1990) and McMichael (1993). Although details are in dispute, the following is generally conceded. Tropical moist forest contains a large proportion of the world's biodiversity (including many species not yet characterised, and many, no doubt, of considerable medicinal importance for humans). It continues to be destroyed at a considerable rate, perhaps as much as 7.5×10^6 hectares yr^{-1}. At such a rate, the forests of certain countries such as Ivory Coast may not survive long into the 21st century. Forest destruction not only reflects the activity of the logging trade, but also felling by subsistence farmers (often those displaced from their own lands), and more organised clearing to provide land for cattle. Once trees are cleared, the soil rapidly degrades. There is therefore a significant loss both of the CO_2-absorbing capacity of the trees, and of the bound carbon of the ecosystem. Forest degradation is therefore believed to contribute significantly to global warming. Furthermore, biodiversity and

Bioethics for Scientists. Edited by John Bryant, Linda Baggott la Velle and John Searle.
© 2002 by John Wiley & Sons Ltd.

nutritious soils are irreversibly lost in the process. Erosion sometimes leads to severe problems from mudslides, or flooding downstream of the felling. To say this is not to say that there might not be ways in which certain types of forest might be sustainably managed, but very little of this sort of management has been tried in practice, partly no doubt because the price of the timber has never been geared to the cost of replacing it.

4.2   WHAT THE PARTIES TO THE DEBATE MIGHT SAY

We can characterise the different positions in the debate about what should happen to the world's tropical moist forest in the form of an imaginary debate before a special session of the UN Conference on Environment and Development. Seven speeches follow, by representatives of the following:

1. the government of country X
2. a logging company
3. the indigenous forest-dwellers
4. the subsistence farmers of the area
5. cattle ranchers
6. a multinational chemical company
7. the UN Commission on Sustainable Development[1].

1. 'Good morning. I represent *the Government of country X*, and I'm here to explain why it is that we've decided to negotiate with the logging company to allow them new licences to fell tropical rainforest in our country. You also need to know why we've decided to ignore the environmental groups who maintain that the forest is a vital reserve of biodiversity, the precious last home of a rare big cat called the white jaguar, and also a globally important sink of atmospheric carbon dioxide.

What's important for you all to realise is that we're facing a bill for $3 billion by the end of this year. That's now the extent of our foreign debt repayment. Everything we or anyone else says about our options is against that background.

What about aid, I hear you say. OK, there is overseas aid, which last year came to $1.8 billion. We're expecting somewhat less this year as the G8 tighten their belts, but it's still a very good start. Though it's also fair to say that the aid

[1] The speeches may seem to treat in a prejudicial way the activities of certain nations and groups, in particular the behaviour of Japanese and American big capital. It happens that it is Japanese companies which are particularly involved in exploitation of tropical hardwoods in South-East Asia, and a major US company has done a deal in Central America similar to that suggested by the chemical company depicted here. The role-play casts doubt on the effectiveness of the declarations emerging from the United Nations Conference on Environment and Development, which met at Rio in 1992. This is not of course to imply that there was no value in the Rio process. The exercise may also be thought to caricature the attitudes of the different elements of 'country X'. However, the exercise has been tested in class over a number of years, and the authors are satisfied that the element of caricature is only that necessary to convey the ethical challenge; it certainly does not set out to give offence.

tends to involve us in hidden costs and dependencies – it's often arranged so that we then have to rely on foreign engineers or foreign equipment for maintenance.

Even with the aid we're left with a big bill to meet, and we dare not default again or we'll never get any inward investment.

Our rainforest is the one asset which can bring us a really big injection of hard currency. Even at today's timber prices, which we regard as artificially low, the logging is probably worth $20 billion over the next ten years. We have no choice but to realise as much as we can of that revenue, and fortunately the logging licences fall due this year. We have a Japanese-owned company offering us $2 billion for a 5-year licence, providing we don't set too many restrictions on them. We have little choice but to go ahead with that.

Unfortunately there have been a lot of political complications. You'll hear from the Abazi, the indigenous people of the forest – worthy people in their way, but not either taxpayers or voters, so not big factors in our calculations were it not for the publicity they get. If we could put a tax on the bleeding-heart liberal articles that are written about the Abazi we'd start to make some economic progress.

You'll hear from the subsistence farmers, the rural poor. They are a problem certainly. Some of them even vote. But the foreign debt position has to be our first concern. They'll probably tell you lies about the ranchers. Take them with a big pinch of the salt they put on their maize meal. The ranchers are friends of ours, good respectable people who pay tax, make political contributions of the right sort, earn us hard currency.

Sure, it's sad to see the forest go, and the beautiful white jaguar, sure we know about biodiversity and species extinction, sure there are sometimes erosion problems later, but our budgetary problems come now.

In conclusion I'd just remind you that it is our decision: Principle 2 of the Rio Declaration emphasises that "States have . . . the sovereign right to exploit their own resources. . .".'

2. 'In this debate I represent *the logging company*. You've just heard that finally the government is coming round to our way of thinking. Though I may say they were difficult in the first instance, going on about their responsibility for a carbon dioxide sink and about those Abazi. Just trying to beat up the price in our opinion.

All that stuff is all very well of course, but our responsibility is first and foremost to our shareholders. Good respectable people, and good solid Tokyo pension funds. People are living a long time in Japan these days and pensions have got to show profits. So we have got to export as much profit as we can to Japan. That means stressing our operating costs to the government of X and declaring as little profit as possible.

As for the Abazi, well, we always take the greatest trouble to compensate them. Even when they have no formal documents we have regarded them as the landowners and entered into profit-sharing arrangements. We routinely write off large sums towards replanting schemes we hope to bring in. We sponsor research into tropical soils not only in Osaka University but at MIT as well, and we have promised, at very considerable expense, a reserve habitat for two hundred white jaguar.

What really damages the environment is not our proper and legal activity but illegal and uncontrolled felling by the drifters who come into the forest along our roads, looking for land to farm. They make it impossible to do much in the way of replanting, because all they do is move in and rip the new trees up. And new trees come expensive. Why are there so many of these people in the forest? Ah well, you must ask the ranchers that.

People sometimes say to us – tropical hardwoods are a scarce resource. They should be much more expensive than they are. The governments who own them should form a cartel, like OPEC. To which our reaction would be that those people could no more form a cartel than they could build the Great Wall of China. But yes, of course it would suit us if the commodity were more expensive, provided of course that profit margins could be retained. No doubt it would be possible to fund much-needed research on how to make these very complex and often almost inaccessible parts of forest loggable on a proper sustainable basis. But we can't help with this. We don't dare raise prices, or cut production; if we do that the next company along the road will undercut us. It's a tough world out there, and we have our employees to think of. Good hardworking folk with families to support.

In conclusion we would like to quote Principle 12 of the Rio Declaration: "Trade policy measures for environmental purposes should not constitute . . . a disguised restriction on international trade".'

3. 'I speak for *the Abazi people* in this matter. We are grateful for a voice in so great a deliberation, but we would like to begin by expressing our reluctance to speak too quickly or simply. It is our custom for the men of the tribe to decide great matters concerning the tribe over a very long period, thinking and singing and smoking over it.

Our people have lived in the forest since time began. The forest is our food, our clothing, our wisdom. No-one could own the forest, any more than they could own the air they breathe.

So we have tried to understand the idea that someone might give us money to move away so that the forest could be cut down, or that someone else might give us money to stay and teach them how to harvest the medicines of the forest. Both ideas are foreign to us. If the forest were cut down we could not see or hear or hunt. If the medicines were taken away the birds and animals of the forest would die.

We have been accused of having a primitive religion, of being a museum piece. The people called anthropologists have studied us for years. They cannot see or hear the forest, but they stay and study, and our laws of hospitality require us to receive them. There was only trouble once, when one of us desired to take a woman anthropologist. Though we lost many women to those tree-slaying subsistence farmers, who cut and burn the trees and steal our places and pollute our river.

We have been asked to speak about our religion as a factor in the discussion, as though any human being could live without religion. We have been told we may not kill the jaguar, as every young man must do to gain manhood. We have deliberated all this in our council, and our wisdom is that we should make no law to please others. If the jaguar were not hunted by day, the boys would go out at night into greater danger to hunt him then, for only thus could they become men.

We wish to live in peace with all others in the forest. We want our women to be safe. A tree or two cut down at the edge of the forest is not murder, for trees fall occasionally of themselves. But if the river is made dirty, or loses its banks, that is a murder and a blasphemy, for the streams of the river are the life-blood of the forest. Our statement concludes now, but we would end by quoting Principle 22 of the Rio Declaration: "Indigenous people . . . have a vital role in environmental management and development . . . States should recognize and duly support their identity, culture and interests. . .".'

4. 'I speak for *the subsistence farmers* and their families. Our case is simple. We mean, we should say at the start, no harm to our brothers and sisters the Abazi. Some of our men have already married Abazi, so there is no prejudice between us. No, our situation is a simple one. We are here in the forest trying to survive. No more, but no less.

We lost our traditional lands and way of life to the ranchers. A few of us received payment; most of us were simply driven north by people they called cowhands, but who had never looked after cattle in their lives. They were mercenaries, pure and simple. So here we live, on the edge of the forest, up and down the loggers' roads, and we try and farm as we did before, though now on the bad land the trees leave. We farm the forest the way the ancient Maya did, moving on before the land is totally depleted, harbouring our stocks of maize seed as we move from place to place. We take down the young trees for fuelwood – the big ones take too much felling, and gasoline for chain-saws is always scarce.

What would make our lives better? The return of our land, but it is hard even to hope for that. Some sort of development on the edge of the forest which would give our young men work and which would pay for the things we can't make or grow. Hospitals for our sick, schools for the brightest of our children. Politicians we can trust.

We know there is likely to be a lot of work with the logging, and the roadbuilding that goes with it. But none of us has ever had a decently paid job from the loggers. Their foremen come from Japan; the men who drag trees to the trucks and the women who cook for them are mainly Filipinos. So our young men hang around on the edge of the logging camps, but hardly ever to any purpose.

Young men? Yes, our population is very young, and the rich countries will say there are too many of us. But the load on the planet is to do with population multiplied by consumption, and consumption in the West is a hundred times ours. Anyway, there is no choice for us but to go on conceiving children. We have no machines to clear the forest or to harvest maize. We have to do most of this by hand. At least we try and give our women a say in our affairs, which is more than the Abazi do theirs. And we kill the jaguar only if we absolutely have to, if one turns child-stealer.

We are a deserving case, urgently in need of national and international help. Principle 5 of the Rio Declaration confirms it: "All States and all people shall cooperate in the essential task of eradicating poverty as an indispensable requirement for sustainable development".'

5. 'Good morning. I have come here to speak for *the ranching community of country X*. Our case is clear enough, and can be stated briefly. We pursue our traditional way of life – always we have owned land and run our cattle over it. Recently there has been a greater demand for our beef for American hamburgers – they pay their own farmers too much, and fatten their animals on grain, and are surprised when they cannot afford their own meat.

So we have expanded, still living simply and hard as we did in the old days, riding round our land to oversee the stock and protect it from rustlers. And it is true that we have had to employ more security-conscious staff – a bigger operation always involves incidental expenses.

The other thing that has changed in the last 20 years is that our country's international debt has spiralled. No longer can we simply do what we like – we borrowed too much in an effort to develop our infrastructure, and must now pay back in hard currency. That means we must supply what the rich world wants. At the moment that is beef, and since our beef has no BSE like the Europeans and we have never had to use growth hormone as the Americans do there is every sign that our market-share will continue to grow. We are the people with the resources and the vision to meet this challenge. We pay our taxes, we pay our wages, the whole country benefits from our wealth.

It is possible that in the future there may be different demands. If the world oil price were more realistic then gasohol would become an important possibility on acreages such as ours. Again, we are the best people to see such a chance and to exploit it. It is to the benefit not just of our country but of the world that we manage the bulk of our country's land.

You will perhaps have heard from former tenants of this land – small-time operators, whinging, lazy and ill educated. Beware their lies. They were all paid the market rate for their land, and if they have squandered the proceeds it is hardly our fault, nor should we be put in the position of paying them again. Nor, incidentally, will we demean ourselves to quote the Rio Declaration – a bunch of pie in the sky which profited no-one but the hotel-keepers of Rio. Our statement ends.'

6. 'I am the representative of *the chemical company*. Ours is a difficult position and we want to try to spell it out kind of carefully, above all because our shareholders could well misunderstand.

We are offering the government of X the sum of $0.5 billion for a 30-year right to extract whatever plants, genetic material or indigenous knowledge we may be able to find within the tropical moist forest of their country.

Now that may seem a reckless, even a quixotic move, and it is true that we have been profoundly moved not only by consideration of the foreign debt situation of the government of X but also of the vital need for conservation of the forest and preservation of the culture of the indigenous people within it. Otherwise it is highly doubtful if we would ever pay such a vast sum for something that may in the end prove valueless.

It is true that such forests have proved very valuable sources of genetic material in the past and particularly of primitive, disease-resistant strains of cereal crops, and they have also given rise to lead compounds in pharmaceutical research. A classic example is the extraction of the antilymphoma agents vinblastine and vincristine from the rosy periwinkle of Madagascar. Incidentally, we deplore the fact that the Malagasy people received so little reward for these valuable drugs. Not only are we making this very large down payment to purchase the rights so firmly assigned to the Government under the UN Convention on Biological Diversity, but we also propose to make *ex gratia* payments to the tribespeople of the area in which we uncover valuable material. If drug manufacture ensues we undertake to carry out at least 10% of it in the country of origin, and to environmental standards higher than those insisted upon by the Government (if not necessarily as high as those that would pertain in the US).

We anticipate that our use of the forest will be low impact. Almost always the material we extract is then prepared synthetically, so there is no need for bulk recovery methods (though we cannot guarantee this, given the recent difficulty in producing a synthesis of Taxol, the anticancer drug from the Pacific yew). There may in some cases have to be harvesting of organic material in the medium term. And in view of the size of our initial outlay we have had to stipulate not only no logging in the relevant areas but also that we hold the rights to all mineral extraction over the first 15 years. There is a small possibility that commercial amounts of gold may be extractable from the sands of the riverbed.

In conclusion we want to restate that this is a speculative investment that may well prove to be a straight transfer of finance and technology to the developing world. As such it is firmly in the spirit of the whole Rio Declaration, and we urge the Government to accept it.'

7. 'I have been asked to brief the meeting on behalf of *the UN Commission on Sustainable Development*. What you have heard from the other parties illustrates the problems of our work in graphic detail. From the Rio documents we are clear about the following.

(i) The right of the Government to sell and profit from natural and biological resources within their sovereign territory.

(ii) The responsibility of the Government to work for sustainable development of these resources.

(iii) The responsibility of foreign companies to consult with and co-operate with the people of the country in working for a pattern of development that will be sustainable, and to take due account of the precautionary principle (Rio Declaration Principle 15) and the polluter-pays principle (Principle 16).

Specifically under the terms of the UN Framework on Climate Change we note the responsibility on the government of X to prepare national inventories of sources and sinks of greenhouse gases, and to move towards sustainable management. We note that developed countries should bear the full incremental costs of preparing these inventories and policies. The rainforest is a sink of global importance, but the developing countries may reasonably submit that the problem arises out of the over-production of carbon dioxide by the industrialised world. Unless offered full compensation the Government cannot be blamed for deriving revenue from its single major commodity.

Under the terms of the UN Convention on Biological Diversity we call for fair sharing of biological resources and the technology with which to exploit them. We call for urgent assessment of the environmental impact of the proposed logging (and possible mineral extraction), in particular on species that are known to be rare such as the white jaguar. There is a responsibility on all parties to conserve such rare fauna both *in situ* and in captivity.

We note in passing that the chemical company has its head office in the USA, a country that declined to sign the Convention on Biological Diversity.

In the spirit of Agenda 21 we call for the widest possible consultation on these matters at local as well as national and international level. It will be important in particular to hear from women's groups, from young people, and from indigenous peoples, also from concerned non-governmental organisations.

Lastly under the terms of the Statement on Forest Principles at Rio we affirm that forests should be sustainably managed to meet social, economic, ecological, cultural and spiritual needs. The full costs of managing forests should be borne by the international community. We note however that this document is in no way legally binding.'

4.3   ETHICAL ANALYSIS

These speeches raise in particular the following questions.

(i)   • What ethical systems were the different parties using?
      • Where would their positions sit on the grid depicted in Figure 3.1?
(ii) Suppose there were a 'Global Ecological Authority' that resolved these disputes. A plausible brief for such an authority would be that it had full international powers to preserve:

- firstly, human survival on the planet
- secondly (where possible), existing human culture
- thirdly (where possible), other species in descending order of importance according to their complexity and rarity.

What would such a body implement in the situation described? What would be the reactions of the different parties to the role-play to such an imposed solution?

(iii) Towards what solution do the Rio documents point? [See Appendix B and Grubb (1993)]

(iv) How should the situation illustrated in the case-study be viewed in terms of an appropriate combination of the categories of value – instrumental, intrinsic and systemic – listed in Chapter 3?

4.4   FUTURE POSSIBILITIES FOR THE RAINFOREST

Ecosystem destruction, and associated human poverty and exploitation, continue to be widespread. This is well illustrated at the time of writing this chapter (summer 2001) by the situation in Brazil. The Brazilian Congress is debating a proposal to raise the proportion of Amazon rainforest that may be felled for farmland from 20% to over 50%. An area twice the size of Portugal may therefore be lost to the forest. The timber will be sold, much of it not as solid wood but as wood chips, and the cleared land will be used for crop growth and pasture (there are clear echoes here of the debate set out above).

Nevertheless, there are certain signs of hope in limited areas. Park (1992) lists the following:

- research, training and education
- land reform
- ecosystem conservation in designated areas
- reforestation
- marketing non-tree products of the forests
- controls on the timber trade
- consumer pressure in the West
- debt-for-nature swaps[2]
- changes in aid, debt relief and investment policies by the First World.

It is also possible that the recent Kyoto protocols on greenhouse gas emissions trading may help the situation. Under this scheme, countries that wish to release more net carbon dioxide than had been agreed internationally would have to buy emissions 'permits' from countries with low net levels of release. Since rainforest-containing countries include large areas that absorb carbon dioxide, and are often low in industrialization, this might direct funds to their govern-

---

[2] These involve a country guaranteeing to preserve a section of its rainforest in return for some relief from its debt.

ments in a way that would provide vigorous incentives to keep forest intact. Though this way forward is subject to some of the criticisms levelled in Section 3.5 against the polluter-pays principle, and is rejected by the current US adminis- tration, it may nevertheless lead to governments at least beginning to confront the implications of the energy-profligacy of the Western lifestyle.

4.5   THE CONCEPT OF SUSTAINABILITY

Perhaps the most important emphasis to emerge from the Rio Declarations was that the concept of sustainability, which had been stressed by the Brundtland Report (World Commission on Environment and Development, 1987), should be central to future environmental policy. The concept implies that no satisfac- tion of present human needs should be at the expense of the needs of future generations. This seemingly unexceptionable concept has however given rise to great debate. As Mary Midgley has pointed out, sustainability can focus too narrowly on human interests (Midgley, 1997). Moreover the word 'sustainable' is all too often coupled with the word 'development'. UN documents insist on the prospect of development, since so many humans live with such limited access to good food, fresh water, healthcare, education and employment. Sustainable development therefore tends to be a coded expression for the hope that con- tinued global economic growth will provide these things.

However, mature ecosystems are not characterised by growth, and economic growth has in the past always been at the expense of the future potential of the environment, as non-renewable sources have been consumed, and habitats destroyed. We referred in Section 3.5 to the difficulty of designing an economics that does justice to environmental 'capital'. Ehrlich and Ehrlich quote Robert Gray to the effect that 'the costs of sustainability must be calculated in terms of the investment required to restore society's natural capital – to undo the ecologi- cal damage created in the course of doing business' (Ehrlich and Ehrlich, 1996).

Daly and Cobb not only insist on the distinction between sustainable growth – ultimately a contradiction in terms – and sustainable development, that is, improvements in infrastructure which are not at the net expense of the environ- ment.[3] They also make an important distinction between 'weak sustainability', where decline in natural capital can be compensated by a rise in humanly created capital, and 'strong sustainability', where the two sorts of capital must be calculated separately[4].

Where a given project must unsustainably use non-renewable resources, advo- cates of strong sustainability urge that it must be paired with a project that rebuilds natural capital. Nations committed to wholesale unsustainable practice,

---

[3] Daly and Cobb (1994) pp. 69 – 76.
[4] A recent restatement of the need for strong sustainability is that of the British environmentalist Jonathon Porritt (Porritt, 2000).

such as the USA, would therefore have to pair projects with nations where natural capital is better preserved and can be meaningfully increased (e.g. by planting and conservation schemes around the rainforests) – hence the 'hope' we see in global emissions trading, which would be a sign of this pairing beginning to take place.

However, the political reality is that any country or non-governmental organisation can insist that conditions and social practices in a particular context be sustainable, but it is the rich and powerful countries that are able to protect their way of life in global negotiation.

4.6   CONCLUSION

In looking at a real situation of progressive environmental destruction, the tropical rainforest, we have seen how many competing human interests there are, and how difficult it is to satisfy human needs and demands while according status to the non-human world. There are certain limited signs of possible hope, mainly around schemes of exchange between poor countries and the very rich. Two examples are debt-for-nature swaps and carbon dioxide emissions trading. Another important principle is that of sustainability, which may be understood in either weak or strong senses.

In the next chapter we look at two further cases where different interest groups advocate radically different ways forward in respect of the competing interests of humans and the non-human world.

REFERENCES AND RELEVANT READING

Those works marked with an asterisk are particularly recommended for further reading.

Adams, W.M. (1990) *Green Development.* Routledge, London, UK.

Attfield, R. (1999)* *The Ethics of the Global Environment.* Edinburgh University Press, Edinburgh, UK (see especially chapter 6).

Daly, H.E. and Cobb, J.B., Jr. (1994)* *For the Common Good.* Beacon Press, Boston, MA, USA (first published 1989).

Ehrlich, P.R. and Ehrlich, A.H. (1996) *Betrayal of Science and Reason.* Island, Washington, DC, USA.

Grubb, M. (1993)* *The Earth Summit Agreements.* Earthscan, London, UK.

McMichael, A.J. (1993)* *Planetary Overload.* Cambridge University Press, Cambridge, UK (see especially chapter 9).

Midgley, M. (1997) Sustainability and moral pluralism. In *The Philosophy of the Environment.* Chappell, T.J.D. (ed), Edinburgh University Press, Edinburgh, UK, pp. 89–101.

Park, C. (1992) *Tropical Rainforests.* Routledge, London, UK (see especially chapter 6).

Porritt, J. (2000) *Playing Safe: Science and the Environment.* Thames and Hudson, London, UK (see especially chapter 7).

Shiva, V. (1998)* *Biopiracy.* Green, Dartington, UK.

Westoby, J. (1989) *Introduction to World Forestry: People and Their Trees.* Blackwell, Oxford, UK.

World Commission on Environment and Development (1987) *Our Common Future* Oxford University Press, Oxford, UK ('The Brundtland Report').

APPENDIX A. COMMENT ON THE ROLE-PLAY (FOR LECTURERS'/TEACHERS' USE)

These comments illustrate how the authors have used the role-play in teaching. If the role-play is to be used, the instructor needs to decide whether it is most effective to use students or members of the teaching team as role-players. Those not taking any of the seven main roles may be assigned additional roles as members of the press, or of concerned non-governmental organisations of their choice. The other students form the representatives of the 178 countries represented at the Rio Summit.

The timings of this exercise are seven 3-minute presentations, plus changeover time, plus 20 minutes' discussion, plus 5 minutes' lecturer's/teacher's summing-up, total 50 minutes.

The role-players should be encouraged to memorise the central details of their brief, and then improvise their actual address, rather than reading straight from the script. As always with role-play it will be important at the end to remind participants that they do not carry the roles away with them, but finish the class as themselves.

Once the role-play is over and the participants have been debriefed, a discussion may then ensue, which might be focused around the questions raised in Section 4.3. Sometimes students become discouraged when they discover the continuing and irreversible damage to tropical moist forest, and experience the intractability of the situation. That is why it is helpful to begin by considering what solution might be imposed by a benevolent outsider, and then to work from there back towards the conflicts of interest that bedevil the real situation.

The exercise may be completed by requiring students to provide written summaries, reflecting on the most ethically desirable ways forward, and the criteria by which they make their choice.

APPENDIX B. THE RIO DECLARATION

THE RIO DECLARATION ON ENVIRONMENT AND DEVELOPMENT (1992)

Preamble The United Nations Conference on Environment and Development, Having met at Rio de Janeiro from 3 to 14 June 1992, Reaffirming the Declaration of the United Nations Conference on the Human Environment, adopted at Stockholm on 16 June 1972, and seeking to build upon it, With the goal of establishing a new and equitable global partnership through the creation of new levels of co-operation among States, key sectors of societies and people, Working towards international agreements which respect the interests of all and protect the integrity of the global environmental and developmental system, Recognizing the integral and interdependent nature of the Earth, our home,

Proclaims that:

Principle 1

Human beings are at the centre of concerns for sustainable development. They are entitled to a healthy and productive life in harmony with nature.

Principle 2

States have, in accordance with the Charter of the United Nations and the principles of international law, the sovereign right to exploit their own resources pursuant to their own environmental and developmental policies, and the responsibility to ensure that activities within their jurisdiction or control do not cause damage to the environment of other States or of areas beyond the limits of national jurisdiction.

Principle 3

The right to development must be fulfilled so as to equitably meet developmental and environmental needs of present and future generations.

Principle 4

In order to achieve sustainable development, environmental protection shall constitute an integral part of the development process and cannot be considered in isolation from it.

Principle 5

All States and all people shall co-operate in the essential task of eradicating poverty as an indispensable requirement for sustainable development, in order to decrease the disparities in standards of living and better meet the needs of the majority of the people of the world.

Principle 6

The special situation and needs of developing countries, particularly the least developed and those most environmentally vulnerable, shall be given special priority. International actions in the field of environment and development should also address the interests and needs of all countries.

Principle 7

States shall co-operate in a spirit of global partnership to conserve, protect and restore the health and integrity of the Earth's ecosystem. In view of the different contributions to global environmental degradation, States have common but differentiated responsibilities. The developed countries acknowledge the responsibility that they bear in the international pursuit of sustainable development in view of the pressures their societies place on the global environment and of the technologies and financial resources they command.

Principle 8

To achieve sustainable development and a higher quality of life for all people, States should reduce and eliminate unsustainable patterns of production and consumption and promote appropriate demographic policies.

Principle 9

States should co-operate to strengthen endogenous capacity-building for sustainable development by improving scientific understanding through exchanges of scientific and technological knowledge, and by enhancing the development, adaptation, diffusion and transfer of technologies, including new and innovative technologies.

Principle 10

Environmental issues are best handled with the participation of all concerned citizens, at the relevant level. At the national level, each individual shall have appropriate access to information concerning the environment that is held by public authorities, including information on hazardous materials and activities in their communities, and the opportunity to participate in decision-making processes. States shall facilitate and encourage public awareness and participation by making information widely available. Effective access to judicial and administrative proceedings, including redress and remedy, shall be provided.

Principle 11

States shall enact effective environmental legislation. Environmental standards, management objectives and priorities should reflect the environmental and developmental context to which they apply. Standards applied by some countries may be inappropriate and of unwarranted economic and social cost to other countries, in particular developing countries.

Principle 12

States should co-operate to promote a supportive and open international economic system that would lead to economic growth and sustainable development in all countries, to better address the problems of environmental degradation. Trade policy measures for environmental purposes should not constitute a means of arbitrary or unjustifiable discrimination or a disguised restriction on international trade. Unilateral actions to deal with environmental challenges outside the jurisdiction of the importing country should be avoided. Environmental measures addressing transboundary or global environmental problems should, as far as possible, be based on an international consensus.

Principle 13

States shall develop national law regarding liability and compensation for the victims of pollution and other environmental damage. States shall also co-operate in an expeditious and more determined manner to develop further international law regarding liability and compensation for adverse effects of environmental damage caused by activities within their jurisdiction or control to areas beyond their jurisdiction.

Principle 14

States should effectively co-operate to discourage or prevent the relocation and transfer to other States of any activities and substances that cause severe environmental degradation or are found to be harmful to human health.

Principle 15

In order to protect the environment, the precautionary approach shall be widely applied by States according to their capabilities. Where there are threats of serious or irreversible damage, lack of full scientific certainty shall not be used as a reason for postponing cost-effective measures to prevent environmental degradation.

Principle 16

National authorities should endeavour to promote the internalization of environmental costs and the use of economic instruments, taking into account the approach that the polluter should, in principle, bear the cost of pollution, with due regard to the public interest and without distorting international trade and investment.

Principle 17

Environmental impact assessment, as a national instrument, shall be undertaken for proposed activities that are likely to have a significant adverse impact on the environment and are subject to a decision of a competent national authority.

Principle 18

States shall immediately notify other States of any natural disasters or other emergencies that are likely to produce sudden harmful effects on the environment of those States. Every effort shall be made by the international community to help States so afflicted.

Principle 19

States shall provide prior and timely notification and relevant information to potentially affected States on activities that may have a significant adverse transboundary environmental effect and shall consult with those States at an early stage and in good faith.

Principle 20

Women have a vital role in environmental management and development. Their full participation is therefore essential to achieve sustainable development.

Principle 21

The creativity, ideals and courage of the youth of the world should be mobilized to forge a global partnership in order to achieve sustainable development and ensure a better future for all.

Principle 22

Indigenous people and their communities, and other local communities, have a vital role in environmental management and development because of their knowledge and traditional practices. States should recognize and duly support their identity, culture and interests and enable their effective participation in the achievement of sustainable development.

Principle 23

The environment and natural resources of people under oppression, domination and occupation shall be protected.

Principle 24

Warfare is inherently destructive of sustainable development. States shall therefore respect international law providing protection for the environment in times of armed conflict and co-operate in its further development, as necessary.

Principle 25

Peace, development and environmental protection are interdependent and indivisible.

Principle 26

States shall resolve all their environmental disputes peacefully and by appropriate means in accordance with the Charter of the United Nations.

Principle 27

States and people shall co-operate in good faith and in a spirit of partnership in the fulfilment of the principles embodied in this Declaration and in the further development of international law in the field of sustainable development.

5 Environmental Ethics: Further Case-Studies

Christopher Southgate and Alex Aylward

5.1 INTRODUCTION

This chapter continues the application of the frameworks of environmental ethics to problems with both local and global implications. The two case-studies summarised in this chapter both concern national policy, but both are examples of how the stance of one nation can have effects in countries thousands of kilometres away. The first concerns the use of organochlorine pesticides, particularly DDT (Section 5.2); in the second (Section 5.3) we go on to investigate the appropriateness of electricity generation from nuclear power. Some readers of this book will consider that neither DDT nor nuclear power could possibly be used in any circumstances. Whatever conclusion you finally come to, we hope to prise you away from such an absolute position.

First, however, we earth our explorations in an everyday example. Imagine a dairy farmer working land at the head of a river catchment. In his rural community the nearest neighbour lives half a mile (800 metres) away. His grown-up children live in a city 60 miles (100 km) away. He has farmed all his life and needs continually to improve productivity to pay for bank loans, feed, equipment, grass seed, pesticides and pollution prevention measures. He knows his land and believes in the idea of sustainable farming.

It is difficult to find farm workers and he only has the help of one herdsman. His wife has cancer and is in and out of hospital. He suffers from a back injury. This year there has been an outbreak of disease in his dairy herd and the weather has been so bad that the cows have had to stay under cover for weeks longer than usual. It seems that every month has been a wet month.

He cares about the environment and has a slurry store, which collects all the animal waste from the cow sheds, yards and milking parlour. These are stored, and then spread onto his fields at appropriate times during the year. The pressures of work and family commitments over the last several years have meant he has not maintained the store. Overnight it undergoes a catastrophic failure and about 100 000 litres of slurry enter a tributary and the main river,

Bioethics for Scientists. Edited by John Bryant, Linda Baggott la Velle and John Searle.
© 2002 by John Wiley & Sons Ltd.

poisoning the water and killing fish. In the morning he milks the cows, unaware of the disaster. Downstream other farmers abstract water for their animals and crops, cattle drink from the river and water is abstracted for a fish farm. The river also provides water for a nearby paper mill and a sizeable town. It is not until midday that our farmer notices the problem. While he thinks about what he should be doing a Government pollution officer turns up at his farm and informs him of the fish kill, the shutting down of the paper mill, the possible contamination of the town's water supply and likely prosecution.

The farmer is in a state of shock. He feels he is a responsible person providing food for the community but now he has caused a devastating incident affecting hundreds of others, threatening their health and businesses.

What kinds of judgement have you, the reader, made of him? How does he value the environment, family, his stock and other people? What does this tell you about his character, personal qualities and beliefs? Was he morally at fault? Should he have spent more time maintaining/monitoring his slurry store than on the welfare of his family and animals?

True, most of us are not in a position to influence a whole river-system so dramatically, but each individual has an 'ecological foot-print'; whatever choices of lifestyle one makes, one has a greater or lesser impact on the environment. And each reader of this book belongs to a nation with policies that, again, have a certain impact on ecosystems both local and global.

We now turn to the consideration of how nations should construct those policies. It is first necessary to be aware that nations tend to prioritise environmental concerns differently depending on their political, economic, cultural and religious character. The principal aim of 'developing' countries (as viewed from the perspective of modern Western economics) tends to be to reduce poverty and to have secure supplies of food. When such a country is undergoing industrialisation problems occur, such as uncontrolled urbanisation and infrastructure development, increased demands for energy and transport, increases in use of raw materials including chemicals and increased waste production. It is the economically most developed nations that tend to be most concerned about the long-term stability of the health and wealth of their populations. An example of the influence of cultural and religious factors is the differing attitude to animals in different countries. In the USA and UK animals such as cattle are handled, in effect, very much as a commodity (as was very apparent in the UK in 1967 and again in 2001 when millions of healthy animals were slaughtered in attempts to limit outbreaks of foot and mouth disease). In Hindu societies, by contrast, the cow (itself a major producer of the greenhouse gas, methane[1]) is sacred and cannot be harmed.

The world's governments, politicians and pressure groups acknowledge that

[1] An individual cow can produce up to 280 litres of methane per day. This comes from the metabolism of methanogenic bacteria that inhabit the rumen and whose activity is essential for the digestion of grass. In Australia it is estimated that sheep and cattle produce 14% of that country's greenhouse gas emissions.

these differences between nations are difficult to manage at an international level. An important attempt to establish guiding principles was made at the Rio Earth Summit in 1992 with the Rio Declaration on Environment and Development (see Chapter 4 and Grubb, 1993).

5.2 CASE-STUDY I: DDT AND CONTROL OF MALARIA

5.2.1 A GLOBAL PROBLEM

Malaria has been killing humans for many thousands of years. Its cause is a protozoan parasite in the genus *Plasmodium*. There are several species of this parasite but the most widespread and dangerous, causing cerebral malaria, is *P. falciparum*. The parasite's lifecycle requires two major hosts, an animal (for our purposes the human), and the female anopheline mosquito. Once the host has been bitten by a malaria-carrying mosquito the parasite makes its way to the liver, multiplies, and 9–16 days afterwards enters the blood stream to attack and break down the red blood cells. This induces bouts of fever. Further feeding of mosquitoes on infected humans enables the parasite to complete its life-cycle.

Estimates of the impact of the disease are difficult to establish but it is believed that up to 500 million people are affected by the disease. There are over one million deaths a year, the majority being children under five years old, but other high risk groups include pregnant women, and those lacking a fully functioning immune system. There are currently more than 90 countries in which malaria is endemic and 90% of recorded cases occur in Africa.

Current counter-measures include

(i) anti-parasitic drugs to treat those affected and
(ii) vector control, attacking the carrying mosquitoes using organochlorine insecticides, such as DDT (dichlorodiphenyltrichloroethane).

Although chemical control of mosquitoes continues to be quite effective, the parasites and mosquitoes are developing resistance to these pesticides. This is partly the reason why, after initial high hopes in the 1950s and 1960s, when inroads were made into eradicating the disease, the World Health Organisation has abandoned its total elimination strategy. Malaria is now on the march again, claiming more lives and spreading to more countries (possibly with help from global warming).

5.2.2 A SECOND GLOBAL PROBLEM, IN TENSION WITH THE FIRST

DDT and its relatives were much celebrated as pesticides when this property was first discovered (indeed the discovery earned Müller the 1948 Nobel Prize for Chemistry). However, it is now understood that compounds with these chemical characteristics, besides being effective at their original uses, can also have a disastrous effect on other systems. Not only does DDT wipe out many other insects, including those which might have beneficial effects in controlling pests, but this type of insecticide is concentrated by the food chain and accumulates in higher predators. In the 1950s, when these effects were first noticed, the impact on the breeding success of birds of prey such as the Peregrine falcon was particularly marked.

The consequences of using these compounds were brought to the public attention by the founders of the modern environmental movement such as Rachel Carson in her book *Silent Spring* (1962). DDT is now recognised as belonging to a category of chemicals which are persistent organic pollutants (POPs) – others include aldrin, dieldrin, polychlorinatedbiphenyls (PCBs), and dioxins. Projects such as Carson's led to the first laws banning the general use of DDT in USA in 1972. Other Western countries followed the American example, banning use of DDT in the 1970s–80s.

Persistent organic pollutants have common properties as follows.

• They are persistent: they resist degradation through physical, chemical or biological processes (when a sample of DDT is released into an ecosystem, typically over 50% of it is still present two years later).
• The compounds are generally semi-volatile and can travel long distances on air currents, returning to earth thousands of kilometres from the original source.
• They generally have low water solubility and high lipid solubility, which leads to bio-accumulation in fatty tissues of living organisms and bio-magnification of their concentrations by thousands or even millions in food chains.

These compounds are disrupting ecosystems in all parts of the world, with observed consequences in populations of many species of reproductive failure, population decline, abnormally functioning hormone systems, androgynous effects, behavioural abnormalities, tumours and gross birth defects.

At low concentrations, such as are found in the environment, POPs are also having similar effects on human health, having been identified as carcinogenic, mutagenic and teratogenic (causing birth defects), as endocrine system disruptors and as causing immune system changes.

5.2.3 A NATIONAL PROBLEM – COUNTRIES WITH ENDEMIC MALARIA

The countries in which malaria is endemic have a tropical environment and are poor. Poverty is always a contributory factor to disease where health services are insufficient or non-existent, and the cost of treatment is too expensive. Their economies are generally dependent on a few commodities – perhaps even on only one cash crop. Populations may also migrate for work in and out of areas where malaria is endemic. These countries are in debt and require international aid.

The options to combat malaria in poor countries are primarily limited to what those countries can afford, and the anti-parasitic drugs and insecticides used need to be cheap. If you want an inexpensive insecticide then DDT is at the top of your list. DDT has proven effectiveness; it is cheap, easy to apply in homes and to clothes and to spread on mosquito habitats, as well as having a multi-purpose role as a general pesticide for crop protection. It is persistent and requires few applications.

However, mosquito populations are an integral part of tropical ecosystems and they have their place in the food chain. Also, DDT is not a species-specific insecticide and affects a range of organisms from insects to mammals. Moreover, organochlorine pesticides are becoming less effective due to over-use and the increased resistance of the parasites.

The potential of these countries will never be realised whilst the vast majority of the population is undernourished and prone to diseases such as malaria. For a country's economy to be viable it must have a sufficiently healthy population to meet its needs. In addition, if the expectations of the population are not achieved in meeting these needs then civil unrest may occur.

In the last few years the World Health Organisation (WHO) and the United Nations Environmental Programme (UNEP) have provided the negotiating forum for an international treaty to ban the production and use of POPs. Most of the 12 most notorious chemicals, 'the dirty dozen', will be subject to immediate ban. The exception is DDT, on the grounds of its usefulness to countries needing to control malarial mosquitoes. They will be permitted to use DDT until a cost-effective and environmentally friendly alternative is available. The treaty, known as the Stockholm Convention, was formally adopted in Stockholm in May 2001 and individual governments must now ratify. The treaty comes into force when 50 nations have ratified it.

5.2.4. CASE-STUDY QUESTION: SHOULD DDT BE BANNED FROM USE AS AN EFFECTIVE INSECTICIDE AGAINST THE MALARIA-CARRYING MOSQUITOES?

The ethical question is best focused by putting oneself in the position of an environmental consultant advising a government of a poor country in a malarial region whether or not it should accede to international pressure to extend the treaty described above to include the banning of DDT.

To formulate a position you will need to consider:

(i) what value you put on the lives of people in this and other nations,
(ii) what values you apply to ecosystems and the health of the biosphere,
(iii) whether you should suggest what is right for the global community and environment or what may be right for the needs of the country in question and

(iv) where you would place your approach on the grid in Figure 3.1 (see Chapter 3).

See Appendix A for discussion of how this exercise might be used and extended in a teaching situation.

5.3 CASE-STUDY II: ANOTHER GLOBAL PROBLEM ON WHICH NATIONS HAVE TO MAKE DECISIONS – THE ADVISABILITY (OR NOT) OF GENERATING ELECTRICITY FROM NUCLEAR POWER

5.3.1 BACKGROUND

The key balance in the previous exercise was that between the interests of humans and of non-human organisms, particularly those at the top of the food chain such as birds of prey. The question as to whether nuclear power is an appropriate energy-generating technology for the 21st century concerns a different balance, between short-term risks and benefits to some humans and their homes and long-term risks and benefits to others.

The problem can be simply stated. Electricity consumption continues to grow and is still principally generated by the combustion of fossil fuels – coal, oil and gas. All fossil fuels, however, give rise to the greenhouse gas carbon dioxide, believed to be the main reason why the mean surface temperature of the planet is rising and the climate therefore changing. These fuels are non-renewable, finite resources, and their use has a big impact on the environment through pollution and global warming.

Alternative sources of energy, from 'renewable' sources, such as wind, tides, solar power and hydroelectricity, cannot in most countries contribute more than a relatively small fraction of the total electricity that is expected to be required. The one means available which is known to generate a large amount of energy without emitting greenhouse gases is nuclear power. Indeed France has committed itself to a very extensive nuclear programme, which by 2000 delivered 75% of the nation's electricity.

There are however three major objections to the proliferation of nuclear plants. The first is the question of *safety* – accidents have been few, but the few have been fairly severe, especially at Three Mile Island in Pennsylvania in 1979, and at Cernobyl in the Ukraine in 1986. The International Nuclear Event Scale, which ranks accidents according to their severity on a scale of 1–7, ranked Cernobyl as the only '7' yet to occur, but as recently as September 1999 there was an accident in a fuel manufacturing plant at Tokai-Mura in Japan, which was ranked '4' on the scale, and involved a chain-reaction which raised radiation levels within the plant to 15000 times their normal levels. The few incidents that have occurred fall within a remarkably positive safety record for nuclear power compared with most other heavy industries, but nevertheless raise grave doubts as to the wisdom of increased commitment to this mode of generating electricity.

The second objection is the long-term problem *of the costs and risks of decommissioning* – making safe worn-out plants, and storing the waste they produce. These issues are yet to be properly resolved, even 50 years after the introduction of the technology. What is certain is that the sites of power-stations, and their waste-products, will remain appreciably radioactive for thousands of years after electricity ceases to be generated there.

A third objection is that there is a risk, hard to quantify or evaluate, that further development of nuclear power could place in the wrong hands the raw materials for *nuclear terrorism* – either for an actual delivered nuclear weapon or for the scattering of highly toxic waste.

These objections, and the power of environmental protest, have persuaded many countries to phase out their nuclear programme over the next two to three decades, but an editorial in the British newspaper *The Financial Times* commented recently that

> Power consumers are in danger of being robbed of a vital alternative to fossil fuels. Properly built and well-run nuclear power plants are a cheap and safe source of energy. In the west, only a handful of people have died in nuclear accidents, in contrast to the many who are killed each year mining fossil fuels.
>
> The 1986 Chernobyl nuclear disaster quite rightly exposed the industry to intense scrutiny. But it passed the test with flying colours. Abandoning nuclear power because of Chernobyl makes no more sense than abandoning coal mining because of the scores of Russian and Ukrainian miners who die each year.[2]

5.3.2 THE ETHICAL TENSION

Again the first task for an ethicist is to identify the competing principles at work. The precautionary principle (see Chapter 3, Section 3.5) could be applied in either direction – either to assert that nuclear power is necessary because it would be unwise to risk the consequences of appreciable climate change, which would carry with it the prospect of the submersion of whole countries such as Bangladesh – or to assert that nuclear power should not be risked because we do not know how to make either its operation or its waste sufficiently safe to avoid major contamination of the planet.

As in the previous case-study, there is an issue of national *versus* international interests, since the effects of a severe nuclear accident would almost inevitably involve more than one country. Climate change might actually benefit some northerly countries, improving the climate and making food easier to grow, but at the cost of catastrophic effects in low-lying lands (and possibly also the spread of tropical diseases such as malaria).

When Dave Toke of Friends of the Earth writes 'We can only solve this environmental dilemma by doing a comparative environmental impact assessment.' he has immediately to add 'In practice such costs are very difficult to work

[2] 16 June, 2000, p. 14. However, see Appendix A for a comment on sources. Very few are totally neutral on these issues.

out.'[3] All the problems we noted in Section 3.5 of Chapter 3 about relating damage to the non-human world to human gains and losses, and about comparing the needs of the present with those of the future, are complicated here by the need to deal with risk – perhaps a nuclear accident will not happen. What level of risk can be factored into an equation in a consequentialist way, and what level is *unacceptable*, and implies that nations have a *duty* not to incur it?

A wider consideration of the issues allows other factors to be considered. *Must* the world use as much electricity as it plans to? *Are* the costs of alternative energy as much greater than those of fossil fuels as is usually calculated? What courses of action are *politically* possible in a democracy?

It is enough here to indicate the range of issues involved, and to invite the reader to research further. Again, we must note that in a polarised debate there are few sources that do not seek to promote one view more than another (see Appendix A). Three web sites which indicate the range of approaches are those of Greenpeace International, The International Atomic Energy Authority and British Nuclear Fuels Ltd (BNFL)[4].

5.4   A COMMON FACTOR

Note also one other factor common to both the case-studies in this chapter. Just as the insecticidal properties of DDT were much celebrated when first discovered, so too the early days of the nuclear programme saw the industry driven forward by 'brilliant scientists who cared deeply about humankind and were intensely proud of the 'incalculable benefits' that their new technology would bring'[5]. In both cases, environmental problems have given rise to a major backlash, at least in certain countries. The environmental ethicist faces the task of grasping what the current scientific data are, since they are likely to constitute the deepest and most honest insight into what the impact of human activity is, but also anyone giving an ethical opinion on an environmental issue must recognise that those same scientific views are themselves provisional and may well shift within the lifetime of the courses of action proposed.

5.5   CONCLUSION

We have shown that environmental ethics is a subject with which everyone is involved, to a greater or lesser extent, both as an individual and as a citizen of a country. We went on to show how two questions about environmental policy – the banning of DDT and the use of nuclear power – involve dilemmas which go beyond the range of gut-reactions and into a complex matrix of factors. In both

[3] Toke, 1990, p. 76.
[4] Respectively www.greenpeace.org, www.iaea.org/worldatom/ and www.bnfl.com/index1.html.
[5] Porritt, 2000, p. 114.

cases ethics is not abstract decision-making about matters of purely theoretical interest, but is the key to making a decision which must be taken wisely, for the sake of the well-being of hundreds of millions of humans, and countless members of other species as well.

As so often in contemporary ethical dilemmas, different ethical principles, each laudable on its own, conflict. Ethics, here as elsewhere in biology and medicine, requires a balanced judgement to be made now but with a recognition that the balanced judgement may change in the light of future scientific discoveries and technological developments.

REFERENCES AND RELATED READING

Carson, R. (1962) *Silent Spring.* Houghton Mifflin, New York, USA.

Ehrlich, P.R. and Ehrlich, A.H. (1996) *Betrayal of Science and Reason.* Island, Washington, DC, USA.

Grubb, M. (1993) *The Earth Summit Agreements.* Earthscan, London, UK.

Holdren, J. (1992) The transition to costlier energy. In *Energy Efficiency and Human Activity.* Schippers, L. and Myers, A.S. (eds), Cambridge University Press, Cambridge, UK, pp. 1–51.

Porritt, J. (2000) *Playing Safe: Science and the Environment.* Thames and Hudson, London, UK.

Toke, D. (1990) *Green Energy: a Non-Nuclear Response to the Greenhouse Effect.* Merlin, London, UK.

World Health Organisation (1993) *Use of DDT in Vector Control – Conclusions of the WHO Study Group on Vector Control for Malaria and other Mosquito-Borne Diseases, Geneva, 16–24 November 1993,* WHO/MAL/95.1071, WHO/CTD/VBC/95.997.

APPENDIX A. SUGGESTED EXERCISES ON MALARIA AND DDT

These appendices are for lecturers'/teachers' use.

The core task of this exercise is for students to conduct the ethical analysis suggested in 5.2.4, points (i) – (iv), and use this to weigh up the pros and cons of retaining DDT for the control of malaria.

EXTENDING THE EXERCISE I – ASSESSING THE AVAILABLE INFORMATION

To give balanced advice a consultant would need a great deal more information than we have provided! However, information of this sort is rarely presented from a strictly neutral point of view. In particular, the World Wide Web provides a rich source of accessible information. Indeed, there is more information than we can possibly handle. It is important for students to be aware that everyone has their own personal 'filtering processes', which can all too often mean focusing on the information they want to find to support a predetermined view.

Organisations, moreover, seek not only to inform but to influence those who visit their sites. Below are the addresses of three web sites which have been selected to illustrate different ethical positions and the way information is used to promote certain views and counter the views of others.

1. The Rachel Carson Council.
http://members.aol.com/rccouncil/ourpage
2. World Wildlife Fund – Global Toxic Pollution Page.
http://www.worldwildlife.org/toxics/progareas/pop
3. American Council on Science and Health.
http://www.acsh.org

The student should visit each site in turn. In each case they should consider

  (i) what point of view the site-owner wishes them to adopt,
 (ii) how the language and images used tend to reinforce that view and
(iii) what hard information is being presented.

Finally, they should ask themselves whether visiting the sites changed their point of view. Again, this question should be focused by referring students back to the ethical questions in 5.2.4, points (i)–(iv) – what was the change in their *valuing* of different elements of the situation?

EXTENDING THE EXERCISE II – GATHERING MORE INFORMATION

For a more thorough study of the dilemma faced by a country where malaria is rife students should work in groups to do the following.

  (i) Compile a summary of the culture, political, social and economic structures of the country.
 (ii) Consider what important changes are happening in the economy.
(iii) Identify any important ecosystems and changes these systems are undergoing.
 (iv) Consider the effect of malaria on the size of the human population.
  (v) Consider the human suffering that malaria inflicts and the suffering inflicted by pesticides on people and ecosystems beyond the borders of the country.[6]
 (vi) How is the country attempting to control malaria at present?
(vii) What alternatives to using DDT might there be?
(viii) Finally, what course of action would they recommend for their chosen country? Would that be different from students' personal preferences?

APPENDIX B. THE PROS AND CONS OF NUCLEAR POWER

The exercise on the assessment of information given in Appendix A can be re-run with the web sites cited in Note 4.

To extract the utmost from the case-study on nuclear power, the students should

---

[6] As was mentioned in Chapter 3, the effects of organochlorine pesticides are astonishingly widespread. Agents used in the tropics even appear in the breast-milk of Inuit mothers.

identify (a) what *factors and principles* are most important in deciding whether it would be right for a particular nation to invest in a programme of nuclear power stations, (b) how those factors/principles should be ranked as to importance and (c) what conclusion the operation of those factors/principles suggests.

They should also reflect on their own views, and *the way those views have been affected* by the source material they consulted.

6 Human Use of Non-Human Animals – a Biologist's View

David de Pomerai

6.1 INTRODUCTION

One currently popular myth portrays scientists as cold, unfeeling, calculating – uninvolved with the objects of their study. There is, of course, a grain of truth in this caricature; when dealing with experimental measurements, scientific method demands both rigour and impartiality in assessing what they mean. But in another sense this is a deeply misleading myth, picturing the scientist as someone devoid of ordinary human sympathies and emotions. Nowhere is this myth more destructive than for biologists whose experiments use animals. In many instances, there is deep involvement with the animal under study, a sense of awe and wonder at what we are discovering, and respect bordering on love. Even for those working at one remove from living organisms, there can be similar involvement with biochemical networks or gene sequences. Alexander von Humboldt described two centuries ago how scientists become passionately involved with what they are studying. Having worked in departments of Animal Genetics and Zoology for a quarter of a century, I believe this to be a truer picture of most of my colleagues than the opening caricature.

For the past decade, I have worked on a tiny free-living roundworm, the geneticists' favourite nematode, *Caenorhabditis elegans*. Most people have never heard of it, and even among biologists it remains the preserve of a limited (but growing) coterie of enthusiasts. Back in the 1960s, Sydney Brenner had the foresight to spot its potential as one of the few animals that could be analysed genetically in the same way as bacteria (see Brenner, 1974). Few took 'the worm' seriously for the next 20 years, yet in the mid-1980s it became the first (and so far only) animal to have its entire cell lineage mapped out – the family tree of cell divisions giving rise to every one of its 959 adult somatic nuclei (Sulston et al, 1983). A decade and a half later, it became the first animal to have its genome sequenced (reviewed by Hodgkin et al, 1998), although within two years it was joined on the podium by the first drafts of the fruit-fly and human genomes. Within the 'worm community' of scientists, there is a real sense of camaraderie,

Bioethics for Scientists. Edited by John Bryant, Linda Baggott la Velle and John Searle.
© 2002 by John Wiley & Sons Ltd.

born initially out of the scepticism of others who could not see what the worm had to offer. From this shared enthusiasm emerged an in-house community newsletter, quirkily entitled the *Worm Breeders' Gazette* (*WBG* to those in the field). The cover cartoons of this august publication are revealing, because they clearly display a genuine affection for our chosen organism: worm larvae skating when their culture plates are put in the cold room, or the two worm sexes entwined in a heart shape for Valentine's Day. I confess that I am a zealous convert to the worm; again and again I am amazed at its resilience, its astonishing intricacy and (yes) elegance, its continuing ability to surprise and confound predictions, even though so much is already known.

And yet, I am scarcely *kind* towards these worms. We expose them to radiation or toxic chemicals in order to monitor their responses to diverse stressors (an early-warning test for environmental pollution). Some we microinject with the aim of introducing foreign genes, or of inactivating existing genes. Others we freeze down in batches so as to monitor their growth or fecundity. I feel no qualms of conscience about any of these. Nevertheless, *C. elegans* is a *bona fide* animal with muscles and gut, reproductive organs and a nervous system comprising precisely 302 neurones. Like any animal, it can respond to external stimuli, moving away from touch or noxious substances, but towards bacterial food and favourable conditions. In the wild, this worm lives opportunistically in the soil – where it is subject to the vagaries of waterlogging, drought and irregular food availability, such that its numbers boom and crash periodically. However, even if laboratory culture conditions afford a worm paradise by comparison, this could hardly provide an ethical justification for our experimental procedures. What we do to worms would scarcely raise an eyebrow if applied to bacteria or plants, yet both of the latter can certainly respond to external stimuli (on a longer time-scale revealed more clearly by time-lapse photography in the case of plants). Should we therefore be concerned because worms are being maltreated in research laboratories like mine around the world? To some purists, perhaps, there is cause for outrage even here. But I suspect far more people would be alarmed if these same treatments were being meted out to fish, say, or to mice. This implies that some animals are more equal – or more worthy of concern – than others. The mere possession of some sort of nervous system may qualify an organism as an animal (though sponges lack even this), but that in itself cannot qualify the organism as an 'other' worthy of serious moral concern. To make the same point in a different way, many of the close relatives of *C. elegans* are parasitic, infesting the tissues of plants, animals and humans. Some are serious pests, causing widespread economic loss and disease. We do not hesitate to kill such worms with powerful anthelminthic drugs.

In many traditions, respect for animals goes hand in hand with (limited) exploitation and even killing of them. A shepherd protecting his flock may even be prepared to lay down his life for the sheep, but in return he must make his living from them, whether from wool or meat or both. Among Native Americans and Aboriginals, respect for prey animals is entirely consistent with hunting, killing and eating them. Similar respect may be discerned in the peculiarly Anglican (i.e. Church of England) tradition of clergymen-naturalists (Armstrong, 2000), among them men who devoted their leisure time to studying obscure groups of animals, often within a limited local area. That such men collected birds' eggs, or pinned out dead insects collected from the wild, may seem at odds with current conservation practice, but their fascination with whatever they were studying shines through their writings. In such cases, 'subjects of study' might be a more appropriate phrase than 'objects of study'. What is inconceivable, for any of the groups so far mentioned, is wholesale slaughter, wilful neglect or deliberate cruelty towards animals. Such balance and respect seem less prevalent today, as opinions become ever more polarised between those who see animals merely as commodity units and those who wish to accord them rights or moral status equivalent to those attributed to human beings (Regan, 1983; Singer, 1986). The aim of this chapter is to explore the possibility of a middle way.

Outrage at the maltreatment and exploitation of animals tends to be focused on relatively recent applications such as intensive livestock rearing (e.g. battery cages) or routine testing of chemicals on laboratory mice. As animals become commodified – mere units of production or data points in a test series – so any sense of human respect or care seems to evaporate. It could be argued that my own experiments treat individual worms in an even more cavalier fashion, yet earlier in the introduction I claimed that respect for 'the worm' (generically rather than individually) was a hallmark of the *C. elegans* research community. Personally, I would feel far less comfortable performing similar work on vertebrates, but other biologists may not necessarily share my scruples on this score. Several wider issues need to be addressed in later sections of this chapter: (i) a widespread loss of respect for the natural world in general and for animals in particular; (ii) the biological diversity of animals and consequent impossibility of lumping them into a single moral category; (iii) the ethical grounds for reevaluating the moral status of animals; (iv) the ways in which our society has become locked into patterns of animal exploitation (and indeed abuse), which cannot be abolished overnight, and (v) the peculiar problems posed by new genetic technologies in relation to animals. As an ordained Church of England priest as well as a biologist, I will naturally root my (vi) closing outline of a 'middle way' in the Christian tradition, but will also explore how similar conclusions can be reached from very different secular starting points in the evolution of behaviour and in jurisprudence, respectively.

6.2   THE ROOTS OF ANIMAL EXPLOITATION

Clearly it is impossible to provide more than a few pointers on this complex topic. In a landmark paper from 1967 (White, 1967), the historian Lynn White traced the roots of our current ecological crisis deep into the Western Judaeo-

Christian tradition (see also Chapter 3). The words of Genesis 1:28 take on an ominous ring when we consider how Western man has responded all too literally to God's commands to 'be fruitful and multiply' and to 'have dominion over . . . every living creature'. Entire ecosystems have been devastated in the name of Western progress, entailing forest destruction, decimation of dominant animals (such as bison) and massacres of indigenous peoples. More recently, we have added industrial pollution and intensive agriculture to the growing degradation, and even medical advances have had the side-effect of supporting unsustainable increases in human population. The original sense of responsible stewardship implicit in the Hebrew word for 'dominion' (Westermann, 1987) has become lost beneath a welter of profiteering and greed. Inevitably, animals have shared in this general demeaning of the natural world into humanly exploitable resources.

Paradoxically, both Renaissance and Reformation traditions in the West have helped to encourage open and independent inquiry, thereby fostering the development of science to an extent unparalleled in any other culture (Hooykas, 1972). With the success of this enterprise has come a reliance on technology and an implicit trust in 'progress' that have only recently been challenged. Technology in its turn has revolutionised agricultural practice, achieving its goal of high productivity (since World War II) only at the cost of dehumanising the process. The consequences of this are equally as apparent in the disappearance of flower-rich meadows as in the rise of the battery hen. Consumers with disposable income have the liberty to opt for organic or free-range produce, but this may not be realistic for those on limited incomes. It is difficult to reconcile respect for animal welfare with the requirements for high-efficiency and low-cost production of livestock. Moreover, the split between urban and rural life has been exacerbated by mutual distrust and even hostility in recent years, particularly in the UK.

A third factor underlying the low moral status accorded to animals arises from Cartesian dualism, reinforced (alas) by certain Christian traditions which emphasise the salvation of human souls but apparently leave little place for animals. Rene Descartes perceived a radical discontinuity between the thinking 'I am' of a human person, and all animal natures – which he viewed as unthinking collections of mechanical reflexes (and was prepared to engage in vivisection on this basis). It would be rash to deny that this attitude still persists and underlies some of the worst excesses of animal experimentation. An unholy collusion of Christian and scientific attitudes has done much to accelerate environmental destruction and further demean the status of animals. However, the UK has a commendable record in pioneering legislation (the 1876 and 1986 acts governing the use of animals in research) aimed at protecting animal interests and preventing such excesses as far as possible.

6.3 THE DIVERSITY OF ANIMALS

For a zoologist, one of the most frustrating aspects of the debate about animals is the widespread refusal to engage with their sheer diversity. Animals are not one category of things, any more than fungi or seaweeds can be lumped together with flowering plants. In common parlance, any discussion of animals in relation to scientific experiments usually means warm-blooded furry mammals – mice and rats, or more emotively cats and dogs and monkeys. These are the animals highlighted in anti-vivisection literature, so to restore the balance somewhat, I began this chapter quite deliberately with an account of animal work using small invertebrates. We need to begin by asking what distinguishes an animal from a fungus or plant (the other great kingdoms of multicellular organisms). As mentioned earlier, the most obvious distinction – namely a nervous system – does not apply in the case of one animal group, the Porifera (sponges). Coelenterates and some sessile groups have only a simple nerve net to communicate between different parts of the animal. In the absence of any co-ordinating centre, it is difficult to see how such animals represent any great advance over plants in terms of their awareness of the environment. In the nematodes and molluscs, there are nerve rings in the head region which serve as foci for sensory input and as command centres controlling movement and behaviour. Such animals both perceive and respond to noxious stimuli, but it seems unlikely that they would feel pain in any sense that we might recognise. With the segmented arthropods and annelids we find an axial nerve cord bearing some homology with the vertebrate spinal cord, such that the anterior ganglia can be crudely described as a brain. However, anatomical comparability should not mislead us into expecting functional similarity; the brain organisation of even the simplest vertebrates is vastly more complex than this. Among vertebrates themselves, brain size and complexity generally increase from fishes up through amphibians and reptiles to birds and mammals, with primates and humans heading the field. There can be little doubt that mammals and probably *all* vertebrates are capable of suffering, not only pain itself, but also deprivation and maybe the anticipation of maltreatment. Whether we can rely on such anthropocentric (human-based) terms as contra-indicators of animal welfare will be discussed in Section 6.7.

However, lest all this seem too convenient a definition of 'sentience' (a modern reworking of the *scala naturae*), we should also be aware of exceptions to the rule. Most blatant among these is the cephalopod group of molluscs, comprising squid, octopus and cuttlefish. The organisational complexity and sensory capacities of the cephalopod nervous system confer a degree of intelligence unrivalled outside the vertebrates, and certainly justify calls for their experimental use to be regulated in the same way as fish (Bateson, 1992). On the basis of these wide differences between animal nervous systems, I believe (*pace* Singer, 1986) that there can be no biological justification for treating all animals (or even all sentient animals) as morally equivalent in terms of the suffering they are capable of experiencing. This fact, it seems to me, undercuts Singer's main accusation of 'speciesism' (cf. sexism, racism). Though sentience provides a useful rule of thumb, it ultimately draws an arbitrary dividing line across a continuum of different kinds of nervous system that respond in different ways. Treatments that would be cruel if applied to vertebrates have much less (but not zero) moral significance when practised on worms or flies. The above argument tends towards the general conclusion that some degree of speciesism is morally defensible (see also Midgley, 1983). Other things being equal, it is both natural and right to place the interests of our own species above those of animals. If indeed some animals are of greater moral significance than others, on account of the higher complexity of their nervous systems (with the correlate of increased awareness), then experimental protocols that would pass without comment among the lower invertebrates will require much closer scrutiny and justification if it is proposed to apply them to vertebrates. This is the practical outworking of the 1986 Animals (Scientific Procedures) Act, which regulates all UK animal experimentation using vertebrates (but not, as yet, any invertebrates). Only the prospect of significant human benefits can be used to justify experiments causing substantial harm to such animals.

6.4   A HIGHER MORAL STATUS FOR ANIMALS?

If Descartes represents a low tide mark in Western thinking about animals, what of more recent attempts to elevate their moral status, or even to claim some kind of equality with humans? This is not the place for a detailed review, but brief outlines of several salient arguments are sketched below.

6.4.1   ANIMAL PAIN?

As Jeremy Bentham pointed out two centuries ago, 'the question is not, Can they reason? nor Can they talk? but, Can they suffer?' (quoted by Singer, 1986, p. 8). This argument has been developed along extrinsic utilitarian lines by Peter Singer, such that pain becomes the only relevant moral criterion, and all pain experienced by sentient animals is morally equivalent. On this argument, it is indefensible to perform experiments that cause suffering to animals, even with the apparently noble aim of developing medical treatments that might alleviate human misery. Such procedures could only be justified if they held out equal or greater prospects of alleviating suffering in the animal species experimented upon. By the same token, if human society wants new drugs or disease therapies, it should be prepared to perform the necessary experiments upon its own members, perhaps on those individuals too severely disabled to survive. This suggestion sparked vehement protests from disabled people during Peter Singer's inaugural lecture as Professor of Bioethics at Princeton in 1999. Although his position has logical and philosophical consistency, many would part company with these implications, even though they may find his views on animal liberation congenial. I have argued above that there is no zoological category of sentience, but any gradation in how pain is perceived among different animal groups would seem to imply a corresponding gradation in moral significance. I should admit a bias here; as an insulin-dependent diabetic, I owe my life to a treatment developed using animals (initially dogs), and for the first 22 years of my disease I depended entirely on insulin derived from bovine slaughterhouse pancreases. Vegetarian diabetics may take comfort from the advent of cloned human insulin (produced in microbes) in the mid-1980s. For my own part, as a UK citizen, I am grateful that this selfsame transition coincidentally averted any (remote?) risk that might have arisen from using insulin obtained from BSE-infected cattle!

6.4.2 ANIMAL RIGHTS?

A complementary system of intrinsic deontological ethics is derived in large measure from Immanuel Kant. His most famous dictum suggests that we should treat others as ends in themselves, and never solely as means to an end. However, Kant himself denied that animals are subjects of moral concern to humans: 'So far as animals are concerned, we have no direct duties. Animals are not self-conscious, and are there merely as means to an end. That end is man' (quoted by Midgley, 1983, p. 51). This seems a breathtaking assertion today, and many (e.g. Regan, 1983) have suggested that the category of morally significant 'others' should be widened to embrace animals as well as humans, in particular those animals which are the 'subject of a life' (c.f. sentience in Section 6.4.1). 'Animal rights' is often used as a catch-all phrase to describe this trend, although it is significant that Kant speaks in terms of *duties* rather than *rights*. These two stand in a reciprocal relationship in most (but not all) situations. As Roger Scruton has pointed out, human rights normally imply duties to society (a social contract), though clearly this cannot be true for young children whose human rights are not diminished by their inability to undertake reciprocal duties. This argument applies *a fortiori* to animals, since very few of them have duties towards humans (an exception might be made for guide dogs). Even so, the legal protection of rights afforded to human minors could be extended to animals, and this argument is winning wider acceptance in the case of the great apes. Extension of this idea to all mammals might run into serious practical difficulties, for instance in culling populations of herbivores whose natural predators are now extinct, or in dealing with carriers of dangerous infectious diseases (such as plague rats). The concept of criminal responsibility is even less applicable to animals than to human minors, so on what moral grounds could one ever justify killing such animals? Even if the concept of animal rights seems unhelpful, this in no way diminishes the converse responsibility of humans towards animals in their care (and even those beyond?). Whenever we exploit animals for our own ends (whether as livestock for meat or mice for drug testing), we have a duty to ensure their welfare (Barclay, 1992). Neglect, starvation, untreated disease or deliberate cruelty involving animals are, and should remain, punishable offences.

6.4.3   ANIMAL UTILITY?

It is often asserted that the differences between animals are so great that drug properties assessed (usually) in mice cannot usefully be extrapolated to humans. Like many claims from the animal liberation movement, there is a grain of truth here, but one that is greatly exaggerated. The similarity of genetic and biochemical systems across the mammals ensures broad comparability with humans; indeed the striking gene homologies emerging from recent genome projects suggest that even lower invertebrates such as *C. elegans* may provide good models for human disease processes and toxicological assessment. In future, the wider use of these lower 'model organisms' may restrict vertebrate testing to the more advanced stages of a research programme, thereby reducing the numbers of mice (etc) used. Human cell and tissue cultures will also find increasing applications, although it is difficult to assess side-effects in whole organisms using this approach. Molecular modelling of target proteins and their interactions already facilitates rational drug design, again reducing the numbers of animals used in routine testing. But at the end of the day, there will remain an irreducible minimum of animal experiments, simply because no partial system can hope to mimic all the subtle complexities of an intact animal. The trends cited above fulfil all of the 'three Rs' (refinement, reduction and replacement) set by Russell and Burch (1959) some 40 years ago as goals for medical experiments using animals. However, it is also true that these trends are driven in part by economic and bureaucratic necessities; keeping breeding populations of laboratory animals is costly, and in the UK, Home Office form-filling is an increasingly onerous task!

6.5   NO ESCAPE?

One aspect of the argument that is frequently forgotten is the fact that medical sciences in general, and the pharmaceutical industry in particular, are locked into an institutionalised pattern of legal requirements for animal testing. No new drug can be approved for human clinical trials without an extensive programme of toxicological and therapeutic testing on animals. Disasters such as the thalidomide scandal in the 1960s have increased the burden of proof required, necessitating the use of even larger numbers of animals. The whole process is immensely expensive, such that it can cost millions of dollars to check out a new drug. Although these costs pale into insignificance when compared with those required to undertake subsequent clinical trials of that same drug, the pharmaceutical industry would dearly love to find cheaper alternatives. A major pitfall is the requirement to demonstrate rigorous equivalence with current testing methods. To date, not even the deeply unpopular Draize eye-irritation test has been replaced completely by a cell-culture alternative (see Knight, 2001). Other such replacements may be a long time coming, and in the meantime large-scale animal testing will continue. Although it is possible to develop cosmetics using traditional natural products that do not require animal testing, this is not an option available for new drug development. As resistance to current antibiotics becomes ever more widespread among pathogenic bacteria, there is an urgent imperative to develop new classes of anti-bacterial agent, otherwise even simple infections may become untreatable. We ignore this at our peril! Similar points could be raised in relation to livestock; there are simply too many farm animals to be phased out quickly, nor would most breeds stand much chance of survival in the wild. Once again, we are locked into a historically determined pattern of animal exploitation from which we can hope to extricate ourselves only gradually, if at all. No short-term solutions are on offer.

6.6   GENETIC MODIFICATION OF ANIMALS

I do not propose to rehearse the details of how or why genetically modified (GM) animals are created (see de Pomerai, 1996, 1997). Suffice it to say that there are two broad categories of GM animals: those whose genomes are supplemented with novel (usually foreign) genes (see also Chapter 15), and those where the function of an existing gene is altered through targeted mutagenesis. The latter include many strains of mutant mice acting as more-or-less accurate models of human diseases, while the former include such diverse applications as growth-enhanced salmon, and sheep secreting valuable human proteins (e.g. α_1-antitrypsin, AAT) in their milk. Ethical problems are posed by the inefficiency of GM procedures (Chapter 15) and by the high frequency of developmental abnormalities (see Mepham et al, 1998), as is also the case with cloning (Chapter 17). For the purposes of this discussion, however, I wish to focus on one unique feature of all such transgenic animals; the fact that their gene pool, and hence their intrinsic nature or *telos* (Rollin, 1986; see also Verhoog, 1992; Vorstenbosch, 1993) has been deliberately manipulated for human ends. Is this an ethically acceptable procedure?

The intrinsic nature (*telos*) of an organism can be viewed as the sum of its realised and potential genetic capabilities, both what it *is* and what it could *become*, its evolutionarily determined 'needs and interests'. At the least, this *telos* is modified – and perhaps violated – by human interventions which add to or delete from the native genes appropriate to the species. Arguably, the AAT sheep are a benign example of modified *telos*, since AAT is extracted non-intrusively from their milk, and affords great therapeutic benefits for patients with cystic fibrosis or emphysaema. At the other extreme, the notorious on-comouse involves a violation of *telos* (Linzey, 1993), since these mice have been genetically programmed to develop tissue-specific cancers through the insertion of an activated oncogene. Whether their use as models for testing anti-cancer drugs can be justified on utilitarian grounds remains a moot point. It is true that breeding only as many oncomice as you need might reduce the wastefulness and suffering associated with conventional routes (involving carcinogens) for generating mice with particular cancers, but on the other hand oncomouse cancers are far more uniform genetically than those resulting from carcinogen treatment. At least initially, the patented oncomouse has not proved a commercial success (Fox, 1993).

Whether modified or violated, it is pertinent to ask whose *telos* is being altered by genetic engineering – that of the individual or that of the species? The former involves a circular argument, since every cell of the transgenic individual is endowed with the new or disabled gene; it is just as much part of the genetic make-up of that individual as are the mutant CF genes in a human patient suffering from cystic fibrosis. The latter is a somewhat evanescent concept; faster than we could list all the allelic variations that comprise the total gene-pool of a species, somewhere – in some individual – natural mutation will be creating an allele combination never seen before. Moreover, *telos* encompasses phenotype as well as genotype – the combined effects of innumerable gene-variants working together. Did the *telos* of the first wild dogs domesticated by man encompass both the poodle and the greyhound? Though derived from the same ancestral gene pool, is not the *telos* of these modern breeds quite distinct, both from each other and from their ancestors? Arguably, the alterations of *telos* involved in long-term selective breeding are far more radical than anything effected by single-gene changes introduced through genetic engineering. We should also recall that several dog and livestock breeds suffer from common congenital defects and deformities; selective breeding of animals for traits favoured by humans can sometimes violate rather than merely modify their *telos*. A notion similar to species *telos* was outlined by C.S. Lewis in answer to the theological question of whether animals have souls. His answer was equivocal; maybe yes for those animals involved in human relationships, but probably no for the vast majority – unless it might be (for lions) in the form of a 'richly Leonine [self] . . . that expresses whatever energy and splendour and exulting power dwelled within the visible lion on this earth' (Lewis, 1940, p. 131). Here, Aristotelian *telos* (Rollin, 1986) meets with Platonic idealism, but its other-worldly frame of reference is hardly helpful in grappling with the realities of manipulating animal genomes. An absolutist view of *telos* would prohibit any interference with animal (plant? microbial?) genomes whatsoever, whether through genetic manipulation or selective breeding, but such a universal ban risks losing the GM baby along with the bathwater. Indeed, as also mentioned in Chapters 8 and 9, some people do have an intrinsic or deontological objection to moving genes between organisms; this view would presumably not even permit human drug proteins to be produced in bacteria. On the other hand, if we accept that limited interference with animal genomes may sometimes be permissible, then the yardstick we use to draw a dividing line between acceptable and unacceptable applications is likely to boil down to a consequentialist cost–benefit analysis of utility versus suffering (Vorstenbosch, 1993).

The ethical questions raised by genetic engineering of animals are by no means trivial. I believe we should not look for universal rights and wrongs here, but should instead consider each case on its merits. Perhaps the most acute problems are raised by mouse gene mutants used as models for many human diseases in which genetic factors play a role. As I have pointed out previously (de Pomerai, 1996, 1997), the initial generation of such knockouts is a long-winded and laborious process, producing very small numbers of homozygous mutant individuals. Major ethical difficulties ensue only when such mutants prove to be useful as disease models. In such cases, large numbers will be bred in order to test novel drugs aimed at appropriate molecular targets associated with the disease process. There is a 'double whammy' here; not only are these mice genetically engineered to develop a particular (probably distressing) disease, but they will also suffer from any unforeseen or toxic side-effects of the untried drugs being tested. Yet, realistically, this may be the only rational way forward to develop effective new drugs to combat some of the most intractable diseases afflicting large numbers of humans. As a society, we have to ask ourselves – at what price medical progress? Already, the 1999 annual UK Home Office returns have shown an upward trend in the numbers of animals being used in experiments, and a major component of this increase results from the adoption of transgenic mouse strains as models for human disease.

6.7   A MIDDLE WAY?

As a Christian, I would begin with the words of Jesus from Luke 12:7 – 'You are worth more than many sparrows'. This well known phrase is a two-edged sword, for on the one hand it acknowledges a moderate form of speciesism (a human life counts for more than a bird's), yet at the same time it points to a real value of the sparrow's life in the sight of God, if not of humans. This is reinforced by the preceding verse: 'Are not five sparrows sold for two pennies? Yet not one of them is forgotten by God'. By extension, we may claim justification for experimenting on animals with the aim of alleviating human diseases and the suffering they entail, but this does not mean that anything goes. Each God-remembered animal life has a moral worth whose sacrifice needs to be weighed up against any benefits that might accrue. Of course, there is no guidance as to relative scales of value; indeed, the human benefits (actual or hoped-for) are strictly incommensurable with the animal suffering involved (Singer, 1986). In no sense is the animal a volunteer, and it cannot know anything of the purposes for which it has suffered. Since our Christian-influenced moral code is biased in favour of the vulnerable (the poor, the foreigner, the orphan and the widow in many Old Testament texts), this hard truth should give us cause for serious reflection before embarking on any experiment likely to inflict harm on 'sentient' higher animals (such as vertebrates). Indeed, our moral concern should reach out beyond vertebrates – beyond even the animal kingdom – to embrace the whole of creation (see e.g. Page, 1996 and Chapter 3 in this book).

This seems to leave us with the *status quo*, where widespread animal suffering continues in the name of science (mostly, but not wholly, in the area of medical research), but is at least scrutinised externally in the UK under the terms of the 1986 Animals (Scientific Procedures) Act. This entails periodic review of animal usage, through annual returns to the Home Office, through visits from Home Office Inspectors, and through internal review mechanisms. These regulations currently apply only to vertebrates, though there is a case for including cephalopods (Bateson, 1992). Concern is focused on the general welfare of the animals used, on the severity of test procedures administered (and whether anaesthesia is used, if appropriate) and on the numbers of animals involved. While this system licences experiments that some people might find unacceptable, it does act to limit the scale of animal suffering and where possible to ameliorate conditions. It certainly keeps the minds of experimenters focused on issues of animal welfare, and encourages moves towards the aforementioned 3 Rs of Russell and Burch (1959). At the core of the process is a utilitarian cost/benefit analysis balancing the likely suffering involved against prospective gains (such as new drugs or insights into disease processes). This is inevitably crude, since neither aspect can be accurately quantified, and there is no scale of equivalence. Some people view the entire enterprise as so flawed as to be worthless, but most such either wish to see all animal experiments banned (Regan, 1983; Singer, 1986), or conversely to have the current restrictions relaxed. A Christian viewpoint, as I see it, would eschew both of these extremes, while at the same time emphasising the God-given value of *all* living things (a concern whose embrace extends far beyond the vertebrates; Page, 1996).

Similar views can also be adduced from a variety of secular starting points, including philosophy (Midgley, 1983), behavioural science and jurisprudence (see below). Concepts such as 'pain', 'suffering' and even 'welfare' are inevitably suffused with anthropocentric thinking, and this is unlikely to be helpful when considering animals (let alone plants or microbes) very different from ourselves. In a detailed critique of welfare criteria, Barnard and Hurst (1996) call for an organism's life history (including survival, growth, reproduction etc) to be considered within its own evolutionary and ecological context. We humans judge animal welfare from our own perspective – that of a long-lived vertebrate which invests much time and effort in looking after juveniles (our children). It follows that individual survival is a major human preoccupation, to the extent that we read a similar priority into animal lives. But is this necessarily the case? Returning briefly to *C. elegans*, this worm develops from fertilised egg to hermaphrodite adult in a mere three days at 25 °C, then produces about 300 self-fertilised offspring over an equally brief reproductive period. Thereafter, the worm can survive for about another 2 weeks, and this post-reproductive period can be extended dramatically by various genetic mutations or by anti-oxidant treatment. But from an evolutionary point of view, that individual worm has already fulfilled its purpose in passing on its genes to the offspring. Its post-reproductive lifespan is essentially redundant and is unlikely to be extended by pressures from natural selection. If anything, long individual lifespans would tend to result in food competition, which might disadvantage the offspring. In this case at least, it is hard to see what real benefit accrues to the worm (either individually or as a population) from prolonging its lifespan. This is a rather stark example of the so-called 'disposable soma' hypothesis, which sees the body as an expendable 'machine' built by the genes for the purpose of reproducing themselves (Kirkwood, 1993). To take another example from the vertebrate world, dominant males are successful at reproduction (and hence at passing on their genes), but often live short and dangerous lives filled with conflicts to maintain their dominant position (which is of course transient, as sooner or later that position will be usurped). The palette of underlying behaviours (aggression etc) is determined by evolution, such that reproductive success constitutes a greater priority for that male than long life or mere survival. Barnard and Hurst (1996) introduce a helpful analogy with motor vehicles. Human welfare priorities (directed towards prolonging life and minimising 'stress') would be analogous to keeping a Ferrari well oiled and beautifully polished in the garage. It will certainly last a long time like that, but sitting in a garage is not what it was actually designed to *do*! Equally, driving an off-road vehicle at top speed round and round a race track is likely to cause damage and malfunction, which would likewise be true if driving a racing car across open rough country! In each case, the treatment received is at variance with the design specification of the car. This begs the obvious question of how we can gain an understanding of the evolutionary and ecological 'design specifications' of an animal – which Barnard and Hurst (1996) interpret as the decision-making rules (and priorities) by which that animal lives its life. However, without some insights of this nature, our welfare criteria will remain hidebound by human preconceptions, which might be totally inappropriate to the animal in question. Furthermore, this argument is equally relevant to other living organisms (all of which have been 'designed' by evolution), and is therefore as applicable to a radish as to a chimpanzee. Nevertheless, the *moral* significance of interfering with that organism's well-being will be far more serious in the latter case than in the former. The rule-of-thumb outlined in Section 6.3 above suggests that our concern should increase in line with the complexity of the nervous system (or more broadly, the ability of an organism to respond to its environment). Human-based welfare criteria will apply more accurately to other long-lived mammals (e.g. elephants) than, say, to short-lived rodents. In cases where evidence is ambivalent, the benefit of the doubt should be given as a matter of course.

A different approach to such questions has been developed from jurisprudence (the philosophy of justice) by Beyleveld and Pattinson (2000), in the form of a critique of the thesis advanced by Gewirth (1978). As human agents, we acknowledge a duty (both legally and morally) *not* to inflict wanton pain or suffering on other human agents. Can this duty be extended to so-called *partial agents* – those who share some but not all of the characteristics of humans (a category which would include human embryos and foetuses as well as non-human animals)? Both Gewirth (1978) and Beyleveld and Pattinson (2000) concur that this duty should be extended to partial agents *in proportion to the degree of approach to being a full (ostensible) agent*. However, whereas Gewirth sees this protection as a question of generic rights, arising from the animal's status as a partial agent, Beyleveld and Pattinson argue that such generic rights are only applicable to full agents, and can only be extended to partial agents under the precautionary principle (rather akin to Barnard and Hurst's (1996) 'benefit of the doubt'). To the extent that an animal exhibits behaviours and responses similar to those of human agents, we have a duty to protect that animal from wanton pain. This can take us beyond the *scala naturae* of nervous-system complexity, since it allows the possibility of considering characteristics that may not be mediated through the nervous system. Again, this should widen the circle of our concern beyond the animal kingdom. It is still true that noxious treatments inflicted on vertebrates would have greater ethical significance than similar treatments applied to worms (which is where we started!), but it would also imply that treating plants or even microbes in the same way would have *some* moral cost greater than zero. By extension, whatever we humans do to harm our natural environment has moral significance, even if there is no discernible effect on larger animals. Both this paragraph and its predecessor demonstrate a striking convergence (in terms of conclusions) with the Christian 'middle way' advocated at the start of this section, despite starting from very different contexts and assumptions.

In a little-quoted passage from the end of his article (White, 1967), Lynn White focuses on St Francis of Assisi as an alternative pattern of Christian relationship with the natural world, standing in sharp contrast to the exploitative domination so characteristic of Western history and theology. The 'middle way' advocated in this final section does not give us clear and unequivocal guidance as to what should or should not be permitted in experiments using animals. It leaves us struggling with many difficult and emotive issues, and consensus will be hard to achieve. Open and rational debate may even prove impossible in the UK context, where animal experimenters who dare to raise their heads above the parapet are demonised and sometimes terrorised by animal-rights extremists. Those who advocate much greater care for the natural world might reflect that much of the information gleaned from laboratory experiments on living organisms (not just animals) provides essential underpinning for the remediation of ecosystems destroyed by human actions. Only thus can we hope to repair some of the damage we have caused to the biosphere as a whole.

REFERENCES

Armstrong, P. (2000). *The English Parson-Naturalist: a Companionship Between Science and Religion*. Cromwell, Trowbridge, UK.

Barclay, O. (1992). Animal Rights: a critique. *Science and Christian Belief,* **4**, 49–61.

Barnard, C.J. and Hurst, J.L. (1996). Welfare by design: the natural selection of welfare criteria. *Animal Welfare,* **5**, 405–433.

Bateson, P. (1992). Do animals feel pain? *New Scientist,* **138**, 30–33.

Beyleveld, D. and Pattinson, S. (2000). Precautionary duty as a link to moral action. In *Medical Ethics.* Boylan, M. (ed), Prentice-Hall, Upper Saddle River, NJ, USA.

Brenner, S. (1974) The genetics of *Caenorhabditis elegans*. *Genetics,* **77**, 71–94.

de Pomerai, D. (1996). Animal transgenesis: a view from within. *Science and Christian Belief,* **8**, 39–60.

de Pomerai, D. (1997). Are there limits to animal transgenesis? *European Journal of Genetics in Society,* **3**, 4–12.

Fox, J.L. (1993). Transgenic mice fall far short. *BioTechnology,* **11**, 663.

Gewirth, A. (1978). *Reason and Morality*. University of Chicago Press, Chicago, IL, USA.

Hodgkin, J., Horvitz, H.R., Jasny, B.R. and Kimble, J. (1998) *C. elegans*: sequence to biology. *Science,* **282**, 2011.

Hooykas, R. (1972). *Religion and the Rise of Modern Science*. Scottish Academic, Edinburgh, UK.

Kirkwood, T.B.L. (1993). The disposable soma theory – evidence and implications. *Netherlands Journal of Zoology,* **43**, 359–363.

Knight, D.J. (2001). New directions in toxicology. *Chemistry and Industry*, 5th March, 140–143.

Lewis, C.S. (1940). *The Problem of Pain*. Fount, Glasgow, UK.

Linzey, A. (1993). Created, not invented: a theological critique of patenting animals. *Crucible*, April–June, 60–67.

Mepham, T.B., Combes, R.D., Balls, M., Barbieri, O., Blokhuis, H.J., Costa, P., Crilly, R.E., de Cock Buning, T., Delpire, V.C., O'Hare, M.J., Houdebine, L.-M., van Kreijl, C.F., van der Meer, M., Reinhardt, C.A., Wolf, E., and van Zeller, A.-M. (1998). The use of transgenic animals in the European Union. *Alternatives to Laboratory Animals*, **26**, 21–43.

Midgley, M. (1983). *Animals and Why They Matter: a Journey Around the Species Barrier*. Penguin, Harmondsworth, UK.

Page, R. (1996). *God and the Web of Creation*. SCM, London, UK.

Regan, T. (1983). *The Case for Animal Rights*. University of California Press, Berkeley, CA, USA.

Rollin, B.E. (1986). On *telos* and species manipulation. *Between the Species,* **2**, 88–89.

Russell, W.M.S. and Burch, R.L. (1959). *The Principles of Humane Experimental Technique*. Methuen, London, UK.

Singer, P. (1986). *Animal Liberation: a New Ethic for our Treatment of Animals*. Cape, London, UK.

Sulston, J.E., Schierenberg, E., White, J.G. and Thomson, J.N. (1983) The embryonic cell lineage of the nematode *Caenorhabditis elegans*. *Developmental Biology,* **100**, 64–119.

Verhoog, H. (1992). The concept of intrinsic value and transgenic animals. *Journal of Agricultural and Environmental Ethics,* **5**, 147–160.

Vorstenbosch, J. (1993). The concept of integrity: its significance for the ethical discussion on biotechnology and animals. *Livestock Production Science,* **36**, 109–112.

Westermann, G. (translated by Green, D.E.) (1987). *Genesis: a Practical Commentary*. Eerdmans, Grand Rapids, USA.

White, L., Jr. (1967). The historical roots of our ecologic crisis. *Science,* **155**, 1203–1207.

REFERENCES

7 Human Use of Non-Human Animals: a Philosopher's Perspective

R.G. Frey

7.1 INTRODUCTION: THE ONGOING DISCUSSION

Much discussion takes place today over the use of animals in medical/scientific research. At least in the public media, the impression one forms is that, unlike such discussion in the past, medical and scientific researchers are very much on the defensive. In the past, simply pointing to the benefits that such research has conferred upon us has often been enough to carry the day. It is the standard line of defence by scientists to ask, of anyone in the 'animal welfare' or 'animal rights' lobbies today, whether they would be prepared, if their child were ill, to forego the very animal research that would make their child whole. If forced to choose between an animal and one's child, then everyone would choose their child. Doubtless this is true; it does not however win the argument any longer. For the mere fact that each of us would strongly desire to save our own child does not show that what we are doing, in the course of trying to save it, is morally right and permissible; it only shows that each of us strongly prefers our child's health and life to the health and life of any particular animal (or group of animals). And this result frustrates working scientists, to say the least, since the implication remains that, at least in very general terms, it is not clear that the worthy aim of benefiting humans justifies sacrificing on any scale, let alone a large one, animal health and lives.

At this juncture, it is tempting for scientists to resort to a number of different arguments. First it is tempting to ask why we are forbidden from doing what we can to improve and extend our lives and to extend our working scientific knowledge of the world around us. The answer, of course, is that we are not forbidden to do these things. The only thing that is presumably forbidden to us is to achieve these things by immoral means and that is what is pressed against animal research. Secondly, it is tempting to have recourse to some biological imperative to survive on our parts (see also Chapter 12) and to chalk up to that

Bioethics for Scientists. Edited by John Bryant, Linda Baggott la Velle and John Searle.
© 2002 by John Wiley & Sons Ltd.

imperative our use of animals in medical and scientific research. Even assuming that there is such an imperative, however, still leaves us with the problem of why, uniquely where animals are concerned, we cannot moderate our behaviour by moral means and so, as the opponents see it, refrain from using animals in this research. But what if there is no other way at present for us to obtain the knowledge in question except through the use of animals and animal models (see also Chapter 18)? We should still have a problem to face. Let us suppose that there is no way at present for us to obtain an important piece of knowledge except through seriously impairing or destroying certain *human* lives: presumably, most people would argue that the pursuit of knowledge, or indeed, enhancement and/or extension of some human lives does not justify impairing or destroying these other human lives. So, at least where other humans are concerned, biological imperatives seem subject to moral constraints. Why then is the same not true where animals are concerned? Thirdly, it is tempting to argue that species makes all the difference here: what is done to a human is one thing, what is done to a non-human animal is another (see Chapter 6). Yet, even here, there is an obvious problem, even in very general terms. Why, if it is wrong to put out the eye of a child, is it not wrong to put out the eye of a dog or a rabbit or a mouse? If we cannot do the former in the pursuit of knowledge how can we do the latter? To say that the difference resides in the species to which the two individuals belong is not yet to say anything very direct, for both the child and the dog will experience the loss of the eye, feel the pain involved, lead diminished lives and so on. If species is to make the difference, one needs to say more about how citing species works in this regard. I cannot rub a compound on to the skin of a child if that compound is toxic and will severely impair the health of the child but I can rub the compound on to skin of an animal who will experience exactly the same effects as the child would and whose life will be severely impaired as a result. How does the difference in species magically transform the morality of rubbing the compound on to the skin of a sensitive, feeling creature?

For these and other reasons, I find that many scientists today are on the defensive about their activities involving animal research (although an alternative perspective is given by other authors in this book: see Chapters 6 and 18). Fear of laboratory break-ins and personal assaults are not the nub of the matter: there is afoot today on the part of many people the idea that what such scientists do is not morally defensible. As this idea spreads so the reputability of medical/scientific research – heretofore a noble activity – is called into question. No working scientist can be pleased about the prospect of being regarded as morally dubious.

We can refine the issue in debate here. That we care for other humans is almost certainly involved in why we engage in animal research in the first place but that care does not in any simple way translate into moral permissibility for infecting healthy animals with terrible diseases, including ones not native to their species, inflicting pain on them and killing them. Put differently, the fact that humans benefit from animal research is certainly the major part of the story for why we engage in that research (as argued in Chapters 6 and 18) and any attempt to justify research on animals will almost certainly go through this argument from benefit. But that argument on its own does not show that it is right to do what we do to animals.

The reason for this is that the benefits in question could be obtained from doing to humans what we currently do to non-human animals. This is undeniable. If benefit alone were what was driving the argument, then using humans could equally as well, if not better, achieve that benefit. (After all, extrapolations from the animal to the human case seem inherently less reliable than extrapolations from human to human.) Yet here, virtually everyone would agree that it would be morally wrong to do to humans what we do to animals. Presumably therefore, the argument from benefit must be accompanied by another argument, one that specifies why, whatever the piece of research in question, it would be wrong to do to humans what it is that one proposes to do to animals. What could be the difference?

The tempting answer here for scientists is to claim that while humans are morally considerable and so are members of the moral community, animals are not members of that community. Therefore what is done to them does not matter morally. If what we do to animals does not count morally, then it does not matter morally what we rub on their skins, what infectious diseases we give them or what illnesses we use to take their lives. This approach is tempting precisely because it makes short work of the argument needed to supplement the argument from benefit: the reason that we cannot do to humans what we currently do to animals is that humans but not animals are morally considerable. Thus, what we do to humans counts morally in a way that what we do to animals does not.

In my view, the temptation to use this argument must be resisted, even though I think that there is a point to the line of argument that can go towards building a defence of animal research. A threefold argument can best capture the point in question (Sections 7.2–7.4).

7.2 MORAL CONSIDERABILITY I

As I have argued elsewhere (Frey, 1988, 1996, 1997), moral considerability turns on whether a creature is an experiential subject with an unfolding series of experiences that, depending upon their quality, can make that creature's life go well or badly. Such a creature has a welfare that can be positively and negatively affected, depending on what we do to it. With a welfare that can be enhanced or diminished, a creature has a quality of life. Even humans that lack agency, such as those in the advanced stages of Alzheimer's disease or those that are severely mentally enfeebled, are experiential creatures and so are beings with a quality of life that can be positively or negatively affected by what we do to them. Equally, however, rodents, pigs and primates are experiential subjects with a welfare and a quality of life that our actions can affect. While there may well be creatures where I am uncertain about where they are experiential (see Chapter 6), rodents, pigs and primates do not seem dubious cases. Thus, in my view, such creatures are morally considerable and so are part of the moral community on the same basis that we are: we are all experiential subjects with a welfare and quality of life that is affected by how life treats us.

One of the things that matters so much to us about pain is how it can substantially diminish the quality of a life; this possibility of diminished quality exists for all those who can experience pain. So, no-one treats massive pain, all other things being equal, as a high or desirable quality of life in either us or animals. Thus, while human lives have a quality that can be increased or diminished, so too do ('higher') animals.

Elsewhere, not surprisingly, I have endorsed a quality of life view of the value of both human and animal life (Frey, 1993, 1995, 1998). According to this view, as a life worsens in quality, its value diminishes; as its quality increases, its value increases. Thus, I do not believe, as many do[1] today in medical ethics, that lives of a radically diminished quality are equally as valuable as those of normal or very high quality (cf. Singer, 1994). Indeed, increasingly today, not even those who have to endure lives of radically diminished quality pretend otherwise, as the incidence of requests for assistance in dying testifies (but see also Chapter 17).

No-one will deny that the Alzheimer's patient or the severely mentally enfeebled are members of the moral community in this first sense of being morally considerable, since they remain, whatever their present state, experiential subjects with a welfare and a quality of life. Our actions with respect to them can augment or diminish that quality. This is true of all kinds of human being who have their quality of life radically diminished, whether from amyotrophic lateral sclerosis, sickle-cell anaemia, cystic fibrosis or whatever. All kinds of human being live lives of significantly reduced quality: reduced, that is, compared with the quality we identify in healthy, 'normal' adult humans. Yet they remain morally considerable.

Those, however, such as anencephalic infants or in conditions such as permanent vegetative state (PVS) are more problematic cases. For although what happens to these individuals may well affect the welfare and quality of life of other people, such as parents and other family members, it is not obvious that they themselves have experiential lives that would include them in the class of morally considerable beings. For it is not obvious how what we do to them could have any effect on *their* welfare or quality of life (even if those charged with care, for example of PVS patients, do all they can to enhance physical comfort). Indeed, in the cases cited it is not obvious that the unfortunate individuals in these conditions have a welfare or quality of life at all. (See also the discussion in Chapter 17, in Singer, 1994 and in Warnock, 1998.)

---

[1] Or perhaps pretend to do.

This first sense of moral considerability then is that in which creatures who figure within it are all those who have experiential lives and so a welfare and a quality of life. Animals as well as humans are in this sense morally considerable. I reject therefore the view that the way to defend animal research is by denying that animals count morally. I can see no more reason for denying them moral considerability or standing than I can for any being who has an unfolding series of experiences that constitute its life.

Notice that in order to be morally considerable in this first sense it is not necessary that one be capable of being a full moral agent and thus be morally responsible for one's actions. People who fall into the late stages of Alzheimer's disease have ceased to be full moral agents but they remain morally considerable in this first sense since what happens to them experientially continues to affect, either one way or the other, their welfare and quality of life. The fact then that animals are not moral agents does not affect their moral considerability in this sense.

7.3 MORAL CONSIDERABILITY II

Creatures that fall within the class of morally considerable beings are however not all alike; some are included as agents, some as 'patients' (or as the objects of moral actions by agents); the former are morally considerable in a sense that the latter are not.

This second sense of moral considerability is one in which morally considerable beings have duties to each other, in which reciprocity of action figures, in which there are standards for the assessment of conduct and in which reasons for action, especially where deviations from standards occur, are appropriately offered. The absence of agency, the absence of offering and receiving standards and reasons, matters because those who cannot do these things are not appropriately regarded as moral beings in the full sense of being held accountable for their actions. To be accountable for what one does, in a community of others who are accountable for what they do, is not the same as being considerable in one's own right. In this second sense, morally considerable beings inhabit a community of similar beings in which they offer and receive standards and reasons for action.

Some human beings are not members of *this* moral community since they are incapable of presenting standards for evaluation of conduct, of conforming their conduct to these standards and of receiving and weighing reasons for action. Disease and illness can undo agency. Also, young children and the severely mentally enfeebled are not members of this moral community. In this sense, many more human beings fall outside the moral community as a community of agents than fall outside the moral community as a community of morally considerable, experiential beings. Indeed, some human beings, such as those in a permanent vegetative state and anencephalic infants, fail to be morally considerable in both senses. (This explains contemporary controversies about whether to withdraw feeding via tubes from the former – see Chapter 17 – and whether the latter may be regarded as organ donors). On the other hand, while most, if not all, animals fall outside the moral community in this second sense, a great many fall inside the moral community in the first sense (as discussed above in Section 7.2). Thus there are going to be some humans outside the moral community altogether, even while some animals are within the moral community in the first sense.

To be morally considerable in this second sense is quite different from being morally considerable in the first sense. Agency, construed as acting on weighing reasons for action in the light proffered standards, is not required in order to be morally considerable in one's own right. So, why does agency matter? In my view, the answer has to do with determining the value of a life.

In the usual human case, if one can be held to be *causally* responsible for an action, we can go to ask whether one is also *morally* responsible for that action. For animals however, while causal responsibility exists, since we can say for example that the cat was the cause of the vase being knocked over and broken, we do not think that we can go on to ask whether they are morally responsible for what occurred. Once we accept that animals are not moral agents, we in effect deny that they act in the appropriate sense for being held morally responsible.

Moral discourse is appropriately addressed to and received by those who act for reasons, and the assessment of those reasons by the rest of us is central, with *causal* responsibility for an outcome established, for establishing *moral* responsibility for an outcome. Over such assessments, all kinds of argument are possible, as different sets of reasons may appeal to all kinds of standards; this kind of debate is an important part of our lives, when we seek to justify what we do or propose to do. Arguing about reasons and the standards they imply, trying to change people's views about the reasons they accept and find compelling and urging some other set of reasons are all things that figure prominently in this second sense of moral considerability. Indeed, it is just such activities that constitute that sense. To creatures who are incapable of acting for reasons, however, none of this applies, and none of the interaction among individuals that is involved in this second sense of moral community holds of our interactions with those incapable of acting for reasons. One can be close to one's cat and deeply concerned about its welfare; one can even hold oneself accountable in some sense, through ownership, for the cat's welfare. But there is no sense in which we hold the cat accountable for our welfare, not because it is like a child, but because, even when it is grown up, having our welfare as a reason for action is not something that we think cats can do. It is precisely this sort of thing that is characteristic of being part of a moral community in a full sense, as an individual who interacts with other individuals in the reciprocal sense I have been describing.

7.4   THE MORAL COMMUNITY

On the quality-of-life view of the value of a life, being a member of the moral community in this second sense can enrich one's life and so enhance its quality. It does this by sculpting the relations in which we stand to each other and so affecting how we live and judge our lives. The moral relations in which we stand to each other are part of the defining characteristics we give of who we are. We are husbands, wives, fathers, mothers, sons, daughters, brothers, sisters, friends and so on. These are important roles we play in life, and they are informed by a view of the moral burdens and duties they impose on us, as well as the opportunities for action they allow us. Seeing ourselves in these relations is often integral to whom we take ourselves to be. In these relations, we come to count on others, to see ourselves as interlocked with the fate of at least some others, to be moved by what befalls these others, and to be motivated to do something about the fate of these others to the extent that we can. Our lives are affected in corresponding ways. Though there is no necessity in any of this, being a functioning member of a unit of this kind can be one of the great goods of life, enriching the very texture of the life one lives.

Binding ourselves to others, pledging ourselves to perform within the moral relations in which we stand to others and holding ourselves responsible for shortcomings in this regard are all part of what we mean by being a functioning member of a moral community, within which we live our lives with other members. We come to count on others and they on us: the reciprocity of action and regard characteristic of fully functioning moral communities find their root in these moral relations. As a result, we come to take certain reasons for action almost for granted. We come to take the standards, which the rules and duties that in part comprise these relations, to be, *prima facie*, ones that it is appropriate by which to judge our own and others' actions. Here, too, there is no actual necessity in the matter: we can come to reject the standards implied in the usual understanding of these moral relations in our societies in favour of others. However, that these standards take a normative form by which we can evaluate reasons and actions, whatever their substance, is the crucial point; for it is a normative understanding of these roles that seems crucial to how we see ourselves within them, to how we live our lives and judge many of our actions within those lives and to how we judge how well or badly those lives are going.

Our participation in such a community enriches our lives. Even in a minimal form, it achieves this merely by enabling us to cooperate over extensive areas of our lives with at least some others to achieve those of our ends that involve others. Put differently, the relations in which we stand to each other aid us in the pursuit of our ends and projects, many of which require the cooperation of others to achieve, and the pursuit of these ends and projects, the pursuit, as some would have it, of one's conception of the good life, adds enormously to how well we take our lives to be going. Since our welfare is to a large extent bound up in these kinds of pursuit, to ignore this fact is to give a very impoverished account of human life. Since all these ends and projects can vary between persons, there is no single way of living to which every one of us is destined or condemned. Agency enables us to select different ends and projects, to mould and shape our lives differently, to achieve and accomplish different things.

Beyond any such minimal form, however, the very way we live our lives as, for example, spouses and parents, in order to fulfil what we see as our obligations within these moral relations, forms part of the texture and richness of our lives. We often cannot explain who we are and what we take our ends in life to be except in terms of these relationships in a moral community, and we often find it difficult to explain why we did something at great cost to ourselves except through citing our links to certain others. Thus, the fact that most humans are members of the moral community in this second sense is a very important feature of their lives: not only can they live out lives of their own choosing, moulded and shaped in ways they desire to reflect the ends and projects they want to pursue, but they can also live out these lives in a normative understanding, for example, of the relationships that characterise both themselves and others, relationships in terms of which they see themselves as linked to these others. Here, also, the normative understanding of these relationships enables us to see our lives as going well or badly depending upon how these relationships are affected by what we do to others and by what they do to us. The reciprocity so characteristic of a functioning moral community is not the mere reciprocity of action; it is also the reciprocity of judging actions from a normative point of view that sees enhancing the welfare of another as a reason for action.

To be sure, there are many things that we do that find analogues in the animal world, but what agency enables us to do is, in the sense intended, to fashion a life for ourselves, to live a life moulded and shaped by choices that are of our own making and so reflect, presumably, how we want to live. Achievement or accomplishment of ends so chosen in this regard is one of the great goods of human life and is one of the factors that can – again, there is no necessity in the matter – enrich individual human lives.

The appeal to species, then, in order to try to account for why we do not do to humans what we currently do to animals involves something deeper than any mere appeal to the fact that different creatures are of different species. In the human case, what the scientist is appealing to is this shared life in a moral community, and what prevents him or her from experimenting upon those that form part of this community is that shared life and the moral relations that he or she sees holding among the members of that community. Nor is there anything 'speciesist' about this particular use of species; for the claim is not that mere membership in different species is all that determines how one may be treated. Rather, what one is trying to do is to show why we think that normal adult human life is richer for being lived in the way I have described and to link this richness to claims about the quality of life of the humans in question. If there is something in animals that enables them to make up for the lack of agency, so be it; but we should need evidence of what this thing is. For its absence seems crucial to our understanding of this second sense of moral community in which we stand to humans, a sense in which the value of the lives of the humans in question is traced to the quality of their lives, where that quality is partly formed by living lives within a moral understanding of relationships and community.

7.5 CHOOSING AMONG VALUABLE LIVES

The point of the above remarks should now be clear. I think the fact that most humans are morally considerable in the second sense is a significant fact about their lives, one that holds the potential for dramatically enriching their lives and so conferring upon them a high quality. Being a functioning member of such a community enriches our lives, and while there are unquestionably many dimensions of richness in our lives that we share with animals, this one we do not. In any argument which tries to make out a difference between the value of human and animal life, therefore, such features as this participation in a community are bound to be important.

Nothing in all this involves our agreeing that one life is superior to another. All that is being suggested is that the lives in question are different and that this difference has impact on the quality and so the value of the life being lived. This is true between humans and animals. But it is also true between humans, and a word on this fact is necessary before this chapter is concluded.

In my view, what underlies scientific appeals to differences between humans and animals, as a justification for why they may not do to humans what they do to animals, are in fact, as I have indicated, appeals to the value of human life. Normal adult human life is more valuable than animal life, and the reason for this lies in the capacities and scope for enrichment that the different lives possess. If there is something in a mouse's life that confers upon it the richness of life of the normal adult human, we shall need to be told what it is; for not all lives, merely because they are, as it were, living things, have the same richness and quality. This difference in value we acknowledge, when, for example, we face the proverbial choice between saving a dog or saving a man and choose to save the man. It is not that the dog may not be good after its kind, not that it may not be in good health and so a dog with a reasonably high quality of life; it is rather that we take the man's life to have a higher value because we take it to have greater richness and so quality[2].

Now it should be obvious that, while this is true for a 'normal' adult human life, it is not true of all human lives. Disease, illness and acts of nature itself can severely impair human lives, so severely in fact, that their quality is radically reduced precisely because their capacities and scope for enrichment have been virtually destroyed. Increasingly, even in the news media, people who live such

[2] Although we might also note that adherents of the 'selfish gene' hypothesis (e.g. Dawkins, 1989) would ascribe this choice to our being evolutionarily programmed to save the genes to which our own are most closely related: we save the man and not the dog.

lives come forward asking for relief from them through assistance in dying. In the past, it was a plank of most religions that all human lives were of equal worth, that no one among us had a life less worthy than that of any other; but, today, when more and more people think about their lives and the lives of others in terms of the quality of life they have, this religious plank begins to appear insecure. Some lives appear so blighted, so devoid of richness and capacity for enrichment that we would not wish them on anyone, and those whose misfortune it is to live them seem increasingly vocal about their desire for options in continuing to be forced to live lives of such radically diminished quality. It is not we, but they themselves, who seek options.

It should be obvious, therefore, that purely on a quality-of-life view, not all human life has the same value, that it will turn out, normally through disease and illness, that some lives can be so blighted and devoid of richness and capacity for enrichment that their quality and so value plummets. No one familiar with our hospitals and nursing homes can fail to appreciate the point (but see also the discussion in Chapter 17). Of course, it will be asked, who am I to judge someone else's life as being of low quality? However, judgements of this kind are made daily in our hospitals in all kinds of context, including resource allocation ones. Again, the old religious imperative to love everyone and love them equally may well still exert influence but there is no reason to think that love can *only* express itself through striving mightily, e.g., to keep alive people who, if given a choice to continue or not in their present state, would choose not to do so. So we need to learn how to cope with the fact that not all human lives have the same richness, quality and value.

And one particular thing with which we will have to cope is this: some human lives may be so blighted by disease and illness that they cease to be members of the moral community in the second sense and, indeed, in the first sense. By contrast, some animal lives, while not members of the moral community in the second sense, are members of the moral community in the first sense. This fact creates a tension in our ethical thinking: a number of people today in medical ethics contexts talk about what it would be right or wrong to do to those in a permanent vegetative state or to anencephalic infants, cases in which humans either fall out of the moral community in the first sense or else never make it into that class. Tension is also created between our views of humans and animals by this fact: if one can save a dog or someone in a permanent vegetative state, and if the former but not the latter is at least a member of the moral community in the first sense, is there not reason to favour the dog? It is perhaps not a conclusive reason; after all, the effect on other human beings of so favouring the dog remains to be considered. But a reason would certainly exist for defending why one chose the dog.

This very fact – that a reason can exist for favouring a dog over a human – may upset some people and be taken to constitute the case for reverting to a view of human life as favoured above *all* other lives, but we operate with quality-of-life views constantly today, in all manner of settings involving medical ethics for humans, and it is hard to see why we should depart from such views here, where the implications for humans may be adverse. In the end, then, the problem that animal experimentation really raises is a problem about humans: how are we to think ethically about those human lives that, through illness, disease or nature, have ceased to be morally considerable in the second sense and, at least for some human lives ultimately, even in the first sense? This is further discussed in Chapter 17.

REFERENCES

Dawkins, R. (1989) *The Selfish Gene* (2nd edn). Oxford University Press, Oxford, UK.

Frey, R.G. (1988) Animal parts, human wholes: on the use of animals as a source of organs for human transplant. In *Biomedical Ethical Reviews*. Humber, J.M. and Almeder, R.F. (eds), Humana, Clifton, NJ, USA, pp. 89–101.

Frey, R.G. (1993) The ethics of the search for benefit: animal experimentation in medicine. In *Principles of Human Health Care Ethics.* Gillon, R. (ed), Wiley, New York, pp. 1238–1248.

Frey, R.G. (1995) The ethics of using animals for human benefit. In *Issues in Agricultural Bioethics*. Mepham, R.B., Tucker, G.A. and Wiseman, J. (eds), Nottingham University Press, Nottingham, UK, pp. 335–344.

Frey, R.G. (1996) Medicine, animal experimentation and the moral problem of unfortunate humans. *Social Philosophy and Policy*, **13**, 181–210.

Frey, R.G. (1997) Moral community and animal research in medicine. *Ethics and Behavior,* **7**, 123–136.

Frey, R.G. (1998) Organs for transplant: animals, moral standing and one view of the ethics of xenotransplantation. In *Animals and Biotechnology*. Holland, A. and Johnson, A. (eds), Chapman & Hall, London, UK, pp. 190–208.

Singer, P. (1994) *Rethinking Life and Death.* Oxford University Press, Oxford, UK.

Warnock, M. (1998) *An Intelligent Person's Guide to Ethics.* Duckworth, London, UK.

III Ethical Issues in Agriculture and Food Production

8 GM Crops and Food: a Scientific Perspective

Steve Hughes and John Bryant

8.1 INTRODUCTION

In Chapter 1 the basis for ethical decision making in biotechnology was discussed. In this chapter we focus on GM crops and food as a particular facet of biotechnology and address the sorts of information needed to make the ethical decisions about its implementation. We also, from a scientific perspective, endeavour to untangle the many convoluted and conflicting issues which have grown up around this technology against the principles of individual rights and social and global responsibilities. In order to do this it is necessary initially to look briefly at the history of genetic modification and its application to food and crops against the established modalities of plant breeding and agronomy.

8.1.1 PLANT BREEDING

It is clear from archaeological records that humans have been collecting, saving and planting seeds for over 12 000 years. It is also clear that, consciously or unconsciously, early farmers made use of genetic variety, saving seeds from plants that had desirable characteristics. This selection by farmers was in effect the start of plant breeding and it rapidly led to significant changes in the crops exposed to this selection.

Archaeological studies have shown that in the 'fertile crescent' in the Middle East selecting for yield led rapidly to significant increases in the number and size of grains per ear in wheat, even in the absence of any deliberate cross-breeding (Hillman and Davies, 1990). Thus, crop improvement, based on making use of the plant's genetic makeup, has been part of agriculture for a very long time. However, the establishment of genetics as a science in the early years of the 20th century under the influence of Bateson and Biffen, and the 'rediscovery' of Mendel's work, led to much more rapid progress in crop breeding, in particular because breeders started to cross-breed between different landraces or varieties

Bioethics for Scientists. Edited by John Bryant, Linda Baggott la Velle and John Searle.
© 2002 by John Wiley & Sons Ltd.

of a crop species. These early days of genetics-based crop breeding led to further significant improvements in crop performance, such as resistance to particular plant diseases.

The significance of genetic variation to crop improvement through breeding was highlighted by the Russian geneticist Vavilov in his exhaustive study of the agro-ecology of crop origins and their centres of diversity (Vavilov, 1960). Thus, the breeder is constantly looking for new genetic variation in a particular crop species in order to generate new varieties that are better in some way, but the 'classical' breeding techniques do not necessarily provide access to the variation or the characteristics needed. In some cases, the crop species in question does not possess the desired character and so no amount of crossing between different varieties of that species will lead to the acquisition of the character. In such cases, hybridisation with closely or distantly related wild species (for example Peruvian potatoes) is attempted in order to introduce the appropriate gene into the crop. However, plant breeders often deliberately induce new genetic variation within a crop species by the use of γ-radiation or of chemical mutagens. Although a high percentage of the seeds that are treated this way actually die, the survivors very often show useful genetic characters. Thus, the variety of barley (Golden Promise) used to make much of the beer in the UK is a mutant generated by radiation, as are several of the high-yielding rice varieties.

We may regard the use of mutagens and radiation as being very far from natural (even though plants may be exposed to such agents in nature, albeit at much lower levels). However, as we argue more fully below, a distinction between natural and un-natural is not an unfailing guide in ethical decision making. Further, many of the breeding methods that do not employ mutagenic agents still actually require extensive human intervention. Plant breeders employ a variety of techniques in order to by-pass inter-species fertilisation barriers and thus to facilitate some of the cross-species hybridisations, for example between a crop species and its wild relatives. A particularly interesting example of this concerns our 'daily bread'. It is ironic in the context of a discussion of GM crops that good bread is often taken as an example of 'natural goodness' when the wheat varieties used for bread-making contain significant amounts of genetic material from other grass species. In the 1960s, scientists at the Plant Breeding Institute, Cambridge, UK, discovered a gene that regulates chromosome pairing in grasses. Use of this in wheat enabled hybridisations to be performed that would hitherto have failed because of a breakdown of chromosome pairing. This in turn facilitated the incorporation of a segment of a rye chromosome containing many genes into one of the wheat chromosomes and a gene conferring resistance to eye-spot from a wild grass species, *Aegilops ventricosa*, into another. Furthermore, the phenomenon of centric fusion was the basis for the construction of an intergeneric wheat/rye hybrid chromosome known as the 1B/1R event, which is widely deployed in current European wheat varieties (Lupton, 1987).

Bread wheat thus contains significant amounts of foreign genetic material. It is interesting to note that had these transfers of genetic material been achieved by GM techniques they would have been subject, in the UK, to all the precautions that are currently applied to GM crops (see next chapter). As it was, the new varieties went through the normal variety trials and registration and were quickly adopted into agriculture in the 1970s, where they gave very large increases in yield compared to the varieties used previously (Lupton, 1987).

Despite the success in improving performance and yield in many crop species it needs to be stated that 'conventional' plant breeding is an imprecise process. Several thousands of crosses have to be made to obtain one or two useful 'new' varieties with the right characteristics; selection has to be carried out over several generations. This is because when crosses are made, an unknown number of genes, typically 1000 to 2000, of largely unknown function will be transferred into the candidate variety with the wanted characteristic (gene), even when the hybridisation is between two varieties of the same species. Some of these unwanted genes may have deleterious side effects in the plant (reduced yield, susceptibility to disease) or possibly in humans. So, once the desired character is in the candidate variety, further selection is needed to eliminate, if possible, these unwanted genes from that variety. If the unwanted characters cannot be eliminated, that variety is of course excluded from further development. All this means that in a classical pedigree breeding system new varieties take at least ten years to reach the market from the moment of the initial cross. This period incorporates many seasons of intense natural selection in the unprotected environment of field trials as well as intensive selection by breeders for yield and agronomic properties.

Perhaps this measured process has contributed to the very safe history and benign (low risk) image of plant breeding. Despite the imprecise nature of breeding technology, and the propensity of plants to elaborate an extensive arsenal of toxic compounds to repel their predators, instances of novel toxicity and mishaps for consumers are rare even in the case of very wide hybridisations. While reassuring, this seeks explanation, just as does the scientist's confidence that GM-derived plants will not pose novel problems of toxicity. An explanation may be framed along the following lines. If a plant derived from hybridisation were to divert its resources to the elaboration of new toxins (which it might well have the genetic potential to do on the basis of its parentage) it would be unlikely to survive the repeated artificial and natural selection of the breeding process. This would be either because it lacked its *normal* profile of toxins and failed to confront its predators and parasites, or because it devoted less resource to seed yield. Thus we would argue that plants producing abnormal profiles of toxins are eliminated in the selective process unless a strong selection is applied for something linked to that trait. This was the case with the high alkaloid potato where the disease resistance trait introduced by crossing with a wild relative was directly linked to the production of an alkaloid, solanine.

We have indulged in this 'polemic' since we believe that it is relevant to the role of scientists in informing the debate over relative risks of new and established technologies. In this case, with knowledge and experience of an outwardly bland process, and a knowledge of the balancing of metabolic processes, we argue that, provided the safeguards of repeated environmental and yield selection are applied, then novel toxicity is unlikely to be the outcome of any breeding technology. At the same time, scientists are not omniscient and cannot provide an absolute assurance that genetic admixture will never give rise to unprecedented events. This admission would, for some, be sufficient to invoke the precautionary principle in relation to GM technology[1]. As we shall argue later the precautionary principle in practice has rather more to do with political agency and endowed power than risk management in society. However, in this case our precise objection to the principle is that it freezes the debate concerning the relative risks of the old and new technologies in relation to their perception and their management.

It is justifiably argued that at least half of the improvements in crop productivity achieved during the 20th century were attributable to breeding success. This has had major social, political and economic consequences in relation to the availability and price of food, and the ability of the UK, for example, to be self-sufficient in grain production since the 1970s. Despite this, in contrast with GM technology, the goals and procedures of plant breeding have rarely been questioned other than by breeders themselves or by their sponsors, the farmers. This is particularly notable given the parallel processes of commercialisation and commodification that the seed development sector has undergone (see Chapter 10 on the patenting of genes for plant biotechnology). The exception perhaps is the study by Jack Kloppenburg (Kloppenburg, 1988) of the consequences of the introduction of F1 hybrid technology to maize breeding in the USA in the 1930s. In this context he identifies significant changes in the selection of breeding goals, in particular the breakage of the informative link between farmers and breeders and its replacement by the drivers and demands of capital replacement and corporate growth. This analysis anticipates much of the criticism levelled latterly at the so-called industrialisation of agriculture and the role played in this by new technologies such as GM. In a farsighted manner it begins to focus attention on the question of how and by whom decisions about the goals of a new technology are taken, rather than on the prosaic notions of points for and against.

It is against this background that we now introduce specifically the topic of genetic modification and its role as a new technology of plant breeding.

---

[1] Developed first in the World Charter for Nature but subsequently enshrined in the 1992 Rio Declaration in principle 15, the precautionary principle states that 'in order to protect the environment, the precautionary approach shall be widely applied by States according to their capability. Where there are threats of serious or irreversible damage, lack of full scientific certainty shall not be used as a reason for postponing cost-effective measures to prevent environmental degradation'.

8.2 GENETIC MODIFICATION

8.2.1 INTRODUCTION

Genetic modification (also known as genetic manipulation or commonly as genetic engineering) was invented in 1973. The technique was based on the natural gene transfer mechanisms that occur in bacteria (and certain other micro-organisms). Micro-organisms can transfer small pieces of DNA called plasmids between cells; gene transfer can also take place when bacteria are infected by viruses (bacteriophage). The invention of different techniques for transferring genes to bacterial cells was based on these processes. The techniques also utilise specific endo-deoxyribonucleases to cut and DNA ligases to re-join DNA molecules in a very precise way. The technique was very rapidly taken up by the pharmaceutical industry. This is exemplified by the deliberate transfer to bacteria of the human gene that encodes insulin, culminating in 1982 in the licensing for use in human therapy of insulin produced by genetically modified bacteria. The bacterium 'reads' the human gene as if it were one of its own and makes a product, i.e. a human hormone, that it has never made before. This success has been followed by the production of many therapeutic proteins and of several vaccines in genetically modified bacteria and other micro-organisms. The technique is used to produce an enzyme that is used in the manufacture of one of the components of aspartame (used to sweeten 'diet' fizzy drinks) and of an enzyme called chymosin for use in cheese making, thereby avoiding the need to extract the enzyme from calves' stomachs[2]. Cheese made this way is sold as 'vegetarian cheese'. It is strange that with all the resistance to GM foods and crops (see next chapter), these applications seem to have escaped attention, especially because, in eating vegetarian cheese, the product of the genetic modification process, i.e. the chymosin, is consumed. Is it the label 'vegetarian' that allays consumer worries?

8.2.2 RESEARCH USES OF GENETIC MODIFICATION

The ability to transfer genes from any living organism into cells of micro-organisms (and later into animal and plant cells) opened up a research use for GM, a use which has grown enormously in scope and sophistication over a quarter of a century. The knowledge about gene structure and function that we now possess could not have been dreamed about prior to the invention of GM. The various genome projects that are sequencing the complete genomes of particular species simply would not be possible without GM techniques. In the context of crops, the recent publication of the complete genome sequences of a 'model' plant species, *Arabidopsis thaliana* (Thale cress) (The Arabidopsis

---

[2] It is interesting to conjecture on the consumer response which might be engendered to the traditional rennet process were it introduced today.

Genome Initiative, 2000), and of a major cereal crop, rice[3], is noted. The information from *Arabidopsis* and from rice, combined with the currently more fragmented genetic information from several other crops, will be of enormous benefit in the identification and tracking of useful genes for the plant breeder. Thus we now have a detailed knowledge of the position of genes along the lengths of the chromosomes and this knowledge is being used to achieve gene transfer, for example between cereal species, both by 'conventional' plant breeding and by GM techniques. So, the contribution of basic GM techniques and the array of methodologies that grew from them has been of huge value in basic and applied research in fields as diverse as medicine and agriculture. Against that background we now consider the genetic modification of plants.

8.2.3   PLANT GENETIC MODIFICATION

The initial development of plant genetic modification also relied on natural mechanisms, namely the transfer of genes from the pathogenic bacterium *Agrobacterium tumefaciens* to the host plant's genome. The genes transferred in this way actually reside on a plasmid, the Ti-plasmid, in the bacterium and only a part of the plasmid, the T-DNA, is transferred (Figure 8.1). The transferred genes become integrated into the plant's DNA and provide the genetic information that causes the plant to form a gall and the cells in the gall to produce nutrients for the bacterium. This is effectively a form of genetic parasitism. Disarming the T-DNA thus led to the development of useful 'vectors' for the genetic modification of plants. In some more recent plant GM techniques, the bacterium has been dispensed with altogether and the gene of interest is either fired into the plant or enters the plant cell through wounds made by very short sharp electric shocks or by slivers of silicon carbide. The array of techniques now available provides a very versatile addition to the 'tool-kit' of plant breeders in their quest to introduce new genetic variation.

Plant genetic modification by gene transfer is both *precise* and *imprecise*. It is precise because, unlike 'conventional' plant breeding techniques, one or a few specific genes, conferring desired characters, are transferred to a plant, the rest of whose genetic characteristics are otherwise unaltered. Thus, single wanted genes can be moved with precision into candidate varieties. However, genetic modification is imprecise because of *position effects*: there is no control over the place within the plant chromosome that the incoming genes are inserted. This causes great variation from plant to plant in the first 'GM generation' in the level of expression of the incoming gene. This means that there must be extensive screening and selection of the first generation of GM plants followed by observation of the stability of inheritance in subsequent generations. However, in practice this phase is shorter than the 'sorting' and evaluation phase in conventional breeding, leading to a faster development of the new varieties. Thus some

[3] http://www.tigr.org/tdb/rice/BACmapping/descriptions.html

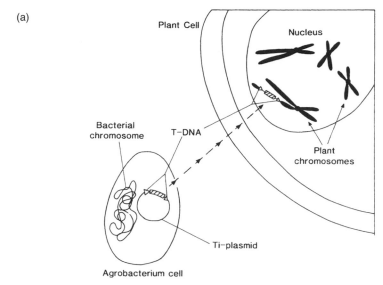

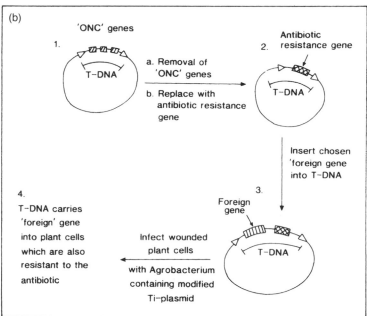

Figure 8.1. (a) Cartoon, not to scale, illustrating the natural gene transfer mechanism mediated by the bacterium *Agrobacterium tumefaciens*. (b) Cartoon, not to scale, showing how the Ti plasmid of *A. tumefaciens* may be used to transfer a selectable marker gene and any other 'foreign' gene into the genome of a recipient plant. Note that other types of selectable marker gene, in addition to those conferring antibiotic resistance, have also been developed.

Both cartoons are reproduced by kind permission of Professors Mike Jones, Keith Lindsey and David Twell and were originally published in Lindsey and Jones (1989).

of the advantages of GM are that it enables the addition of specific genes to well characterised varieties, that those genes can come from widely separated species and that new varieties may more quickly enter the testing and trialling procedures required of all new plant varieties.

However, the main advantage of this technique is that it increases hugely the genetic variety available to the plant breeder whilst avoiding the problem of bringing in unwanted genes. Further, the ability to handle individual genes in the 'test tube' means that direct biochemical modification may be used to generate new variants in existing genes within a crop species. Thus we need to emphasise that although much has been made in some sections of the media of some examples of inter-species gene transfer, many applications of GM will involve transfer of individual genes between varieties of the crop species, or increasing the number of copies of a gene within a crop variety, or switching off genes (as with the Flav^{r}Sav^r® tomato) or even putting back a gene after it has been biochemically modified in the test tube. The distinction between 'conventional' plant breeding and GM techniques is thus very dependent upon the context of the technical debate rather than upon the goal or outcome. Two further examples will help to illustrate this. Firstly, short-stalked wheat was obtained by conventional breeding in the 1970s (Lupton, 1987). However, the allele that causes the short stalk has now been identified, isolated and transferred to rice by GM techniques without the problem of bringing in unwanted characters. Secondly, herbicide tolerant oil-seed rape has been obtained both by mutagenesis in conventional breeding programmes and by GM techniques. This example will be looked at again when we examine some of the objections to GM that have been raised.

8.2.4   APPLICATIONS OF PLANT GENETIC MODIFICATION

In dealing with the actual and potential applications of plant GM we need to emphasise again a point made in the previous section, namely that many of the individual genes that will be transferred in these applications will not cross wide species barriers. This will have implications in the ethical debate for those concerned about inter-species transfer. We also need to distinguish clearly here between GM crops and GM foods, a distinction that is often overlooked in discussion. The 'vegetarian' cheese referred to earlier is clearly a food that is produced directly by a process that involves GM. Further, as already noted, the component produced by GM, i.e. the chymosin enzyme, although used in the cheese production process, is to a large extent trapped in the cheese and thus also becomes part of the product.

Turning to consider crop modification, we can separate those modifications that directly affect the food component of the crop (and thus may come into the category of GM foods) and those that affect plant characteristics that have nothing to do with the food component (but see Chapter 9 for a critique of this position). In the GM food category, the slow-softening tomato, the first GM crop to come to market, is a clear example: the GM process affects the quality of the consumed product. There are other applications under development that also directly affect the food product. However, it is very likely that only a relatively small proportion of the modifications will affect directly the food component of the crop; many applications will enable the crop to withstand environmental and biological stresses while others will affect the way that a crop is grown. There are also several current applications to non-food products (e.g. biodegradable plastics) from crop plants, including some crops that are grown only for their non-food component(s), and the number of examples is likely to grow. The following list is not exhaustive but gives examples of current applications and near-market research. It should be noted that several of the applications have the potential to reduce markedly the use of agro-chemicals. The applications that are asterisked are actually in commercial use (although not in the UK).

Modifications directly affecting the food content of crop plants

1. Slow-ripening fruit with longer shelf-life.*
2. Changes in fat, starch or protein content to increase nutritional value and/or improve baking and food-processing qualities.
3. Increased vitamin, nutrient or anti-oxidant content. The most interesting example of this is 'Golden Rice', which has greatly enhanced vitamin A content and thus has the potential to save the sight of millions of children in Asia. This was developed by Professor Potrykus and his collaborators (Ye et al, 2000) in Switzerland with the financial support of a philanthropic organisation, the Rockefeller Foundation (Kryder et al, 2000). Another example is the possibility of increased vitamin C content, thus providing increased levels of anti-oxidants (protection against cancer etc) (see Muir et al, 2001).
4. Modifications to post-harvest behaviour, e.g. prevention of sprouting in potato thereby avoiding the need to treat with chemical anti-sprouting agents.
5. Removal of allergens and other potentially toxic substances from certain foods (e.g. removing cyanide-generating chemicals from cassava).

Modifications affecting crop 'performance'

1. Increased resistance to viral, bacterial and fungal diseases.*
2. Increased resistance to environmental stress, especially cold, drought and salinity.
3. Increased resistance to predation by pests, especially insect pests.*
4. Early production of flowers (and hence seeds), leading to a shorter growing season.

5. Prevention of premature shedding of seeds.
6. Diversion of plant resources from general biomass to specific production of the desired product (usually seeds/fruit). This includes the development of short-stalked cereal varieties.

Modifications affecting crop husbandry

1. Herbicide-tolerant crops leading to improved weed control with less use of weedkiller.*
2. Short-stalked varieties (as already mentioned above).

Crops for non-food use

1. Enhancement of plants' ability to remove toxic chemicals from soils.
2. Production of pharmaceutical products in plants, including edible vaccines.*
3. Production of industrial and research chemicals in plants.*
4. Modification of oils, fibres and other polymers to meet new industrial requirements.
5. Production of bio-degradable plastics, detergents etc. in a sustainable production system.
6. Many of the applications that affect crop performance and husbandry are equally applicable to ornamentals and other plants grown for non-food uses, e.g. expression of chitinase in roses to provide resistance to fungal pathogens.

8.2.5   PRIORITISING THE GOALS

The list of goals above is a consequence of distinct avenues of prioritisation both in the research and the implementation phases of the technology. For instance, vitamin A-enhanced rice is a goal prioritised by a charitable foundation on behalf of the rural poor and disadvantaged. Insect-resistant cotton was prioritised on the basis of proprietary technology that had the attraction of reducing the need for insecticides and the winning of market share for cotton seed. Herbicide tolerance was prioritised on the basis of proprietary technology that had the potential to control weed competition and enhance yields, with financial return to farmer and breeder and herbicide manufacturer, while avoiding the negative environmental effects of weed control via deep tillage.

The latter two examples represent the more generalised position where priorities are set by narrow sectoral interests coupled to the need for the owners of the technology to recoup their investment. While these are reasonable goals they comprise single relatively crude solutions to single agronomic problems. This assumes that a single solution is appropriate in all contexts, which is to say that they are 'narrowly informed'. At this stage in the development and deployment of the technology this is not unexpected or unreasonable, but for the future a large test of the technology will be the extent to which its goals can be democratised. How can we ensure that social needs as well as issues of environmental impact are factored into *decision-making* so that local goals can be informed by local knowledge and deployed as diverse solutions to match diverse problems.

One solution is to widen the consultative process beyond the closed rooms of institutions. Another is to broaden access to the technology by relaxing the control of ownership of the technology and/or by making the tools more applicable at a small and local scale. One compelling example of the latter approach which we can quote would be the development of *apomixis* as a tool for breeding. Apomixis is a mode of asexual reproduction through seed which is widespread in the plant kingdom though uncommon among crop species. For practical breeding purposes it could offer the possibility of instant fixation of hybrids as *true-breeding* F1 lines without the need for parental inbreeding that is normal in F1 hybrid systems. Apomixis could, for the small local breeder, provide a mechanism for making hybrids, for example between local landraces and high-yielding varieties. In this case many diverse lines from the cross could be evaluated locally for local performance and the best performers maintained as asexually reproducing lines.

The development of such a tool, available for important species, especially those which are difficult to breed, is dependent upon GM technologies, but would be of inestimable value for *local* breeding. The tool has the potential to re-establish small-scale, local breeding initiatives but could equally be valuable to larger-scale corporate breeders for trait introgression and the maintenance of otherwise transient and unstable gene combinations. Apomixis, if constructed so as to be compatible with male sterility, might also find application as a means of preventing gene flow (see below) from transgenic lines (Coghlan, 2000; Masood and Jefferson, 2000). The Bellagio Declaration represents a stand taken on the ethical development of this technology by a group of practitioner scientists[4] in the light of their concern that the technology should be developed in a way that makes it broadly available.

8.3 OBJECTIONS TO GM TECHNOLOGY AS APPLIED TO CROPS

8.3.1 INTRODUCTION

This section focuses on the objections that have been raised against GM *per se* rather than those raised against intensive or industrial agriculture or against the ways in which GM is being used at the social and financial levels. Nevertheless, we acknowledge the importance of general social issues and of issues in global justice that apply to the use of all technologies (and not just to GM) and note

---

[4] See the Bellagio Apomixis Declaration at http://billie.harvard.edu/apomixis

that crop GM has proved a good starting point for the discussion of these issues (Anderson, 1999; Monbiot, 2000). Indeed, these issues are raised by Sue Mayer in the next chapter. However, we note that many campaigners conflate these issues in a confusing way, intermingling their objections to the applications of GM with criticisms of the science and the scientists, almost along the lines of 'The behaviour of company X in the area of GM food/crops is unethical, therefore the science must be "wrong"'. Here we attempt to separate out the issues and thus focus on the objections that have been made to the technology itself.

8.3.2   IT IS NOT NATURAL (THE 'YUK FACTOR')

Although gene transfer mechanisms do occur in nature and indeed have been made use of in the development of GM techniques, it is obvious that use of GM requires extensive intervention. It is, in that sense, not natural. However, as we have been at pains to point out above, it is no more 'un-natural' than many practices used in conventional breeding and it lacks many of the side-effects of conventional breeding (unwanted genes for example). Furthermore, so much that we rely on for 'normal' life is not natural that we wonder how much validity this argument carries. Indeed, Reiss and Straughan (1996) have argued that the natural versus un-natural distinction has no place in this ethical debate. Others have a more pragmatic view, defining natural as what we have done and now do, and un-natural as what we might do in the future or have yet to do. In this regard the 'yuk' response can emerge as an aversion to change.

However, we recognise that there are some with a deontological objection (see Chapter 1) to gene transfer (because of their belief or value systems it is regarded as intrinsically wrong and therefore abhorrent to them). We might presume that such people will avoid diet drinks, vegetarian cheese, recombinant vaccines and other pharmaceuticals produced by GM techniques, in addition to ensuring that they reject all products of GM crops (for a discussion of consumer choice, see the next chapter). Nevertheless, the natural/un-natural boundary, however nonsensical and inconsistent in logical argument, does highlight the need for ethical principles to support the drawing of lines between socially acceptable and non-acceptable intervention, and between reasonable and overbearing regulatory imposition. In dealing with this issue the Nuffield Council on Bioethics Working Group on GM Crops (Nuffield Council on Bioethics, 1999)[5] was confident that the technology of GM *per se* is not morally abhorrent, as indicated by the universal acceptance of life-saving and enhancing pharmaceuticals. Nevertheless, the working group recognised that within a pluralistic, multicultural society exemplified by the UK, the use of cultural value systems[6] as the basis for drawing lines and setting boundaries, is problematic. Consequently they concluded that the rights of those who wished to avoid GM products on the

[5] *Genetically Modified Crops: the Ethical and Social Issues*, www.nuffieldfoundation.org
[6] In this context we regard religion as part of culture.

basis of conscientious objection should not be over-ridden[7]. This implies the requirement for the labelling of GM-derived food products. However, the report suggested that the requirement was not absolutist. It was tempered and lines were drawn on the basis of justice and the balancing of rights in terms of the precise labelling provision and thresholds, in order not to generate unreasonable costs which would tend to fall disproportionately on the less well off.

8.3.3   IT IS TOO NEW/NOT ENOUGH IS KNOWN/MORE EXPERIMENTS ARE NEEDED (IT IS BEYOND RISK ASSESSMENT)

This objection is frequently raised in the media. The argument then continues that it is unethical to produce GM foods/crops because we are firstly experimenting on humans (as consumers) and secondly exposing the natural environment to unknown hazards. In the British newspaper *The Guardian* of 9 March 1999, for example, Don Atkinson, a financial page journalist, wrote that 'the supermarkets . . . have been . . . force-feeding the public with the results of revolting genetic experiments.' For the scientist it is difficult to know how much work is necessary or desirable to ensure that 'enough is known'. In this context, the time-scale to the commercial production by GM techniques of human insulin is very interesting. GM techniques were first developed in 1973. The cloning of the human insulin gene and its use to produce insulin in bacterial cells was reported in 1977 (although it was probably actually achieved in the previous year). The product, recombinant human insulin, was licensed for therapeutic use in 1982.

Now let us compare that time-scale to Monsanto's herbicide-tolerant soya beans. Plant GM was first developed in 1983 and by 1985 could be routinely achieved with several different plant species. By the end of 1985, plants resistant to the herbicide glyphosate (Roundup® or Tumbleweed®) had been produced by GM techniques and had even been subjected to small-scale field trials. Glyphosate-tolerant soya beans were first grown commercially on a large scale in the USA in 1996 and the amount grown has increased annually since then. However, if soya beans were a UK crop they still would not be permitted to be grown here commercially (this is being written early in the year 2001). It is our contention that so many billions of GM plants have been grown in labs, glasshouses, in field trials and on a commercial scale that we do not need more experimentation to prove the validity of GM *per se*.

8.3.4   HOW DO WE KNOW IT IS SAFE?

This is a good question and is asked mainly in the context of food safety. How do we know that the genetic modification, whether or not it directly affects the food

---

[7] Would we not all these days wish to avoid the consumption of food derived from plants the growth of which had been 'encouraged' by the sacrifice of virgins, as in the earlier forms of organic agriculture?!

content of the crop, does not have unexpected side effects? It is for this reason that all new food crop varieties, whether conventionally bred or produced by GM techniques, are subject to food safety evaluation and institutional approval. However, one of the problems here is that the evaluation may not be in the public domain and thus the public are unaware that it takes place. It is ironic that one of the most detailed papers describing appropriate techniques for food safety testing of GM crops (Martens, 2000) has been written by a scientist working for one of the companies[8] that has come in for most criticism of its behaviour in respect of these crops! Further, the public's confidence is not helped by the media portrayal of preliminary developments as if they were near-market products. Indeed, we believe that media 'hype' has played a major role in shaping public attitudes to GM in the UK (however, see the next chapter for a contrasting view). Nevertheless we recognise that, rightly or wrongly, attitudes to science in the UK have been affected by the BSE debacle. Food is an absolute necessity and thus is an emotive topic. People need to be sure that their food is safe (and this goes for 'organic' foods as much as for those grown in conventional agriculture).

One example that is often quoted by campaigners in order to support their case that GM is inherently unsafe is that of the amino acid tryptophan produced by genetically engineered bacteria. Very few people suffer from tryptophan deficiency but nevertheless significant numbers of people take tryptophan as a dietary supplement. As such, it escapes the rules that govern the safety of both foods and drugs (at least under current UK and USA rules), and here the problems start. Tryptophan may be produced commercially by bacteria that have been genetically modified to over-produce it (i.e. to make much more than they need for their own metabolism). One problem with this is that some of the excess is converted by the bacteria into a by-product that is harmless to the bacterial cells but poisonous, even lethal, to mammals. The tryptophan must therefore be properly purified. Several batches of tryptophan were delivered to the USA from a company in Japan that, in order to save money, had omitted some steps from the purification procedure and, unbeknown to them (because they also failed to carry out proper quality control on the product), the harmful by-product remained with the tryptophan. The results were horrifying: over 30 people died and many more suffered from a crippling muscular disease (Gershon, 1990; Hill et al, 1993).

We would wish to emphasise that this is a tragic case but it has nothing to do with the ethics or safety of foods produced from GM crops despite its citation by campaigners. Nevertheless we must not be complacent and it is necessary that the safety testing of all new foods, not just those from GM crops, is both adequate and in the public domain. Finally we note that some of the groups that have urged caution in the application of GM technology to crops, such as the Union of Concerned Scientists in the USA, state that there is nothing inherently unsafe about GM techniques (Rissler and Mellon, 1996).

---

[8] Dr Martens wrote the paper whilst at Monsanto Europe (but also see the review by Kaeppler, 2000).

8.3.5 GM CROPS WILL DAMAGE THE ENVIRONMENT

As pointed out by Reiss in Chapter 1 and by Southgate in Chapter 3, the ethical issues concerning the natural environment transcend generation and species. Environmental harm may not only change the ecological balance by changing the frequency and/or distribution of species but in doing so may also affect the welfare ('happiness' in utilitarian terms: Chapter 1) of future generations of humans. Are there environmental hazards particularly associated with GM crops? Should the precautionary principle (see the next chapter) be applied? The Nuffield working group (Nuffield Council on Bioethics, 1999) took the position that regardless of our personal view of the environment, whether it has the qualities of an organism, whether it has rights, the duty of care placed upon us as temporary custodians is overwhelming. This position can be seen in ethical terms as respect for the rights of future generations, which, for many, extend beyond stewardship of the productive natural resource to the preservation of the sense of wellbeing that the forms of nature induce in us. This reason is sufficient to trigger a cautious and considered approach to the widespread and rapid changes in agronomic practice that are likely as a consequence of the implementation of the new technologies. While recognising that changes in farming practice are not unique to GM crops, the working group argued the need for institutional oversight, in terms of regulations ensuring best practice, and continuing surveillance. They recommended to the UK government the setting up of an 'overarching committee' for this purpose with a broad ranging brief to consider the environmental, social and ethical implications of agricultural biotechnology.[9] Though size is not everything, the group were concerned that scale effects and cumulative effects should be prioritised as one of the key steps of going forward from small and farm-scale trials to commercial deployment.

The working group did not support the notion of moratoria or outright bans on the technology, but at the same time recognised that there is a window of opportunity as the trialling of transgenic crops is progressively undertaken, to reappraise the impacts of all modern agriculture on the environment as well as the questions which are specific to GM crops.

We deal with these latter questions under two headings, firstly gene flow and secondly more general environmental effects, including effects on biodiversity.

8.3.6 WILL THE GENES ESCAPE?

Genes inserted by GM techniques are not autonomous entities. The new genes are integrated stably into the crop plant chromosomes and form part of its genetic make-up. They cannot thus 'escape' as individual genes and then invade other living organisms. They will behave no differently when ingested by humans

---

[9] In the event a committee was set up under the title Agricultural and Environmental Biotechnology Council in parallel to the Food Standards Agency. In its initial consultative phase www.AEBC.gov has proposed study projects on gene flow and the consequences of scale-up.

or other animals than the thousands of other genes which plants contain. This simple fact should allay the fear imparted to many that transgenes themselves pose a threat to health. Checking the stability of integration takes place over several years during the evaluation of GM varieties, just as the stability of inheritance is checked for crop varieties produced by conventional breeding (with the exception of F1 hybrids). So the question should rather be *Will the crop plant hybridise readily with wild species and if so what will the consequences be?* Furthermore, this question must be applied to both conventionally bred and GM crops because there is *no difference* between the hybridisation potential of a conventionally bred variety and a GM variety of the same crop species (some writing on this subject has been most misleading in this respect). For most crop plants such out-crossing is not a possibility. For example, for maize grown in the UK or in mainland Europe there is not a wild relative within several thousand miles. Even in the USA it is unlikely that modern maize could ever hybridise with its presumed wild progenitors. However, using UK crops as our examples, there is for sugar beet a possibility of crossing with wild beet and for oil-seed rape a chance of hybridisation with wild radish and with wild *Brassica* species. Taking the latter example, rape and its wild relatives are insect-pollinated and the pollinators certainly fly far enough to distribute pollen (whether from genetically modified or conventionally bred plants) over several hundred metres. Even so, cross-hybridisation between these species must be considered a very rare event and further there is no certainty that a hybrid will be fertile. The rarity of this event is such that, despite the vast hectarage of rape grown in the UK, the *establishment* of hybrids between rape and its wild relatives has not been reported although they certainly occur. The possible reasons for failure of such hybrids to establish themselves are that firstly they may be less 'fit' than the parental species (Hauser et al, 1998a, 1998b) and secondly that they are in any case very rare. For example, in populations of wild radish growing in the vicinity of oil-seed rape crops, the frequency of hybrids was less than one in one hundred thousand (Chevre et al, 2000); in populations of wild *Brassica* species in areas where oil-seed rape is extensively grown, one hybrid plant was detected in $16\,000\,\mathrm{km}^2$ (Davenport et al, 2000). However, even the rare occurrence of such hybrids between rape and its wild relatives means that oil-seed rape genes can find their way into the wild plant genomes if the hybrids themselves then back-cross with the wild parent. This can of course only happen if the hybrids are fertile. So the consequences of such introgression should be considered, but we reject the idea that 'superweeds' could arise this way, in so much as the concept of a superweed is linked to the vision of plants with exaggerated potential for invasion of the environment. Generally the acquisition of a transgene by a weed would tend to make it less fit, competitive or invasive. In the case of the acquisition of herbicide tolerance weeds would become more competitive than their neighbouring weedy cousins when treated with the herbicide. Since herbicides are not used in the unmanaged environment the consequences of out-crossing would only ever be seen within crop. The consequence here would be loss of utility of the herbicide-tolerant crop type and the corresponding herbicide for that application. It should be sufficiently within the interests of farmers, breeders and agrochemical manufacturers for this not to happen that we might expect to see avoidance and surveillance of such effects to be high on the agenda.

Long-term experiments have been carried out, by default, which can inform this issue. For instance, some 30 years ago a new allele of the gene responsible for the synthesis of a characteristic seed fat (the *low erucate* allele) was introduced into the environment (i.e. into agricultural production) in oil seed rape (canola) varieties. Since that time the allele has progressively become a standard in rape varieties, and pollen containing it must have reached all parts of the temperate agro-ecological zone. All the opportunities for 'pollution' of other rape varieties and out-crossing to wild relatives have been available to this gene for over 30 years, yet no consequences have been reported. This could be because farmers and seed producers have been very effective in managing diverse crop varieties, or because the trait in question reduced the fitness of the recipient wild species, or because the frequencies of out-crossing are negligible (as we have argued above). Even if we ignore the final explanation the scenario is still reassuring and informs our risk assessment. Thus it is wise to conduct large-scale field trials of any GM crop that has the potential to out-cross, and to assess whether there are any circumstances in which the introgressed transgene could increase the fitness of a recipient weed. However, in this context, in relation to herbicide tolerance it seems strange that a conventionally bred herbicide-tolerant rape may be introduced without environmental evaluation whereas its GM equivalent is subject to several years pre-commercial environmental evaluation.

A related question is whether GM crops themselves will become established in the non-agricultural environment as 'superweeds'. Experience of agricultural crops in general is that the strong genetic selection for use in agriculture has resulted in their being very poorly competitive in the wild. This has been confirmed in a ten-year study of five cereal and non-cereal crops in which none of the crop plants established themselves in the wild (Crawley et al, 2001). Furthermore, and crucial to this debate, the GM crops were no more invasive or competitive than their non-GM counterparts.

8.3.7 GM CROPS WILL REDUCE BIODIVERSITY

Concerning the effects of the crops themselves on the environment, there is no doubt that intensive farming *per se* has led to a loss of biodiversity. This is often overlooked in the GM debate although the effects of changes in farming practice in the UK have been highlighted by recent literature[10]. Genetically modified crops are no worse than conventionally bred crops in respect of their use in intensive agriculture. We must add that some actual and potential uses of genetically modified crops will have the potential to reduce the use of agrochemi-

---

[10] For example, see reports by the UK's Royal Society for the Protection of Birds on the decline of certain farmland bird species: http://www.rspb.org.uk

cals, such as pesticides, themselves a major cause of the loss of local biodiversity through impacts on non-target organisms. However, even this raises concerns. If a crop plant is modified to be resistant to insect pests (essentially by producing its own insecticides) will beneficial insects also be harmed? There is indeed some evidence of increased mortality amongst lacewing flies, ladybirds, and in the USA, monarch butterflies (Jesse and Obrycki, 2000, and the next chapter) but this evidence must be taken firstly alongside the fact that agricultural insecticides kill all insects and secondly that the use of insect-resistant crops will reduce significantly the use of these insecticides. Furthermore, a recent report from the USA (ASPB, 2001) suggests that the experiments on monarch butterflies were misleading in respect of what actually happens in the field. In fact, GM maize with its own insecticide is unlikely to have any impact on the monarch butterfly population: caterpillars feeding on plants 'naturally' dusted with pollen of maize genetically modified with the Bt toxin (i.e. natural insecticide) gene had a 17% higher mortality rate than caterpillars feeding on milkweed plants dusted with non-GM pollen. The effect was most marked within 3 metres down-wind of the edge of the field although could be detected up to 10 metres from the field edge; this represents only a very small proportion of the caterpillars' available food sources (Jesse and Obrycki, 2000). Once again however, it seems wise to conduct farm trials with such GM crops before adopting them for commercial use as indeed is suggested by Jesse and Obrycki.

In relation to the wide-scale deployment of herbicide-tolerant crops, concern has been expressed by environmentalists that highly effective weed control, as promised by post-emergence use of a broad-spectrum herbicide such as Roundup®, will lead to loss of biodiversity at a number of levels. The concern is linked to the depletion of the weed seed bank, which may be a direct food source for farmland birds but more important may be the loss of the weed plants themselves as host to insect larvae, which fulfil the dietary needs of some birds. Loss of hedgerows and field margin habitats has compounded this threat, which should be examined carefully. We would contend (as interested naturalists and birdwatchers), that the threat, if real, substantiates our point that agronomic practice is the issue, not transgenic plants *per se*. Any agronomic weed control system, be it selective herbicides, broad-spectrum herbicides used in conjunction with herbicide-tolerant crops or organic deep tillage[11] will, if effectively deployed, reduce biodiversity. That is the whole rationale! What is required is sensible discussion of strategies for minimising consequences not only for the visible members of the biota recognised by the public as 'the countryside' or as 'wildlife', but also for the invisible microorganisms which sustain the essential processes of chemical recycling below ground.

Various suggestions have been put forward to ameliorate effects on non-target organisms. Of these, proposals for re-establishment of field margins with 'herbicide free cordons', and the more systematic creation of refuges and corridors for wildlife, are worthy of close attention.

---

[11] Deep tillage has the additional effect of killing insect larvae directly.

Finally we note that, although organisations campaigning against GM crops have often done so in the name of environmental concerns, some of the actions of certain campaigners bear no relation to these issues. For example in the UK in 1999 there was a wanton destruction of female (and therefore non-pollen-producing) transgenic poplar trees. The trees had been modified so that the wood needed less harsh chemical treatment than the normal poplar wood in the manufacture of paper. This in turn would reduce the chemical outputs into the environment from paper mills. It can only be presumed that the destruction of such trees arose either from a deeply held deontological objection to GM technology *per se* (see Chapters 1 and 2) or from a desire to act against a trans-national company that thus represented 'big business' (cf. Monbiot, 2000).

8.3.8 GM CROPS WILL CONTAMINATE 'ORGANIC' CROPS

Organic farmers have expressed concern that genetically modified crops may contaminate their crops. A number of organisations, but in the UK primarily the Soil Association, are involved in validation of organic crops. Until the advent of GM crops, the particular variety of a crop species grown was of no concern. What mattered was the husbandry of the crop and good stewardship of the soil. However, the Soil Association has outlawed varieties generated by genetic modification despite the real possibility that some of these varieties may be particularly suitable for organic husbandry. The situation in the USA is some-what more complex because of the greater number of validating organisations and the disagreement amongst those organisations about GM crops. Focusing on the UK, the Soil Association has previously recognised that organic farmers cannot be responsible for the actions of nearby 'conventional' farmers and so neither the drift of pesticide or fungicide from a neighbouring farm nor the possibility of fertilisation by pollen from the same crop species, conventionally grown on the neighbouring farm, led to withdrawal of validation. Indeed, dilution of the organic crop by a small percentage of non-organic material was accepted in the validation process[12]. This leads us to a specific case that illustrates the general point very well. The claim has been made that validation of organi-cally grown sweetcorn would be threatened by pollination from a nearby field of genetically modified maize (maize and sweetcorn are different varieties of the same species as has already been mentioned). This claim needs looking at in some detail. Any planting of genetically modified crops in the UK must be approved by ACRE, the UK Government's advisory committee on release to the environment. The committee observes the rules on closeness of planting of varieties based on knowledge of the rate and distance of pollen spread. Maize is wind pollinated; the wind can carry pollen grains considerable distances. How-ever, the pollen dehydrates and becomes non-viable very quickly (and the pollen of genetically modified plants is no different in this respect from the pollen of conventionally bred plants) and so the ability to bring about a successful

---

[12] Recent labelling requirements imposed by FDA allow up to 5% of non-organic crop to be included in organic-labelled products.

fertilisation falls off very quickly with distance. It is such data that both ACRE and plant breeders use to define inter-varietal planting distances whether the varieties are 'conventionally' bred or genetically modified. So, there is a possibility that a few seeds in a few cobs of 'organic' 'sweetcorn' will have arisen from fertilisation by pollen from a GM maize. Under the normal terms of validation, the presence of seeds arising from pollination from a conventionally bred but non-organically grown neighbouring fodder maize crop would not result in loss of the right to use the term 'organic', even though, by the same standard, the crop would no longer be describable as sweetcorn either. The debate is not helped by use of such double standards.

The ethical principle at stake is the right to welfare. It is not in the interest of our welfare to be unnecessarily subjected to uncertainty, fear and alarm by exposure to misinformation. The words *contamination* and *pollution* have been widely used in a manner akin to shouting 'fire' in a crowded room, as witnessed by the unnecessary wearing of protective suits by those taking highly publicised direct action against GM crop trials. However strongly the opponents of the technology feel about their views it is questionable whether the deliberate generation of fear and alarm are acceptable methods of voicing them.

8.4   THE ADEQUACY OF THE REGULATORY ARRANGEMENTS IN RELATION TO PUBLIC WELFARE

The regulation of the technologies of GM is unique in human endeavour in that it proceeded from a recognition among *scientists* that here was a technology which had the capacity to ramify faster than their individual ability to consider its consequences, rather than as a reaction (contrast with the panic reaction to the recent railway accidents in the UK) to an incident or an actuarial prediction of some tangible damaging consequence.

To date there has been no foreseen or unforeseen damaging outcome that can be used as a yardstick for regulatory safeguards. Regulation has had therefore to be generous on the side of caution and to assume worst case in the framing of precautionary measures. Examples of this can be found in the regulations for contained use of microorganisms, where guidelines for disposal of waste from GM experiments has required routine sterilisation conditions well in excess of those required to kill the host organism most usually used for GM work.

The history of the imposition of regulation should reinforce public confidence in the concern of scientists for public welfare and safety. The defining moment in 1974 was the publication in the prestigious journal *Science* of a letter from the distinguished biologist, Paul Berg. This was followed closely by the Asilomar Conference in 1975, at which scientists entered into a voluntary moratorium on the use of the recombinant DNA technology while its possible risks and applications were studied and a set of guidelines developed to ensure its development for human benefit. Guidelines have since been translated into regulations, which have been progressively modified as knowledge has grown and the technology has become increasingly sophisticated, and can be practised in a wider range of hosts. Along with the regulations a cadre of specialist scientists has emerged within research institutions, commercial organisations and public (government) supervisory and surveillance bodies who carry responsibility for oversight of regulatory compliance and the reporting of incidents. Nevertheless, after almost 30 years of activity diffused into almost every biological research institute and university, and of large-scale commercial production and widespread use of products in medicine, we have no instance of actual harm to report. (The reader will recall that the 'tryptophan tragedy', Section 8.3.4, was caused by a failure to purify the product rather than arising from the GM *per se*.) This speaks highly both for the intrinsic safety of the technology and the adequacy of public safeguards.

The counter-argument is that most of this history of safe practice comes from the contained use of GM organisms and that only recently have we started to populate the world significantly with GMOs. The further argument is then that there is no history or basis for informed assessment of the risks of eating or releasing GMOs into the environment, and consequently the precautionary principle should be applied, or a moratorium on field trials and commercial planting should be imposed. At this point the discussion becomes highly politicised and polarised. The precautionary principle becomes exposed not as a guide for ethical decision-making but as an instrument of political agency as dissenting voices struggle to influence policy. We note here the words of the science journalist Matt Ridley: '. . . the precautionary principle . . . ignores the risk of doing nothing . . .. Standing still is not a risk-free option.'[13] As scientists we would argue that we have sufficient knowledge and experience to inform risk assessment sufficiently to avoid the need for moratoria or the precautionary principle. Within Europe and the USA some 30 000 field trials have been conducted and some trillions of transgenic plants have been grown in the field. At the same time we accept that risk assessment is an imperfect tool and that best practice in the scale-up from small plots to field trials to farm-scale trials to commercial planting should be moderated by surveillance and critical review by a diversity of informed observers. As was articulated by the Nuffield Working Group, the transparency of this process is particularly important to demonstrating to the public that their welfare is a priority. This approach corresponds to the consequentialist ethic of 'proceed with caution'.

On the global stage there is less cause for reassurance. The issues of GM have been conflated with those of the global economy such as commodification, import substitution, technological colonisation and non-tariff barriers to trade. There has been a history of major confrontation between developed and developing countries in the negotiation of the Biosafety Protocol within the frame-

---

[13] Ridley, M. *Technology and the Environment: the Case for Optimism.* Prince Philip Lecture, London, UK, May 8, 2001.

work of the Convention on Biodiversity. This confrontation has been at least in part articulated by campaigning groups anxious to capture moral high ground on behalf of developing countries. These issues are further explored in Chapters 9, 10 and 11.

There is much ground still to be covered in the negotiation of an international provision that is not overburdened by political considerations. It is to be hoped that the Conference of the Parties to the Convention on Biodiversity discussions will be able to move on to engage with the actual biosafety provision and that the ethical principles of justice and welfare will be able to reappear.

8.5   THE ROLE OF SCIENTISTS

As scientists and citizens we have several roles in the ethical debate and future of the technology. It is our responsibility to disentangle and articulate the complexities of the issues raised in the light of our knowledge. At the same time it is our responsibility to point out where there are gaps in our knowledge that are relevant to decision-making based on the balancing of risks. In this light it is our duty to review and inform policy on continuing surveillance especially in relation to novel agronomy and its environmental impact. Support for these institutional processes takes account of the key issue of environmental protection, which is that 'markets have failed to make activities internalise their full costs'[14] or even recognise them in relation to the consumption of environmental resources (renewable and non-renewable), and environmental impact. It is further our responsibility to support the informing of decision-making where we have special knowledge, as it is also to resist disinformation where this is propagated by the media. Finally, it is our responsibility to ensure that our own science remains widely accessible. The Nuffield Council on Bioethics Working Group specifically recommended that scientists and their host institutions avoid exclusive licensing arrangements for their intellectual property so as to ensure broad and equitable access.

8.6   CONCLUSIONS

We have attempted to deal with a complex array of issues, some of which we as scientists regard as specious, that have sprung up around the GM crop debate in order to create final space for a consideration of the key challenge, which is the following: How do we apply ethical principles to the decision-making processes which will govern the future deployment of GM technology in order firstly to ensure that its benefits can be delivered where they are most needed, and secondly to ensure that any risks that might be associated with rapid changes in

[14] James Boyle, *A Politics of Intellectual Property: Environmentalism for the Net?* http://james-boyle.com/

agronomic practice are minimised? We dissect this requirement into a set of three issues, illustrated by analogies, onto which we need to apply our ethical principles.

8.6.1   HAMMERS AND NAILS

Hammers represent the current set of tools, and nails the diverse set of applications. How do we select the nails and who wields the hammers? As we have seen and shall see in Chapter 10, the hammers are currently in the hands of a small number of commercial players and hence they select the nails. If one has a hammer one is tempted to bash with it to justify having it; that is to say, that with a hammer in your hand everything may look like a nail. Added to this, the hammers are as yet quite crude, and bear the assumption that small genetic additions will solve all problems. We have argued both for tools to be made more widely available and to be made more sophisticated and applicable at local scale (less like hammers and more like surgical forceps). We have further argued for the choice of goals to be more democratic in terms of the social, agronomic and economic outcomes which they support. Local information is very important to the choice of 'nails'. Broad consultation is a critical component for ensuring both justice and fairness in the distribution of the technology either as finished products or bare tools, as it is also for the support of informed decision-making. This is particularly relevant to the fair distribution of both benefit and risk. If as society we are expected to underwrite the risks of changes to agronomy and the environment then we should be heard. The better informed choice of nails should feed back onto the better design of tools.

In general there is reason for concern in that the current choice of goals is very heavily balanced towards agronomy in the developed world. Although the point is often made that the technology has much to offer for agriculture and food security in the developing world, this is poorly reflected in the levels of research recorded for poor peoples' crops (with the possible exception of rice). As can be seen by analysing the abstracts of international conferences on plant molecular biology[15], the relative effort deployed on poor peoples' crops and on agronomic problems that overwhelm resource-poor agriculture is depressingly small. (see also Chapter 11).

8.6.2   SHARKS AND RED HERRINGS[16]

Red herrings represent some of the issues raised, as seen above in opposition to GM technology, that over-emphasise risk concerning food safety or the environ-

---

[15] For example, the sixth International Plant Molecular Biology Congress, held in Québec in June 2000.
[16] For readers unused to UK idiom, a red herring is an irrelevant (and probably distracting) item introduced into debate.

ment and are not really specific to GM in any case. The problem with red herrings is that they take our eyes off the sharks, which, in this case, represent the narrow focus of decision-making. Sharks also appear in the guise of restrictive environments for innovation and creative solutions. These are manifest either as oppressive regulatory strictures that act as cost disincentives, especially against the small innovator, or as exclusive cartels based on intellectual property or vertical integration (the control of the food supply chain for an individual crop from top to bottom). This analogy thus enables us to focus on the issue of justice and fair distribution of risk and benefit.

8.6.3   BABIES AND BATHWATER

As we have seen, careful study of the relationship between the monarch butterfly and 'bt corn' (i.e. corn expressing the *Bacillus thuringiensis* insecticidal toxin) has shown this issue to be a red herring; these studies in fact put bt technology in a good light in comparison with the non-selective spray-applied insecticides. However, had the initial surmise been correct, would that have implied that GM *per se* was a bad thing? No, certainly not. It would have shown that this particular application was inappropriate for widespread use and should either not be used at all or not used in locations where there was a possibility of non-target insect mortality from pollen consumption. We have argued that institutional oversight will be necessary to ensure that the rights of future generations to a sustainable and enriching natural environment are protected, and have generally aligned ourselves with the recommendation of the Nuffield Report concerning surveillance of the consequences of scale-up.

Undoubtedly, there will be mistakes made in the choice of goals for the technology just as there have been for all new technologies, but these should be seen in the light of the decision-making process and its deficiencies, not as a characteristic of the technology itself. These mistaken goals will be, as they deserve to be, thrown out of the bath when subjected to informed scrutiny. Conversely, there will be applications, among which we anticipate will be such crops as Golden Rice, that will in time be regarded as social goods and direct contributors to human welfare. Within the constraints of 'proceed with caution' these are the 'babies' to be nurtured and kept well clear of the plug hole.

The issue of babies and bathwater is relevant to the rights of individuals who have belief- or values-based objection to the technology, to be able to avoid it or at least avoid eating it. The simplest way to respect their rights is to throw out both baby and bathwater. However, this ignores the rights of those who might cherish the baby.

The ethical challenge is to find a solution that takes account of the legitimate interests of all parties but does not assign a veto to any one of them. The challenge is also to find ways of differentiating between baby and bathwater without becoming proscriptive and censorious.

ACKNOWLEDGEMENT

We are indebted to Dr Richard Jefferson of the Centre for Application of Molecular Biology in Agriculture for introducing us to his valuable metaphors of babies, sharks and hammers and nails.

REFERENCES

Anderson, L. (1999) *Genetic Engineering, Food and Our Environment*. Green Books, Dartington, UK.

The Arabidopsis Genome Initiative (2000) Analysis of the genome sequence of the flowering plant *Arabidopsis thaliana*. *Nature*, **408**, 796–826.

ASPB (2001) EPA, ARS, PNAS show monarch caterpillars can live in harmony with bt corn. *ASPB News*, **28**, 14. (See also www.ars.usda.gov/is/br/btcorn).

Chevre, A.M., Eber, F., Darmency, H., Fleury, A., Picault, H., Letanneur, J.C. and Renard, M. (2000) Assessment of interspecific hybridization between transgenic oilseed rape and wild radish under normal agronomic conditions. *Theoretical and Applied Genetics*, **100**, 1233–1239.

Coghlan, A. (2000) The next revolution: dispensing with sex could transform agriculture. *New Scientist*, **168**, 5.

Crawley, M.J., Brown, S.L., Hails, R.S., Kohn, D.D. and Rees, M. (2001) Biotechnology: transgenic crops in natural habitats. *Nature*, **409**, 682–683.

Davenport, I.J., Wilkinson, M.J., Mason, D.C., Charters, Y.M., Jones, A.F., Allainguillaume, J., Butler, H.T. and Raybould, A.F. (2000) Quantifying gene movement from oilseed rape to its wild relatives using remote sensing. *International Journal of Remote Sensing*, **21**, 3567–3573.

Gershon, D. (1990) Genetic engineering: tryptophan under suspicion. *Nature*, **346**, 786.

Hauser, T.P., Jorgensen R.B. and Ostergard, H. (1998a) Fitness of backcross and F–2 hybrids between weedy *Brassica rapa* and oilseed rape (*Brassica napus*). *Heredity*, **81**, 436–443.

Hauser, T.P., Shaw, R.G. and Ostergard, H. (1998b) Fitness of F–1 hybrids between weedy *Brassica rapa* and oilseed rape (*Brassica napus*). *Heredity*, **81**, 429–435.

Hill, R.H., Caudill, S.P., Philen, R.M., Bailey, S.L., Flanders, W.D., Driskell, W., Kamb, M., Needham, L.L. and Sampson, E.J. (1993) Contaminants in L-tryptophan associated with eosinophilia-myalgia-syndrome. *Archives of Environmental Contamination and Toxicology*, **25**, 134–142.

Hillman, G.C. and Davies, M.S. (1990) Domestication rates in wild-type wheats and barley under primitive cultivation. *Biological Journal of the Linnean Society*, **39**, 39–78.

Jesse, L.C.H. and Obrycki, J.J. (2000) Field deposition of Bt transgenic corn pollen: lethal effects on the monarch butterfly. *Oecologia*, **125**, 241–248.

Kaeppler, H.F. (2000) Food safety assessment of genetically modified crops. *Agronomy Journal*, **92**, 793–797.

Kloppenburg, J.R.(1988) *First the Seed: the Political Economy of Plant Biotechnology 1492–2000*. Cambridge University Press, Cambridge, UK.

Kryder, D.R., Kowalski, S.P. and Krattinger, A.F. (2000) *The Intellectual and Technical Property Components of pro-Vitamin A Rice: a Preliminary Freedom to Operate Review*, ISAAA Briefs No. 20. ISAAA, Ithaca, NY, USA.

Lindsey, K. and Jones, M.G.K. (1989). *Plant Biotechnology in Agriculture*. Open University Press, Milton Keynes, UK/John Wiley & Sons, Chichester, UK.

Lupton, F.G.H. (ed) (1987) *Wheat Breeding: Its Scientific Basis.* Chapman and Hall, London, UK.

Martens, M.A. (2000) Safety evaluation of genetically modified food. *International Archives of Occupational and Environmental Health,* **73 (suppl. 1)**, S14–S18.

Masood, E. and Jefferson, R. (2000) Seeds of dissent. *New Scientist,* **168**, 66–69.

Monbiot, G. (2000) *Captive State: the Corporate Takeover of Britain.* Macmillan, London, UK.

Muir, S.R., Collins, G.J., Robinson, S., Hughes, S., Bovy, A., deVos, C.H.R., van Tunen, A.J. and Verhoeyen, M.R. (2001) Overexpression of petunia chalcone isomerase in tomato results in fruits containing increased levels of flavanols. *Nature Biotechnology,* **19**, 470–474.

Nuffield Council on Bioethics (1999) *Genetically Modified Crops: the Social and Ethical Issues.* Nuffield Council on Bioethics, London, UK.

Reiss, M.J. and Straughan, R. (1996) *Improving Nature? The Science and Ethics of Genetic Engineering.* Cambridge University Press, Cambridge, UK.

Rissler, J. and Mellon, M. (1996) *The Ecological Risks of Engineered Crops.* MIT Press, Cambridge, MA, USA.

Vavilov, N.I. (1940; English translation 1960) *World Resources of Cereals, Leguminous Seed Crops and Flax, and their Utilisation in Breeding.* Israel Programme for Scientific Translations, Jerusalem.

Ye, X., Al-Babili, S., Kloti, A., Zhang, J., Lucca, P., Berger, P. and Potrykus, I. (2000) Engineering the pro-vitamin A (beta-carotene) biosynthetic pathway into (carotenoid-free) rice endosperm. *Science,* **287**, 303–305.

9 Questioning GM Foods

Sue Mayer

9.1 INTRODUCTION

In recent years, especially but not exclusively in the UK, there has been a public uproar over genetically modified (GM) foods. This has been characterised as being a hysterical reaction, whipped up by irresponsible newspapers and pressure groups. Others, including the British Prime Minister, Tony Blair, in a speech to the European Bioscience conference in November 2000[1], have suggested that there is an anti-science sentiment growing in the UK, which threatens progress and business opportunities. There is also a strand of thought that argues that questioning GM foods in the West is morally indefensible, as GM crops will be essential to feed a growing world population (Wambugu, 1999).

This chapter examines what is known about the basis of public opposition to GM foods in the UK and whether there is any evidence of anti-science sentiment. It also considers how well founded the concerns expressed by people are. Finally, this chapter asks whether being critical of GM foods in the developed world will contribute to an inability to feed the world and whether the decisions to be taken about the technology are a matter for science alone.

9.2 PUBLIC ATTITUDES TO GM FOOD

There has been considerable research on public attitudes to GM food before and during the current furore which started in the UK in 1997. This is a combination of quantitative opinion poll data, which can show the extent to which certain views are held and qualitative data, which, through the use of interviews and focus group techniques, can help understand the underlying basis of concerns.

It is clear that concerns about GM foods are widespread in Britain. For example, a MORI/GeneWatch poll in 1998 (MORI/GeneWatch, 1998) showed that 77% of the British public want a ban on the growing of GM crops until their

[1] For text see www.number-10.gov.uk

Bioethics for Scientists. Edited by John Bryant, Linda Baggott la Velle and John Searle.
© 2002 by John Wiley & Sons Ltd.

impact has been more fully assessed. A similar number (73%) are concerned that GM crops could interbreed with natural, wild plants and cause genetic pollution. The poll also reveals that 61% of the public do not want to eat GM foods (an 8% increase since a similar MORI poll was conducted in December 1996) and 58% of the public oppose the use of genetic engineering in the development of food (a 7% increase on 1996).

Although such public opposition to GM crops and foods has been said to be due to a lack of knowledge about the technology, the evidence does not bear this out. Comparing the results of European surveys in 1991, 1993 and 1996 shows that knowledge about the technology has increased in Europe but optimism about its ability to improve the quality of life has decreased (Biotechnology and the European Public Concerted Action Group, 1997). The 1996 results also demonstrated that 74% of the European public support labelling of GM foods; 60% believe there should be public consultation about new developments and just over half (53%) feel that current regulations are insufficient to protect people from the risks of the technology. The most recent Eurobarometer results from 1999 confirm the trend of decreasing optimism about the ability of biotechnology to improve life (Gaskell et al, 2000). Such concern is not confined to Europe; opposition to GM foods is rising in other countries including the United States (Hornig Priest, 2000).

Qualitative research explores public attitudes in more detail and has shown that public concerns about the technology are complex (Grove-White et al, 1997) and include

- fears that the risks are likely to be unpredictable,
- scepticism about scientific assurances of safety in the light of the BSE ('mad cow disease') crisis,
- moral unease about transferring genes between species in a way that is unnatural, i.e. an *intrinsic* or deontological objection to the methodology itself,
- a feeling that the technology is being driven by profit and not the public interest,
- a lack of control about where the technology is going and
- doubts that GM foods are necessary because 'food is fine as it is'.

So it can be seen that anxieties include not only the environmental and human health risks of GM crops and foods but also the very process itself – the transfer of genes between species – which raises serious moral concerns. Other issues such as the patenting of genes, cells and plants – which are fundamental to the commercialisation of the technology – also raise questions of morality. Underlying these anxieties is a lack of trust in the institutions supposedly managing risk and ensuring safety.

However, whilst this evidence shows that people are indeed concerned about GM crops and foods and that this increases the more they know, it does not answer the question about whether people are right to lack confidence in the regulation of GM crops and the trajectory of the technology and whether their moral concerns are being respected.

9.3 REGULATING GM CROPS AND FOODS

One of the underlying factors in public disquiet over GM crops and foods is a lack of trust in institutions to predict and manage the risks involved in their use (Grove-White et al, 1997). Trust, in terms of food safety, has been hard to gain since the advent of 'mad cow disease' and new variant CJD (Creutzfeldt–Jakob disease) in the UK, where firm assurances of safety and 'no evidence of harm' were made for several years before an embarrassing about-turn was made and a link between consuming affected meat and developing new variant CJD was acknowledged.

The release and consumption of GM organisms (GMOs) may pose new risks because the genetic modification and the introduction of novel genetic material may affect an organism in ways not possible through conventional breeding techniques.

For example,

- other genes may be disrupted,
- the genetic change may not be stable and
- biochemical pathways may be disrupted.

Because GMOs are living, the nature of the risks is different from other technologies (such as the use of chemicals) as they have the potential to multiply and cross with other related species, making the risks not only irreversible but multiplying. Whilst many of the dangers remain theoretical, and some concerns may prove unfounded, there is little hard evidence in either direction on which to base decisions, particularly at such an early stage of the commercialisation and wide scale use of the technology.

The question then is 'How well does the regulatory process deal with this scientific uncertainty?' Because there has been no actual documented harm from the use of GMOs, the regulations are intended to be precautionary. The precautionary principle (PP) has become an important approach to environmental and human health protection where scientific uncertainty exists (see Chapter 3). In situations where there is the possibility of serious harm, the PP states that lack of scientific proof of harm should not be allowed to prevent action to limit or contain the harm (see O'Riordan and Cameron, 1994).

The implementation of a precautionary approach adopted in Europe is one of risk assessment and risk management. First, in risk assessment, the potential hazards are identified and then their probability determined. Risk management is then used to prevent any remaining dangers if it is agreed that it is reasonable to proceed and any remaining risks are minimal (but see also Chapter 8).

In the case of releasing GM crops to the environment, the risk assessment process involves considering whether the gene(s) introduced into the GM crop could spread to related wild plants and what consequences this might have on an ecosystem. It also includes predicting whether the GM crop itself could become a troublesome weed. Risk management measures then include things such as separation distances between the GM crop and non-GM crops to reduce the chance of pollen transfer, cleaning farming equipment between GM and non-GM crops to reduce spread and the use of a herbicide to destroy any problem weeds that might emerge.

To investigate the risks, a step-by-step, case-by-case approach is taken. Experiments are conducted in the laboratory and then in the field with gradually declining containment if it is deemed safe to do so. The presumption is that, at each stage, hazards will be accurately identified and that management techniques used will prevent any harm arising.

Whilst this process sounds straightforward and sensible, it is inevitably riddled with scientific uncertainty and subjective judgements. For example, what is included in the scope of harm is a matter of debate. Until 1998, the wider impacts of growing a GM crop on biodiversity if it alters agricultural practice (such as with herbicide-tolerant crops altering the pattern of herbicide use towards broader spectrum chemicals) were not considered in the risk assessment. They are only now being included as a result of external pressure from non-governmental organisations (Levidow and Carr, 2000a).

When potential adverse effects are included in an assessment, knowledge may be incomplete and contradictory. Secondary impacts on insects consuming so-called 'Bt crops', which have a gene transferred from the bacterium *Bacillus thuringiensis* coding for an insecticidal toxin, have been contentious. One study indicated that monarch butterfly larvae could be harmed by pollen from Bt crops falling on the milkweed plants on which they depend (Losey et al, 1999). This has been criticised as not being representative of the real world, where lower levels of Bt may be encountered than in the laboratory studies (Beringer, 1999; see also Chapter 8). A report in *Nature* (Schuler et al, 1999) showed that beneficial parasitic wasps were attracted to plants as a result of chemical release when leaves were eaten by pest larvae. Because Bt-sensitive larvae ate only small quantities before they succumbed to the effects of the toxin, the wasps were attracted to leaves inhabited by Bt-resistant larvae. Such complex behavioural effects may mitigate against effects on beneficial species. However, laboratory studies showed that Bt can leach into soil from the roots of Bt maize (when the Bt in the crop is in its activated form), where it remains toxic for at least three weeks – whether this will improve pest control or raise further concerns about non-target species remains uncertain (Saxena et al, 1999). What these studies expose is how limited knowledge is about the behaviour of GM Bt crops in the environment, how complex the interactions may be and the problems of replicating the real-life complexities in necessarily contained experiments.

It is not only with such complex indirect effects that knowledge is incomplete; even in what might be considered the more direct effect of gene flow there are no simple answers. As more information is gathered about gene flow to wild related plants, this is considered inevitable for some crops in Europe such as oilseed rape and sugar beet (Lutman, 1999), although currently it is not possible to determine its frequency (see Chapter 8). The complexities of the environment mean that a host of factors, including geography and weather, will influence how far pollen moves and thus prediction becomes extremely difficult and the focus of the question has to turn to whether gene flow matters and whether any adverse effects could be controlled by risk management measures. These demand subjective judgements to be made about the acceptability of a risk and whether risk management is likely to be effective. For example, whilst sounding effective on paper, risk management plans may also not be easily followed in a farm situation. Separation distances between crops to reduce gene flow may not be observed and cleaning of the equipment may not be a plausible option on busy farms with uncontrollable factors such as the weather to contend with.

Similarly, the process in place to determine the safety of consuming GM foods is also contestable. The approach to assessing the safety of GM ingredients is based on the principle of 'substantial equivalence' – if a food is substantially equivalent to the conventional counterpart it is deemed to be safe. This approach evolved from international discussions notably in the OECD in the late 1980s and early 1990s and involves a comparison of various agronomic, biochemical, chemical and nutritional parameters of the GM food relative to existing conventionally produced foods. The composition of macro- and micro-nutrients, known toxins and other anti-nutritional factors are all measured. For example, in potato the macro-nutrients include carbohydrate and protein; the micro-nutrients are any vitamins or minerals and known toxins would include solanine (a glyco-alkaloid). However, there is no standard list of what components must be measured and crucially the approach relies on chemical composition being an accurate predictor of biological activity, an assumption which has been questioned (Millstone et al, 1999). Furthermore, the system will struggle to identify unexpected changes or be able accurately to determine the allergic potential of the introduction of novel proteins into the diet from sources that have not been part of the human diet before. The Medical Research Council has recognised these issues, the difficulty of monitoring for health effects of consuming GM foods and the need for further research (MRC, 2000).

Therefore, there are legitimate questions to be asked about the risk assessment systems that are intended to ensure the environmental and human safety of GM crops and foods. The scope of assessments is a matter of judgement as are the decisions about the acceptability of any risks. Because there are no systems to test the practicality of risk management procedures, these may not be 100% effective.

Of course, risk assessment and management procedures will never be fool-proof whoever is undertaking them, but the public have a right to expect that judgements take the public interest seriously and that the potential for unintended effects is not discounted. That government and its institutions have become too close to industry and its interests is one strand of the public concern about judgements over safety.

9.4   THE PRESSURES OF COMMERCIALISATION

Since the mid-1990s, the emphasis in science policy for publicly funded research in the UK has been on wealth generation and improving the quality of life. The direct potential for scientific knowledge to be useful to industry has, therefore, been an important determinant of how money is spent. Research institutes and universities have been encouraged to build links with, and seek funding from, industry. Areas of science with obvious potential for commercial exploitation, such as genetic modification and genomics, have been supported at the expense of other areas such as ecology or integrated farming methods.

A recent review of public spending on GM food and agriculture biotechnology research concluded that only around 11–16% was on 'safety' of GM crops and foods, the remainder being on research more relevant to development (Barling and Henderson, 2000). Whilst this is changing now, as a result of questions being asked during the GM furore (announcements have been made in 1999 and 2000 of major research programmes to look at gene flow and impacts of GM crops on biodiversity) (Levidow and Carr, 2000b), it is clear that tensions exist in the division of research spending intended to support industry and that to answer questions of safety – a division that may have been too generous to industry and neglected public concerns. Therefore, the feeling that there has been a rush to commercialise GM crops and foods so that research and development costs can be recouped may not be misplaced and may indeed, as the commissioning of new research now suggests, have meant that safety was given a back seat in the process, leaving risk assessors and managers ill equipped to make difficult judgements.

Another dimension of the debate where commercial imperatives have been contentious is in the domain of patenting. A patent gives the inventor a monopoly on the commercial exploitation of the invention for 17–20 years. As part of the revolution in genetic technologies, companies have demanded patent protection for genes, cells, seeds, plants and GM techniques (see Chapter 10). In the past, plant varieties and biological processes have not been patentable because patents were intended to be reserved for inventions, not discoveries, but the biotechnology industry has argued that the scope of patentability must be extended if they are to recoup their research and development costs. This has raised concerns that farmers may have to pay royalties if they keep seed to re-sow – something that could cause particular problems for poor farmers in developing countries and rule them out of any advantages that GM crops might bring. Some research shows that restricting access to genetic material *via* patents may also hinder research in public plant breeding. A 1999 survey of public sector plant breeders in the USA showed that intellectual property protection was hindering their work (Price, 1999). Of those replying to the questionnaire 48% had had difficulties obtaining genetic stocks from private breeders, 45% said this had interfered with their research and 23% said it had interfered with the training of students.

Therefore, not only are commercial imperatives placing pressure on the research agenda and moving the focus from safety to development, but commercial control, in the form of patents, is also being demanded to facilitate the technology. Not only may accidents happen as a result of such haste, and a lack of investment in alternatives restrict options in the future, but the course of technology may not be open to public influence because patents will be used to shape the development of the technology.

9.5 ALLOWING FOR PUBLIC CHOICE

At a time when public confidence in institutions is low, people feel increasingly that they have to rely on their own judgements about safety (see also Chapter 2). People who have moral concerns about products on various grounds such as animal welfare want to be able to act according to their beliefs. In a market economy, one way judgements can be exercised and moral beliefs followed is in what a person buys. In the realm of GM foods, where important questions of human and environmental safety are raised, and deep moral anxieties exist, being able to make choices at the point of purchase is of great importance to many people. However, the biotechnology industry does not wish to have stigma attached to its products.

Therefore, whether and under what conditions GM foods should be labelled has been one of the most contentious issues in their introduction and a strange compromise has been struck. Whilst almost every opinion poll across the world has indicated that people (including many of the scientists involved in GM research) wish to have information about the means of production using GM, labelling is confined to situations where there are measurable changes in DNA and protein in the final food. The labelling of GM foods is covered by the EU's Novel Food Regulation (258/97) and includes foods sold in shops, restaurants, cafés and bakeries. The European Commission has proposed a threshold level of 1% contamination as the trigger level for labelling.

Although the European labelling regulations go further than those of the US, because they are based not on the means of production but on the chemical composition of the final product, many derivatives of GM commodity crops escape labelling. Consequently we have the following.

• Products containing GM soy flour (which may be found in foods such as bread or baby foods) or whole GM soybeans must be labelled with phrases such as 'contains genetically modified soya' because foreign protein and DNA are present.

- Products containing derivatives of GM maize (e.g. starch) or GM soybean (e.g. oil or lecithin) will *not* be labelled, because protein and DNA are removed during their production. These products are often found in a whole array of foods including vegetable oils, prepared meals and chocolate.

This position is justified on grounds that there has to be something measurable (i.e. foreign DNA or protein) in the final food for labelling to be enforceable and that it is only in this situation that there could conceivably be some health risk that justifies labelling. However, this restriction on the scope of labelling clearly favours the interests of the industry and leaves consumers wishing to make their own ethical decisions disadvantaged. The European Commission has recently proposed changes to labelling laws to address consumer needs but the USA authorities are arguing that these changes contravene World Trade regulations.

9.6 THE RATIONALITY AND MORALITY OF QUESTIONING GM FOODS

Whilst there are clearly questions remaining over the safety assessment of GM crops and foods and consumer choice is restricted in the developed world, are these concerns little more than the self-indulgent grumblings of well fed people in the rich countries? The population, which is currently about 6 billion, is expected to reach 8 billion by 2020 and 11 billion by 2050 (Kendall et al, 1997; Vasil, 1998). The advocates of genetic engineering believe that the increasing demand for food must be met without expanding the amount of land used for agricultural purposes (to protect biodiversity) and by addressing issues of soil erosion, salinisation, overgrazing and pollution of water supplies (Vasil, 1998; Monsanto, 1998). However, many organisations in less developed countries, aid agencies and environmental groups are less positive about the role genetic engineering can play in solving problems of hunger and tackling environmental degradation; see discussion in Chapter 11.

The new genetic technologies are already transforming agriculture in the developed world. The entry of the agrochemical companies into GM crops coupled with the acquisition of seed companies and mergers between companies has led to new relationships. These corporations undertake 80% of all GM crop research and dominate the trajectory of the technology. How well do the requirements of such companies sit with the needs of those poor people who cannot afford an adequate diet (in 1994, food production could have supplied 6.4 billion people – more than the actual population – with an adequate 2350 calories per day, yet more than 1 billion people do not get enough to eat (Kendall et al, 1997))?

Increasing control of staple crops may lead to increased dependency on just a few companies, with potentially higher costs as a result. By 1997, in the US,

Pioneer (now owned by Du Pont) had 42% of the maize seed market and Monsanto 14%. The concentration of such seed markets and in field trials with GM crops has reached levels where the US Department of Justice could consider it worthy of consideration for anti-trust action (Lichtenberg, 2000).

Companies' demands for patent protection on genetic resources also threaten public research efforts in developing countries that develop the seeds for poor farmers (Conway and Toenniessen, 1999). Moreover, because patent protection may be difficult to enforce for seeds that can be reproduced naturally, companies have developed 'genetic use restriction technologies' (GURTs) to limit access to new GM crops. This has led to a furore over so-called 'terminator technology', a type of GURT that results in sterile seed being produced. This would prevent farmers from keeping seed for re-sowing – a practice that is vital for 1.4 billion of the world's poorest farmers. Companies have now promised not to commercialise this technology (Conway and Toenniessen, 1999), but this is only because its exposure in the developed world led to a debate involving both the developed and developing world that put the intentions of those involved in GM crop production under scrutiny (see also Chapter 11).

The questioning of the environmental and human health systems for GM crops and foods is as relevant to the developing as it is to the developed world. The successful negotiation of the Cartagena Protocol on Biosafety in 1999, covering the trans-boundary movement, export and import of GMOs, came about partly as a result of the pressure from the debate in the developed world. Without it, there would be no requirement for prior informed agreement to import GMOs and no system for capacity building among developing countries to facilitate their own safety assessments.

So rather than threatening food security in the developing world, the debate about GM crops in the developed world has opened up space for a full debate about the possible social, environmental and human consequences of the use of GMOs around the world. What is important is that this involves all people, especially the poorest in developing countries, rather than the poor in the developing world being used as pawns in a distant political debate. With the introduction of such a powerful new technology, debate and scrutiny should form an important part of its introduction if the widest benefits are to be reaped and hazards avoided. Attempts to stifle such critical inquiry should always be questioned.

9.7   CONCLUSIONS

Recent research shows that negative responses to GM foods in the UK are a result of the social shaping and consequences of the technology rather than any non-specific anti-science sentiment (Grove-White et al, 2000). Thus GM is seen as being remote, with questionable benefits, developed by an industry that is oligopolistic (i.e. under the control of very few 'players') and bringing risks that are undefined, under-acknowledged and long term. In contrast, information technology is seen much more positively as being more visible, interactive and empowering.

A look at how GM crops have been introduced and their evaluation systems reveals that this is not an irrational perspective. The risk assessment system, although intended to be precautionary in nature, has been built on a policy that has had an underlying assumption that the development of GM crops and food is essential for the competitiveness of European industry. Whilst the regulations attempt to put in place safeguards, scientific knowledge is uncertain and ignorance exists and thus what constitutes an acceptable risk is contestable. The narrow boundaries of the risk assessment system have excluded many areas of public concern and choice in the market place has been denied. A lack of any clear public benefits and a bias across the research agenda and patenting towards industry has led to a corrosive lack of trust in the institutions supposed to manage safety.

How can a way forward be found? In a plural democracy, it is essential that differing views and cultural values are respected and taken into account (Bruce and Bruce, 1998). A report from the National Research Council (1996) has said how, in the important step of characterising what should be considered in a risk evaluation system,

> In addition to the biological and physical outcomes that are typically covered, decision makers and interested and affected parties often need to know about the significant economic costs and benefits of alternatives, secondary effects of hazard events, or the efficacy of alternative regulatory mechanisms.
>
> A risk characterisation will fail to be useful if the underlying analysis addresses questions and issues that are different from those of concern to the decision makers and affected parties.

Conventional risk assessment using expert-centred 'scientistic' techniques cannot cope alone with the difficulties of characterising and prioritising risks or deciding how risks should be distributed, or their cultural and ethical implications. A requirement is to find new approaches to developing new technologies and products that are sensitive to social, economic and cultural settings and allow a range of different options (and combinations of options) to be compared. Such approaches need to be flexible, open to divergent values, frank about scientific uncertainties, transparent and rigorous, open to participation and practicably feasible (Stirling and Mayer, 1999). One possible technique to add to the deliberative public debate that is so clearly needed is multi-criteria mapping (Stirling and Mayer, 2000). Techniques such as this are not prescriptive but map out the scientific, social, economic and other issues to allow informed political decisions to be made about the optimum choices and what is an acceptable risk. Science, whilst necessary to inform such decisions, is not a sufficient basis on which to make them (National Research Council, 1996).

REFERENCES

Barling, D and Henderson, R. (2000) *Safety First? A Map of Public Sector Research into GM Food and Food Crops in the UK,* Discussion paper 12. Centre for Food Policy, Thames Valley University, London, UK.

Beringer, J.E. (1999) Cautionary tale on safety of GM crops. *Nature,* **399,** 405.

Biotechnology and the European Public Concerted Action Group (1997) Europe ambivalent on biotechnology. *Nature,* **387,** 845–847.

Bruce, D. and Bruce, A. (1998) *Engineering Genesis. The Ethics of Genetic Engineering in Non-Human Species.* Earthscan, London, UK.

Conway, G. and Toenniessen, G. (1999) Feeding the world in the twenty-first century. *Nature,* **402,** C55–C58

Gaskell, G., Allum, N., Bauer, M., Durant, J., Allansdottir, A., Bonfadelli, H., Boy, D., de Cheveigné, S., Fjaestad, B., Gutteling, J.M., Hampel., J., Jelsoe, E., Correia Jesuino, J., Kohring, M., Kronenberger, N., Midden, C., Hviid Nielsen, T., Przestalski, A., Rusanen, T., Skellaris, G., Torgersen., H., Twardowski, T. and Wagner, W. (2000) Biotechnology and the European public. *Nature Biotechnology,* **18,** 935–938.

Grove-White, R., Macnaghton, P., Mayer, S. and Wynne, B. (1997) *Uncertain World. Genetically Modified Organisms, Food and Public Attitudes in Britain.* Centre for the Study of Environmental Change, Lancaster University, Lancaster, UK.

Grove-White, R., Macnaghton, P. and Wynne, B. (2000) *Wising Up. The Public and New Technologies.* Study of Environmental Change. Lancaster University, Lancaster, UK.

Hornig Priest, S. (2000) US public opinion divided over biotechnology? *Nature Biotechnology,* **18,** 939–942.

Kendall, H.W., Beachy, R., Eisner, T., Gould, F., Herdt, R., Raven, P.H., Schell, J.S. and Swaminathan, M.S. (1997) *Bioengineering of Crops: Report of the World Bank Panel on Transgenic Crops.* International Bank for Reconstruction and Development/World Bank, Washington, DC, USA.

Levidow, L. and Carr, S. (2000a) UK: precautionary commercialization? *Journal of Risk Research,* **3,** 261–270.

Levidow, L. and Carr, S. (2000b) Environmental precaution as learning: GM crops in the UK. In *Cow Up a Tree: Learning and Knowledge for Change in Agriculture: Case Studies from Industrialised Countries.* Cerf, M., Gibbob, C., Hubert, B., Ison, R., Jiggins, J., Paine, M., Proost, J. and Röling, N. (eds), INRA, Versailles, France, pp. 323–335.

Lichtenberg, E. (2000) Costs of regulating transgenic pest-protected plants. In *Genetically Modified Pest-Protected Plants. Science and Regulation.* National Academy of Sciences, National Academy Press, Washington, DC, USA, Appendix A.

Losey, J.J.E., Rayor, L.S. and Carter, M.E. (1999) Transgenic pollen harms monarch larvae. *Nature,* **399,** 214.

Lutman, P.J.W. (1999) *BCPC Symposium Proceedings No. 72: Gene Flow and its Consequences for GM Oilseed Rape.* British Crop Protection Council, Farnham, UK.

Medical Research Council (MRC) (2000) Report of a MRC Working Group on Genetically Modified (GM) Foods. MRC, London, UK.

Millstone, E., Brunner E. and Mayer, S. (1999) Beyond 'substantial equivalence'. *Nature,* **401,** 525–526.

Monsanto (1998) *1997 Report on Sustainable Development including Environmental, Safety and Health Performance.* Monsanto, St Louis, MO, USA.

MORI/GeneWatch (1998) Interviews conducted with 950 adults aged 15+. Interviewed face-to-face, in-home, using CAPI (computer assisted personal interviewing) technology on 6–8 June 1998 at 84 sampling points throughout Great Britain. Data have been weighted to reflect the national profile. Trend information has been included from a

MORI/Greenpeace poll: 1003 interviews among adults aged 15+ were conducted by telephone on 13–15 December 1996. Data have been weighted to reflect the national profile.

National Research Council (1996) *Understanding Risk: Informing Decisions in a Democratic Society*. National Academy Press, Washington, DC, USA.

O'Riordan, T. and Cameron, J. (eds) (1994) *Interpreting the Precautionary Principle*. Earthscan, London, UK.

Price, S.C. (1999) Public and private plant breeding. *Nature Biotechnology*, **17**, 938.

Saxena, D., Flores, S. and Stotzky, G. (1999) Insecticidal toxin in root exudates from Bt corn. *Nature,* **402**, 480.

Schuler, T.H., Potting, R.P.J., Denholm, I. and Poppy, G.M. (1999) Parasitoid behaviour and Bt plants. *Nature*, **400**, 825–826.

Stirling, A. and Mayer, S. (1999) *Rethinking Risk. A Pilot Multi-Criteria Mapping of a Genetically Modified Crop in Agricultural Systems in the UK*. SPRU, University of Sussex, Brighton, UK.

Stirling, A. and Mayer, S. (2000) Precautionary approaches to the appraisal of risk: a case study of a genetically modified crop. *International Journal of Occupational Health and Environmental Medicine*, **6**, 296–311.

Vasil, I.K. (1998) Biotechnology and food security for the 21st century: a real-world perspective. *Nature Biotechnology*, **16**, 399–400.

Wambugu, F. (1999) Why Africa needs agricultural biotech. *Nature*, **400**, 15–16.

10 The Patenting of Genes for Agricultural Biotechnology

Steve Hughes

10.1 PREFACE

Genes are pre-existing entities, products of nature. Plant genes in particular, as components of the whole genome, can float on the wind in seeds or pollen, resisting constraint or ownership, and would seem to fall outside the scope of provisions for patenting as historically interpreted. Yet, at the same time individual genes and their sequences are part of biological and chemical discovery. They represent new knowledge, and can be precisely described as 'new' molecules. They can acquire specific utility. They can indeed fall into a form of ownership as intellectual property, and have done so. The enabling tools and techniques of genetic modification catalysed this, while themselves being the subject of patents, weaving a web of dependence, alliance, litigation and acquisition, which has shaped a new agricultural biotechnology industry. The consolidation and centralisation of power in this industry raises concerns for many, particularly in relation to justice and access to the technology and the fair distribution of risk and benefit. The focus of decision making tends to overlook the role of public-funded science in the growth of the technology. Not surprisingly, the sense of exclusion and scant empowerment experienced by some, particularly in relation to the negotiation of socially relevant goals for agriculture, especially in the developing world, leads them to question the ethical basis of gene patenting. However, the broader and arguably more constructive question is whether, in order to promote justice and fair opportunity, we should seek to restrict and redefine the subject matter which is legitimate for patenting, or seek instead to moderate the scope or the consequences of the enforcement of patent rights.

10.2 INTRODUCTION

Patents are just one of the means by which a form of ownership of biological knowledge and discovery can be assigned to individuals or controlled by institu-

Bioethics for Scientists. Edited by John Bryant, Linda Baggott la Velle and John Searle.
© 2002 by John Wiley & Sons Ltd.

tions. Others include trade secrets, plant variety rights (under the UPOV convention)[1] and, to some degree, requisite skill and 'know-how'. Collectively, these fall into the category of intellectual property, represented socially as intellectual property rights (IPRs), which form the basis for the trading of knowledge, just like any other commodity. Apart from trade secrets, IPRs are not absolute rights. As we shall see in Section 10.3, rights are assigned by a formal institutional process, which has its own set of constraints and requirements.

Property rights are an essential concept and component of any system of exchange. Few cultures have evolved without them, and although many of our cultural/religious value systems ascribe merit to the divestment of property, rights to property remain central to a functional society. However, property rights become problematic when they support a monopoly position for a supplier of goods or services. Monopolies are generally regarded as dangerous in market economies since they restrict competition (and by implication self-regulated price control) and choice. Institutional mechanisms (e.g. anti-trust legislation, Monopolies Commission) have been evolved to restrict the emergence and impact of monopolies. Ironically, the institutional provision for patenting itself evolved from a system assigning actual monopolies ('Letters Patent') in the reign of the British king, James I, and even now its rationale is the promotion of monopoly rights to the exploitation of discrete areas of invention.

The consequences of narrow control of whole fields of technology and the effective monopolies that can result will feature in this chapter. In this regard, it is worthwhile first to examine the precise nature of the rights as well as the obligations conferred by patents.

The immediate discussion makes the assumption that public good and welfare are achievable through commercial activities, and that concepts of the justice and fairness of the international intellectual property rights system can be addressed within this context. Within this utilitarian and consequentialist position lies the further assumption that plant gene sequences, where utility can be demonstrated, are legitimate subject matter for patents. It is recognised that there are counter-positions held by groups who, often on deontological grounds, founded in cultural history and value systems, subscribe to the view that the principle of assignment of genes and their sequences, as well as new plant varieties which contain them, to the public domain, is not negotiable. For them the discussion ends here: genes are not patentable. In the ethical context, this position is both worthy and valid. It will be revisited later in the chapter in relation to the issue of the legitimacy of patenting life forms (see also Chapter 11). It should also be noted that, with respect to the individual case of human gene sequences, there is pressure from the executive branch of the UK and US governments for public access[2].

---

[1] UPOV – Union pour Obtentions Végètales.
[2] The Blair–Clinton statement of 14th March 2000 states 'To realise the full promise of this research, new fundamental data on the human genome, including the human DNA sequence and its variation, should be made freely available to scientists everywhere'.

10.3 PATENT ASSIGNMENT

10.3.1 THE NATURE OF THE MONOPOLY

Granting of a patent implies that formal requirements for originality (i.e. an inventive step), non-obviousness, utility and reduction to practice[3] have been met in the opinion of the official patent examiner assigned by the national patent jurisdiction to the case. This reinforces the point that this form of property right is not an absolute right, but one which is institutionally constrained within requirements for disclosure and conformity.

The grant confers rights in the form of a temporary monopoly on exploitation of the patented invention, coupled to an obligation of disclosure of the nature of the invention as well as the means of practising it. There is no obligation to exploit or license a patented invention and compulsory licensing tends only to happen in times of national emergency. The property right therefore only extends to the exclusion of others from practising the invention. In this light it is important to appreciate that the property right which is granted does *not* assign ownership to the subject matter itself.

The intent of the patent jurisdiction is to reward inventors while stimulating further invention by requiring full disclosure of the intellectual property on which the patent is based. Reward to the inventor or assignee of the patent is furthermore both a means of recompense for risk and upstream research expenditure, and a means of encouraging further investigation of the field. In this context patents and income derived from them can become a powerful element in the overall reward system for scientists even in the public sector[4]. At this level, the patenting system does seem to codify a practice that achieves a balance which is fair and just to both inventors and society in general. However, there is an additional element, which ensures that society does not pay too high a price. This is the principle of exhaustion, which provides that the payment of license fees expires at the point of sale of product that embodies the invention. This means that as consumers we are not required to pay, for instance, each time we use our mobile music player. In effect we pay a one-off fee (built into the selling price) to the manufacturer, which is returned to the inventor depending on the deal struck between the inventor and manufacturer. For consumer goods the inventor would obviously prefer to have royalties linked to sales volume. This principle is clearly not compatible with self-replicating entities such as plants, animals and microorganisms, since the inventor would only ever be able to sell one seed, which then would be freely propagated by others. The principle of exhaustion thus necessitates changes to patent conventions in relation to biotechnology inventions. Perhaps this was recognised in the need for a separate

---

[3] For an interpretation of these requirements in relation to different patent jurisdictions and for a general primer on IP provisions see http://www.cambiaIP.org
[4] In the UK this privatisation of IP became key to institutional science policy as the Research Councils were driven successively by customer contractor and wealth creation requirements by successive governments.

plant variety rights provision (PVR, see UPOV convention discussed below in Section 10.7).

10.3.2 THE SCOPE OF THE MONOPOLY

The extent of monopolies assigned by the grant of a patent is very much a function both of the patenting system itself and of the interpretation of the patent application in the light of contemporary science. In terms of patenting procedure and practice it is important to distinguish between patent applications and granted patents. The filing of patent applications follows certain conventions. The filing is generally expected to contain a description of the invention which demonstrates originality and inventiveness, a set of examples of practical applications and a set of claims. Patent applications submitted through the Patent Cooperation Treaty (PCT) and the World Intellectual Property Organisation (WIPO)[5] or the European Patent Office[6] and more recently through the US Patents and Trademarks Office are published after preliminary examination. There is a tendency to mistake these publications for patents. They are not: in fact patents are only granted and published after further protracted negotiation between the applicant and the patent examiner. An important part of this negotiation involves *the set of claims*. Claims define the field of utility and thus the boundary of the territory dominated by the patent. They set the scale of the monopoly which can derive from the patent. It is generally in the interest of the inventor to frame this as widely as possible in order to maximise the scope and value of the patent. The role of the examiner is to allow only those claims which are truly supported by the description of the invention and the utility which is demonstrated. The allowance of *broad claims* (excessively broad or sufficiently broad, depending on one's viewpoint) has become one of the contested issues in patents in the agricultural biotechnology arena. The issue revolves around the principle of *justice*, and *fair treatment* for inventors, for competing inventors and for potential users of the technology. An example of this is in the initial grant by the US Patents and Trademarks Office of a patent covering all methods of genetic transformation of the cotton plant, to the company Agracetus (US patent 5 159 135) which was contested by competitors, who felt that this assigned too broad a monopoly and that it provided a disincentive for others to seek improvements to the breeding of cotton. The latter objection represents a major point of internal conflict in the patenting system. The allowance of excessively broad claims militates against the intent of the system to encourage further invention. It may have the additional effect of diminishing further research efforts which may act against *general welfare*. On the other hand, if claims are insufficiently broad the inventor may decide to keep the invention secret and not to share it in the patent arena, which again militates against the intent of the system to stimulate further invention. Moreover, there is

[5] http://www.wipo.org/
[6] http://www.european-patent-office.org/index.htm

also the need to reward the inventors of truly ground-breaking inventions with the allowance of some broad claims.

Thus, the need for careful balance in the allowance and rejection of claims is driven by a combination of pragmatic consideration as well as by the ethical principles of *justice* and *fairness*. The Nuffield Council on Bioethics, in its study of GM crops (Nuffield Council on Bioethics, 1999)[7] observed and recommended that 'on balance the broad claims within a patent are only justified where the invention is truly supported by correspondingly broad examples and deserves the award of broad claims . . . (We) recommend that national patent offices, the EPO and the WIPO draw up guidelines for patent offices to discourage the over-generous granting of patents with broad claims that have become a feature of both plant and other areas of biotechnology'. However, despite such recommendations, the allowance of broad claims will remain problematic and prone to the differing interpretations of the different patenting authorities.

10.3.3 CONSTRAINTS ON THE MONOPOLY

Several exceptions, arising from considerations of equity, allow for unlicensed practice of an invention without legal infringement (White, 1999). One of these, the experimental exemption, or research exemption, is relevant to advances in plant science and agricultural biotechnology. The purpose of this exemption is reasoned, on the basis of past judgements, to be the encouragement of innovation and as such it is consistent with the intent of patenting discussed above. Within the European Union this exemption includes acts performed for experimental purposes in relation to the subject matter of the patent though it could also be interpreted as having some overlap with the exemption for acts performed privately for non-commercial purposes. Interpretation of the exemption in the context of commercial research in plant science is complex, in particular the scope for performing (experimental) plant variety trials prior to commercialisation. However, it is clear in providing for the use of patented inventions in research by non-commercial institutions.

10.4 PLANT GENES AS SUBJECT MATTER FOR PATENTS

10.4.1 HOW DID PLANT GENES COME TO BE REGARDED AS PATENTABLE?

In philosophical terms[8] some have argued that historical considerations are not germane to ethical consideration and decisions to be made here and now. However, before discussing plant genes and their utility as commercial tools, or

---

[7] www.nuffieldfoundation.org
[8] The genetic fallacy (see Bradley, 1998).

sources of IP, we need briefly to reflect on the technological developments of the three decades of investigations in biology that spawned this new opportunity. I hope thus to illuminate some of the tensions that lie behind the differing stances taken with respect to the legitimacy of patent rights in relation to living systems.

In the late 1960s, it was publicly funded research into an apparent transient immunity of bacteria to their viruses that led to the discovery and description of a new family of enzymes, the restriction endonucleases, which have become such valuable tools for informative dissection of the genetic material, DNA. Within a very short time of their discovery, these enzymes were in widespread use and were freely exchanged by researchers as their potential utility became clear. This was quite a contrary outcome to the predictions of those who had regarded the restriction and modification mechanism of transient virus immunity as a bit of a side-show to the 'real' genetics. In this phase, collaboration and the sharing of materials and information were the norm, and a vacuum flask of small plastic tubes in ice the routine travelling companion.

It was also publicly funded research that led to the discovery of transmissible bacterial drug resistance and of the mobile genetic elements, plasmids and viruses, that mediated the transmission. These became the core of gene cloning vectors when exploited in concert with restriction nucleases. The availability of cloned genes and the ability to isolate them from complex organisms and thus to study them in simpler environments brought a major discontinuity in biology and medical science. Recombinant DNA, as the technology was known, diffused rapidly into most fields of fundamental investigation, accelerating beyond imagination the rate at which impressive data and information could be extracted. This faculty became a dominant paradigm of biological investigation, which required that in order to understand any phenomenon we must first clone the gene(s) that determine or modulate it. This reductionist paradigm put great weight on gene isolation and characterisation, and much creative energy and laboratory resources were channelled down this avenue.

The potential for using defined gene sequences in the design of novel organisms was also quickly appreciated, as were the commercial opportunities which might stem from this. Not surprisingly, the technologies of recombinant DNA became the subject of patents, held by the key inventors or their host institutions. There then flourished a period of entrepreneurship in which scientists floated commercial ventures on the strength of the promise of the technology and the possession of information about some gene or other. In this context, a means of staking one's claim to the knowledge potentially recoverable from a gene sequence acquired key significance. This raised a difficulty with respect to patents since, as noted earlier, genes themselves are pre-existing entities and not inventions in their own right and were therefore not obviously patentable. A gene sequence *per se* may be necessary, but it is not sufficient to meet the formal requirements of any patent jurisdiction. Nevertheless, such was the pressure to protect this form of intellectual property that strategies emerged for redefining the gene as part of a vector construct or for cloaking gene sequences in invention and utility sufficient to support patent filings and assignation of the intellectual property embodied in the gene (Doll, 1998).

10.4.2   THE PATENTABILITY OF PLANT GENES

The application of the standard criteria of patentability in relation to the granting of patents on DNA fragments within the USA, European and Japanese jurisdictions has been compared in a trilateral study by the relevant patent offices[9]. They conclude that 'Isolated and purified nucleic acid molecule-related inventions, including full-length cDNAs and SNPs[10], of which function or specific, substantial and credible utility is disclosed, which satisfy industrial applicability, enablement, definiteness and written description requirements, *would be patentable* as long as there is no prior art (novelty and inventive step) or other reasons for rejection (such as, where appropriate, best mode (US) or ethical grounds (European Patent Convention/Japan)).'

The mention of ethical grounds in this context brings us back to the question of the legitimacy of patenting in relation to living organisms that are or else contain embodiments of inventions, since this is the means by which plant genes most often generate their utility to become patentable. The ethics of patents on life should be examined in the context of other aspects of ownership of living organisms. Many of us are owners of companion or working or recreational (including fashion accessory) animals, cats, dogs, livestock, horses and so on. Ownership is recognised in law in terms of our responsibilities for and to these animals as well as our rights (ownership) not to have someone else remove them from us. In the case of pedigree animals we also have rights to their genetic material as contained in their gametes or progeny; in fact we can sell it or, alternatively, charge stud fees. Of course, the price we can charge reflects the superiority of the genes and gene combinations within our animals. This is based on our knowledge of their performance and lineage (these factors are part of the conventional breeder's computation of 'breeding value'). The construction now starts to sound rather similar to the circumstances that prevail in the case of gene patents, in terms of the knowledge associated with a gene that gives it value for the production of a new genotype, coupled to rights over exploitation. In fact, if we put the two scenarios side by side, ownership of the animal itself seems more draconian and ethically questionable than the ownership of a patent. The only substantive difference is that patents govern the use of a few genes in many breeding opportunities whereas animal ownership governs the use of many genes in few breeding opportunities. So, unless we find a way to reconfigure our ownership rights over animals, and for that matter plants, it seems difficult to challenge patents on ethical grounds, especially since they do not confer owner-

---

[9] http://www.uspto.gov/web/tws/sr-3-b3b-ad.htm
[10] Sequences containing single nucleotide polymorphisms, which may be used to distinguish an allele in one individual from one in another individual (see also Section 10.6).

ship of subject matter but only ownership of rights to block the exploitation of one's invention by others.

10.5 ENABLING PATENTS AND POWER

10.5.1 INFORMATION, POWER AND PATENTS IN PLANT BIOTECHNOLOGY

Generalised acceptance within the patent jurisdiction of patents on gene discovery began the continuing saga of the commodification or 'enclosure' of the genetic material in which latterly, in the eyes of some, bioinformatics and functional genomics have become willing conspirators. These are the tools that enable the conversion of data (nucleic acid sequences) into knowledge. One of the characters in Tom Stoppard's play *Rosenkrantz and Guildenstern are Dead* reminds us 'Information is power, all information is power' (Stoppard, 1967). This observation is reflected clearly in the contribution of formalised (IP) information about genes to the endowment of power to private institutions.

As we shall see repeatedly, general enabling inventions which provide new tools bring the possibility of adding value to discovery by offering a direct route of exploitation. They are the agency by which genes become inventions. A good example of this is the technology of genetic manipulation of the Cohen–Boyer patent[11], which offered the possibility of transferring any gene into a microorganism wherein there was the opportunity to produce the corresponding gene product or some derivative of it, in large amounts. An example of this type of exploitation would be human insulin. Knowledge of the sequence of the human insulin gene instantly becomes IP and has to be protected either by secrecy or by writing a patent around it. This achieved, the patent protecting the insulin gene becomes dependent on the Cohen–Boyer patent, access to which, happily, was freely negotiable as a one-off fully paid up licence at reasonable cost.

In the case of plants a set of major enabling patents was developed by the construction of a vector system for plant transformation, based on the *Agrobacterium tumefaciens* tumour-inducing plasmid (see Chapter 8). In the course of causing a localised neoplastic growth or gall *Agrobacterium* acts as a natural genetic engineer, transferring a segment of DNA from its Ti plasmid to the plant host. The enabling technology which stemmed from this discovery represents many years of hard work by many scientists. This started in the 1970s when the plasmid was discovered to be the agent of tumour formation via the transfer of DNA, through to the re-engineering and implementation of the system in the mid-1980s, and its subsequent refinement. Much of the work was funded by grants from the public purse or charitable institutions until the final phase, when large commercial investment was injected. The effect of this on the enclosure of

---

[11] The Cohen–Boyer patent dominated all aspects of recombinant DNA technology.

the technology and on the centralisation of control of the technology is discussed later.

However, it is clear from reading the patent literature in this area that a large number of patents have been filed. The precise pattern of dependency and dominance is still unclear but it is clear that the essential enabling patents for plant gene transformation are narrowly assigned. In its study of the field in 1999 the Nuffield Council on Bioethics, looking broadly at plant transformation and its essential associated tools, concluded that five major companies effectively controlled the technologies as a consequence of a combination of skilfully directed research and of commercial acquisition (see also Chapter 11). This represents a focus of power and decision making which is at odds with the diversity of the agricultural context, at least with respect to the requirements of crop variety performance and the local needs of agronomy (see Chapter 8, Section 8.6.2).

Plant gene transformation provides the means by which genes of diverse function may be introduced into a plant so as to generate new traits; some of these novel traits may be useful in agriculture. In this sense the enabling patents may be viewed as a type of cassette into which new gene sequences can be dropped in order to generate utility. In this context it is easy to see how enabling patents add value to knowledge and drive the act of patenting from being a means of protecting invention to being for the protection of discovery, with the enabling patents providing the framework of the invention. The latter are the 'crown jewels' that dominate the deployment of patented genes in agricultural biotechnology. They are the vehicles that establish the utility of genes via their deployment in transgenic plants.

However, there is another side to this coin. While the enabling technology provides a route to utility, and therefore to invention, for a newly discovered gene, the gene patent holder is unable to practice the invention without a licence to the enabling patent. This represents a major entry barrier to the practice of the technology and thus raises the issues of dependency and freedom. In this light it is quite natural for some to reflect on the power these patents have assigned to private corporations on the basis of discovery, much of which was in the public sector. Before discussing these issues (Section 10.5.3) we should note that other forms of entry barrier are also significant. These include the costs of regulatory compliance for safety assessment and the costs and scale demands of performing the technology effectively. It is ironic that while the costs of the technology have come down, pressures exercised by campaigning groups for punitively costly regulatory provision have forced the entry price up, biasing the activity in favour of the largest players.

10.5.2 THE ISSUE OF FREEDOM TO OPERATE

As suggested above, the interdependence of inventions and patents that govern the exploitation of newly discovered genes does place some serious practical constraints on the ability of inventors to practice or sublicense their technologies. These limitations come under the collective title of *freedom to operate*. In order to generate freedom to operate, those who wish to practice an invention are obliged to negotiate licences from all holders of patents on which the invention is dependent. This raises a particular horror not only because of the transaction costs of doing so, but also because it is often difficult to be sure that one knows all about all of the claims of potentially dominant patents, or even who the current assignees are. Thus strategies to secure freedom to operate are developed in a climate of uncertainty and represent a prioritisation of risk management options. In an article entitled 'Can patents deter innovation? The anti-commons in biomedical research', Heller and Eisenberg (1998) drew attention to the impact of transaction costs within a fragmented field of patents and inventions as a disincentive to further invention, contrary to the intent of the patenting system. The ability to navigate this space will breed a new cadre of professionals in the plant biotechnology arena.

The Nuffield Council on Bioethics working group on GM crops recognised the complexity of this issue, especially in relation to the relative negotiating positions of large companies and developing countries. They challenged the Co-Ordinating Group of the International Agricultural Research Centres (CGIAR) to act as a skilled broker in ensuring that the opportunity to participate in breeding and crop research utilising GM technologies did not pass poor countries by as a consequence of IP-based restrictions[12].

10.5.3   CASE STUDY: PRO-VITAMIN-A-ENHANCED RICE (GOLDEN RICE[13])

Many of the issues surrounding freedom to operate are illustrated in the case history of Golden Rice. Golden Rice is a type of rice that has been genetically modified by the addition of four functional genes from other species that endow it with the capacity to over-produce pro-vitamin A. Vitamin A is poorly available in the diets of a number of traditional rice-dependent societies. It is estimated that world-wide some 180 million people in about 60 less-developed countries suffer vitamin A deficiency, which can cause children to lose their sight and risk further debilitation and mortality.

Some argue that this dietary limitation can be addressed by the distribution of vitamin supplements, which could have the negative consequence of generating dependency. Others argue that the short-coming is more readily addressed by persuading rice growers to adopt new crop species, requiring new approaches to agronomy and modified social organisation amounting to a wholesale import of alien culture. In comparison, the introduction of a new rice variety, provided that it is compatible with the local agro-ecology, agricultural practice, taste, cooking and storage requirements of the recipient consumers, is of low impact.

---

[12] Nuffield Council on Bioethics, Chapter 3 paragraph 51.
[13] Golden Rice is a registered trademark.

Furthermore, seed, self-replicating and easy to store and transport, is a very efficient way of distributing a benefit, which, if fairly managed does not generate dependence.

So we can see a persuasive line of argument in support of the European Union, the Swiss Federal Institute of Technology and Rockefeller Foundation in their espousal and funding of this humanitarian goal. The engineering of Golden Rice required the use of genes and technologies that are the subject of patents held by private institutions (Kryder et al, 2000). Although it was the wish of the Rockefeller Foundation and the inventors that the rice itself should not be restricted in its use by patents, it was clear that these other patents on which the Golden Rice was dependent could in principle restrict its use.

The International Service for the Acquisition of Agri-Biotech Applications (ISAAA) were sponsored by the International Rice Research Institute to perform an audit of the intellectual property components of Golden Rice in order to review the *freedom to operate* issues[14]. They identified about 40 patents plus material transfer agreements (MTAs[15]) which, in the worst case scenario, governed freedom to operate. This number is alarming in terms of the potential transaction costs of negotiating licenses with a diversity of patent holders. However, the number reduces dramatically when seen in the light of rice-growing countries, many of which either do not recognise the patents or were not seen by the inventors as sufficiently important for the patents to be registered there. Some are also lost from the list as a consequence of the specific claims they contain, for instance where claimed procedures do not extend to the products thereof. In this case the production of rice for home consumption would not be an infringement in those countries that had not granted such claims. RAFI[16] has made the case that only a handful of the patents are actually an impediment and that the freedom to operate issue has been overplayed. They have argued that the IP issue has become an unfortunate rationale for the transfer of the technology to the private sector.

However, we must now be concerned that Golden Rice and its future development will be narrowly informed by the commercial criteria of the company that controls it. This tends to undermine some of the supportive arguments for Golden Rice made above. Nevertheless, the hope for Golden Rice is that it will provide a means of balancing the rights of children and the yet unborn to a better diet, against the responsibility of avoiding the erosion and devaluation of the cultural values and ethics of the society they will grow up into.

---

[14] See http://www.isaaa.org/briefs/Goldenrice.htm
[15] Material transfer agreements (MTAs) in contrast to patents appear to respect ownership of subject matter. Researchers enter into MTAs as a means of avoiding having to repeat complex and expensive procedures, but in doing so signal subject matter ownership, which can be used to block commercial exploitation subject to the terms agreed. They should be regarded as an alternative and complicating form of IP protection.
[16] RAFI http://www.rafi.org/web/allnews-one.shtml?dfl=allnews.db&tfl=allnews-one-frag.ptml& operation=display&ro1=recNo&rfl=118&rt1=118&usebrs=true

10.5.4 THE INFLUENCE OF PLANT PATENTS ON GLOBAL TRADE IN AGRICULTURAL PRODUCTS

The foregoing discussion of Golden Rice has focused on the IP issues as they are moderated in local production/consumption systems. This raises a key issue in relation to patents and international markets, and the role of the World Trade Organisation as an enforcer of intellectual property rights in the guise of the TRIPS (Trade Related Aspects of Intellectual Property Rights) agreement. TRIPS seeks to engage all trading countries in the observance of patent rights. It is clear in supporting the patent holder in preventing the import of commodities that are subject to the patent into those countries in which the patent right is granted, regardless of the fact that production may have been in a country where the patent is not granted. Freedom to operate thus becomes important to the ability to trade.

A comprehensive study has been carried out by Binenbaum et al, (2000) in order to estimate the realistic consequences of freedom to operate issues in relation to trade in agricultural commodities between the developing and the developed states. Their conclusion was that the issue is less imposing than had been expected owing to the small volumes traded in relation to what is consumed domestically. Only a small number of instances were identified where patent rights might impede significant volumes of trade, e.g. bananas and rice. However reassuring a scenario this analysis paints, it does not take account of future opportunities for increased trade and the pressure on less developed countries that comes from TRIPS to move to the recognition of patent rights or plant variety rights.

Somewhat contrasting with TRIPs, the Convention on Biodiversity (CBD), product of the Rio Conference, seeks to assign national sovereignty to genetic resources whatever the motive for their exploitation (see Chapter 3). This instrument includes the potentially useful genes that might find utility in plant improvement via conventional or transgenic breeding. Physical sovereignty has been complicated by the fact that much crop genetic diversity is now held in *ex situ* collections, which are not necessarily complete in historical provenance. However, this is less problematic than the issue of sovereignty over the knowledge indigenous people possess concerning the characteristics and performance of their local crops (land races). It is such knowledge that gives real value to the genetic resource. Such knowledge is not protected by any formal property right regardless of its value in pointing the way to useful genes which may ultimately be patented as inventions by others. This contrast in the treatment of formal and informal knowledge, which it is compelling to see as a reflection of the North–South divide, does not appear just. Pragmatically, it is difficult to define a system for the protection of informal knowledge, particularly where it has traditionally been freely exchanged between farmers and breeders[17]. Neverthe-

---

[17] A study of the relationship between traditional knowledge and IP is being undertaken by WIPO: see http://www.wipo.int/traditionalknowledge/introduction/index.html

less the spectre[18] of less developed countries having to pay for licenses to exploit genes of which they were long-term custodians and informants, helps to emphasise the disparity in the treatment of formal and informal knowledge and the power this disparity assigns to those who can acquire formal IP rights.

10.6 PRACTICAL CONSIDERATIONS OF GENE DISCOVERY

10.6.1 INTRODUCTION

Gene discovery has been translated into the more profitable concept of 'gene prospecting'. As suggested above, gene prospecting is a knowledge-driven process, which can incorporate indigenous crop knowledge. It also accommodates knowledge derived from the modern tools of plant biology and information technology as a means of mining the data compiled in massive genome sequencing efforts in the developed world.

10.6.2 GENOMICS

Along with organisms ranging from microbes to mammals, plants have been included in the genome sequencing programmes. Sequencing of two of the smaller plant genomes, those of *Arabidopsis thaliana*, a 'model' dicotyledonous species (The Arabidopsis Genome Initiative, 2000) and of *Oryza sativa*, rice, a model monocotyledonous species, has been completed.[19] Databases are filling up with strings of sequence, which hold little value until configured and aligned with maps of chromosomal organisation. This type of organised data, however, acquires its full value as intellectual property when it can be exploited in the discovery of functional genes. Several complementary genomics tools support this activity. Databases of expressed sequences translated from random sequencing of messenger RNA populations are a direct line to many of the functional genes. Comparison of these sequences, known generally as ESTs (expressed sequence tags) with known gene sequences from other organisms, can bridge knowledge into the plant genome and often give clues to the genetic identities of whole families of sequences. The tools for making these inferences belong to the toolkit known as *bioinformatics* and are based on sophisticated statistical methods. However, bioinformatics software is freely accessible via the internet and are easy to use.

Those with access to private gene sequence databases for particular crops, whether these be full sequences or collections of ESTs (Adam, 2000), are thus

---

[18] The term *biopiracy* has been used in this context, see Chapter 11.

[19] The rice genome sequence has been completed in draft form by Monsanto as a contribution to the International Rice Genome Sequencing Project. Monsanto will also share rice sequencing data with others in the international research community subject to conditions including
    'If researchers patent inventions based on direct use of Monsanto's sequence data, the company is given an early opportunity to negotiate a non-exclusive license to such patents'.

enabled to find and patent genes in their single species. Again this generates power for those with sequencing resources who can take on the larger genomes of the crop species.

Genetic maps derived by conventional genetic linkage analysis or via the molecular mapping of DNA sequence polymorphisms have supported the alignment of genetic and sequence data, sometimes giving clues to the functional identity of anonymous genes. In this context it is important to realise that the order in which genes are distributed along the genome map is conserved within many groups of organisms. This phenomenon is known as synteny and is valuable to gene prospecting since it allows gene discovery in one organism to inform discovery in others (Moore et al, 1995).

Despite these aids, which are dependent on existing genetic diversity, for the Arabidopsis genome, so far only 69% of its 26 000 potential genes have been classified by similarity to proteins of known function in other organisms, and as few as 9% have been characterised experimentally. Other approaches will be required to help to assign functions to the remaining genes and to complete the annotation (i.e. the full description) of the genome.

Newer tools rely on modified plants themselves to inform us. These tools exploit the possibility of producing populations of mutant plants via the random introduction of chosen transgenes throughout the genome. 'Functional transgenomics' or 'Mutagenesis with a flag' (or tag), as this is known, enables us to screen for altered phenotypes in terms of the loss or gain of functions (depending on the configuration of the transgene construct). The phenotype can then be related directly to the gene sequences surrounding the tag. Alternatively the process can be reversed. We can start with a gene of unknown function and find a plant within the population that has a tag associated with that gene, and then examine the new phenotype. The prizes in this case will go to those with the ability and specialised knowledge to recognise complicated phenotypic changes which are relevant to agronomic performance. Here we can establish the link to genetic diversity and traditional knowledge, which will be critically informative for this route to gene discovery. This inevitably brings us back to the issue of justice in relation to the limited rights accorded to the custodians of traditional knowledge.

As the remaining genes are classified, identified and characterised, patents will surely follow. There will be public good in this in so much as the information disclosed will assist in completing our annotation and understanding of the genome. Nevertheless, this will have to be set against the complexity and transaction costs that the thousands of individual gene patents will place on practical deployment of the knowledge.

10.7 PLANT VARIETY RIGHTS AND PATENTING

10.7.1 THE UPOV CONVENTION

The assignment of property rights to those plant breeders who produce distinctly new and improved plant varieties was formally established under the UPOV Convention[20].

The convention was initially drawn up in 1961 and became operational on the international stage in 1970. This form of intellectual property right differs somewhat from that granted under patent law although the intent, of providing a system of rewards to inventors via temporary monopoly rights and of stimulating further breeding initiative, is similar. Plant variety rights can be assigned to a plant breeder upon demonstration that a newly bred plant variety is distinct, uniform and stable (DUS) in its characteristics. These requirements are similar to the principles of originality and disclosure in patent law. Further requirements relate to value for cultivation and use (VCU), which is established by competitive testing against standard varieties. This requirement is in the practical sense equivalent to the requirements for utility and reduction to practice in patent law. In order for the breeder to start collecting royalties from seed agents or farmers against a grant of variety rights, the said variety must be made available to all who request it. At this point the variety becomes legitimately available as a parent for further breeding. The monopoly over the collection of royalties is temporary, lasting 25 to 30 years dependent on the particular crop.

The effective obligation placed on breeders to make their varieties available for further breeding, like the requirement for disclosure in patent law, undoubtedly fuels further creativity and initiative, and has supported a spirit of sharing and exchange of germplasm (parental breeding material) between breeders. However, at the practical level, as we see below, it does give rise to some difficulties in relation to the patenting of plant genes.

10.7.2 DOUBLE PROTECTION AND CONFLICT WITH OTHER IP RIGHTS

It is, in practice and principle, possible to take an established elite variety and modify it by the insertion, using GM techniques, of a specific and useful plant gene. The resulting variety would be distinct, uniform and stable and could qualify as a new variety. Now, if one breeder did this to another breeder's variety, a shortcut to competitive advantage could be gained. In this case the restrictive principle within the UPOV Convention of 'essentially derived', which requires a minimum level of breeding activity between successive varieties, would ensure justice to the original breeder.

Conversely if a breeder used a single patented transgene to add a characteristic to his/her own variety and chose to register only the transgenic version, then the

[20] http://www.upov.int/

opportunity of other breeders to further develop the variety could be compromised by the risk of patent infringement. This form of 'double protection', while at one time resisted in the European jurisdiction and by UPOV, is now perfectly legitimate.

Alternatively the breeder could ignore the UPOV system and license the variety under terms that specifically prevented further multiplication. Whether IP protection is supported by a combination of Plant Variety Rights combined with gene patents, or by gene patents alone, the principle of exchange of material between breeders has been lost and with it a key avenue to innovation.

Of course, this outcome is not totally novel. F1 hybrid systems also act as a barrier to the flow of material between breeders, and have done so since the inception of F1 hybrids for maize in the 1930s (Kloppenburg, 1988)[21]. This has been reinforced by patent protection taken out on the parental lines, in the US jurisdiction, where patents have been used routinely to protect plant varieties with special traits. The use of transgenic technologies to generate novel forms of male sterility, which can be activated when required, offers a means of extending F1 hybrid possibilities to many crop species. This further provides the means of avoidance in the sharing of germplasm.

An additional means of preventing further breeding from finished varieties has been developed recently but is not at this stage in use. This is known as Gene/Variety Use Restriction Technology (GURT or VURT) or, more colloquially by its detractors, as 'Terminator'. GURT and VURT depend on an engineered gene switching system, which when activated prevents the maturation of embryos in the next generation of grain[22]. That is to say that the grain is sterile. Not only can it not be used as a seed crop but it is also terminated for the purpose of further breeding. Thus GURT and VURT offers a means of restricting the use of the intellectual property embedded in genetically improved seed to those who pay royalties through purchase of the seed. This obviates the need for surveillance of patent infringement.

There has been considerable reaction to the possible implementation of GURT and VURT (see Chapter 11). It has been widely articulated that there is something counter-aesthetic in both the concept and reality of sterile seed. There is something profoundly disturbing in the sight of a golden ripe field crop, the seed of which is barren. It seems to be an infringement of the received wisdom that in the farmer's care seed begets seed[23]. Regardless of this, F1 hybrids secure precisely the same outcomes as GURT and VURT in commercial practice, and

---

[21] The rationale for this is that seed produced by the F1 hybrid gives rise to F2 progenies, which represent the segregation (randomisation) of the parental traits uniformly represented in the F1. Such progeny are thus highly variable and are not valuable for further breeding in relation to what is predictable from the performance of the F1.

[22] VURT and GURT technology was developed and patented by the USDA in one of its research centres. It was subsequently licensed to a company, Delta and Pine Land, which may or may not at the time of publishing have become a part of the Monsanto portfolio of agricultural companies.

[23] '. . . a grain of wheat remains a solitary grain unless it falls into the ground and dies; but if it dies it bears a rich harvest.' St John's gospel, Ch. 12 v24 *New English Bible.*

have done so for three-quarters of a century. This highlights the potential conflict of cultural aesthetics[24] and consequentialist ethics.

Pragmatists and consequentialists would argue at this point that, provided farmers are not misled into believing that the purchase and repurchase of GURT or VURT seed is unavoidable, and provided that alternative and effective seed sources are available, so that the farmer has a choice, GURT and VURT do not pose an ethical problem. Conversely, others, both international institutions and NGOs campaigning on behalf of the rural poor (see RAFI above)[25] have declaimed their opposition to the technology as the defining 'bathwater' of GM (see Chapter 8), which should, in their view, lead to the whole gamut of GM technologies being thrown out (see also Chapter 11).

Each of the foregoing technologies is applicable both for and against the spirit of the UPOV Convention. On the one hand they do provide, like patents, an opportunity for the originators of new varieties to collect their reward and to overcome the difficulties of policing unauthorised use. Given the small operating margins of breeding companies there does seem to be an element of justice in this.

10.8   CODA

Let us return to our initial question. Do the key problems relate to plant genes as subject matter of patents or is it the nature and enforcement of patent rights which is more problematic? Institutional policy within the major patent jurisdictions clearly supports the legitimacy of plant genes as subject matter provided that they are part of a truly inventive disclosure. The benefits to society of disclosure, as an incentive to further discovery and invention, do seem to stack reasonably *against* the reward of a temporary monopoly to inventors, that is, until we look at the international corporate monopolies and the centralisation of power and decision making that can flow from this. However, this caveat relates more to market trends and globalisation than anything ethically unsound in genes *per se* as subject matter.

Concerning the ethics of the patentability of genes, given that patents do not assign ownership of genes, there does not appear to be a sustainable or convincing argument that patents are abhorrent when compared with other forms of control of genetic material, as in the ownership of whole breeding animals and plants. However keen some might be to read gene patents their last property rites, the overturning of the patenting system in this regard would lead to serious consequences for the flow of knowledge into the public domain and also for innovation.

The balancing of intellectual property rights between new scientific knowledge and indigenous knowledge, given the role that both can play in understanding

---

[24] Often powerful reinforcing elements of deontological ethics.
[25] http://www.rafi.org/web/publications.shtml

and exploiting genomes, remains problematic. The continuing discord between CBD and TRIPS is particularly poignant for those who wish to see those principles of fair recognition that are encompassed by patent law extended to all who contribute their knowledge to the public good.

REFERENCES

Adam, D. (2000) Now for the hard ones. *Nature*, **408**, 792–793.

The Arabidopsis Genome Initiative (2000) Analysis of the genome sequence of the flowering plant *Arabidopsis thaliana*. *Nature*, **408**, 796–826.

Binenbaum, E., Nottenberg, C., Pardey, P.G., Wright, B.D. and Zambrano, P. (2000) South–North trade, intellectual property jurisdictions, and freedom to operate in agricultural research on staple crops. *IFPRI Discussion Paper No. 70.* International Food Policy Research Institute, Washington, DC, USA.

Bradley B.M. (1998) In *Restructuring the Psychological Subject: Bodies, Practices and Technologies,* Bayer, B.M. and Shotten, J. (eds), Sage, London, UK, pp. 74–81.

Doll, J.J. (1998) The patenting of DNA. *Science*, **280**, 689–690.

Heller, M. and Eisenberg, R. (1998) Can patents deter innovation? The anti-commons in biomedical research. *Science*, **280**, 698–701

Kloppenburg, J.R. (1988) *First the Seed: the Political Economy of Plant Biotechnology.* Cambridge University Press, Cambridge, UK.

Kryder, D.R., Kowalski, S.P. and Krattinger, A.F. (2000) *The Intellectual and Technical Property Components of Pro-Vitamin A Rice: a Preliminary* Freedom to Operate *Review,* ISAAA Briefs No. 20. ISAAA, Ithaca, NY, USA.

Moore, G., Devos, K.M., Wang, Z. and Gale, M.D. (1995) Grasses line up and form a circle. *Current Biology*, **5**, 737–739.

Nuffield Council on Bioethics (1999) *Genetically Modified Crops: the Ethical and Social Issues.* Nuffield Council, UK.

Stoppard, T. (1967) *Rosenkrantz and Guildenstern are Dead.* Faber, London.

White, A. (1999) Experimental exemption of patent infringement. *Journal of Commercial Biotechnology*, **5**, 208–217.

11 Crop Biotechnology and Developing Countries

Geeta Bharathan, Shanti Chandrashekaran, Tony May and John Bryant

11.1 INTRODUCTION

The application of biotechnology to agriculture has raised many questions. Some of these relate to the science behind biotechnology but many more are concerned with policies governing the use of the technology and the implementation of those policies. A further key question that emerges is whether there are inherent differences between developing countries and developed countries in this regard. If so, where do those differences lie and how do they affect the application of biotechnology to agriculture in the less developed countries? Is there an appropriate model for the transfer of technology or is it simply a case of imposition by developed countries of perhaps inappropriate technology on less developed countries, based on the economic strength of the former and the economic weakness of the latter? These are important questions and we address them in subsequent sections of this chapter, accepting in doing so that there may be no unanimously accepted answers. We acknowledge that there is some overlap between this chapter and certain others (particularly 8, 9 and 10), but have maintained the overlap for the reader who wishes to access within one chapter all the information relevant to this topic.

11.2 FOOD SECURITY

11.2.1 THE CURRENT SITUATION

According to the World Food Summit, 'food security exists when all people, at all times, have physical and economic access to sufficient, safe and nutritious food to meet their dietary needs and food preferences for an active and healthy life'. However, the reality is very different from this ideal. The following

Bioethics for Scientists. Edited by John Bryant, Linda Baggott la Velle and John Searle.
© 2002 by John Wiley & Sons Ltd.

summary spells out the situation in the less developed countries against which the debate on the application of crop biotechnology is partly set.

'Today, almost a billion people live in absolute poverty and suffer chronic hunger. Seventy percent of these individuals are farmers – men, women and children – who eke out a living from small plots of poor soils, mainly in tropical environments that are increasingly prone to drought, flood, bush fires, and hurricanes. Crop yields in these areas are stagnant and epidemics of pests and weeds often ruin crops. Livestock suffer from parasitic diseases, some of which also affect humans. Inputs such as chemical fertilisers and pesticides are expensive, and the latter can affect the health of farm families, destroy wildlife, and contaminate water courses when used in excess. (Often, pesticides banned in the West are used abundantly without protection or proper precautions in place in Third World countries.) The only way families can grow more food and have a surplus for sale seems to be to clear more forest. (A short-term solution, which leads to greater desertification or soil degradation, leading to the need to clear yet more forest – a vicious circle; see also Chapter 4.) Moreover, analysts agree that land is expected to become scarcer, therefore, any processes that exacerbate this situation need to be stopped. Older children move to the city, where they, too, find it difficult to earn enough money to buy the food and medicine they need for themselves and their young children' (Persley and Doyle, 1999).

This quotation suggests that food security is not just a matter of food production but also of socio-economic circumstances, especially poverty. On this basis it is often argued that the 'food problem' today is one of food distribution, rather than food production (for example, see Rissler and Mellon, 1996; Anderson, 1999). More specifically, it is the lack of purchasing power on the part of those that need food the most (Sen, 1981). India, for example, has a surplus food grain stock of 60 million tonnes, in spite of which 320 million Indians go hungry daily because they are too poor to buy food (Sharma, 2001). A CornerHouse briefing asserts that 'a whole range of unjust and inequitable political and economic structures, especially those relating to land and trade, in combination with ecological degradation, marginalise poorer people and deprive them of the means to eat'. The briefing also adds that 'they starve because they do not have the money to buy food, or do not live in a country with a state welfare system' (The CornerHouse, 1998). Current data lend some support to this view. If we were to take the 1994 food production figures and divide the amount up among the year 2000's world population of 6 billion (6×10^9) there would be 2500 calories/day/person, which is an adequate calorie intake. Further, over the past 30 years, in which the global population has risen by 70%, global food production has more than kept up with population growth (FAO[1], 2000). However, the counter-argument is that redistribution of wealth is a very long-term political problem that does not look like being solved. According to the proponents of the latter argument, what is required therefore is an increased

[1] FAO: the Food and Agriculture Organisation of the United Nations. The web-site – www.fao.org – is very useful.

production of affordable food in the countries where the food is actually needed (Conway, 2000; Conway and Toenniessen, 1999; Wambugu, 1998). They further argue that even though in theory the world could be fed on current levels of production, further increases in population coupled with losses of agricultural land mean that there will be a shortfall in global food production capacity. The only way to deal with this, it is argued, is to increase food production in the less-developed countries where productivity per hectare is much lower than in the intensive agricultural systems of the developed nations. Thus Conway (2000) states that he is far from convinced that the powerful developed nations are ready to deal with the underlying issues in any serious way. In his view the world is not about to engage in a massive redistribution of wealth and we must recognise, he maintains, that the only way to help the majority of the world's poor, who tend to live in rural areas, is by increasing their income and food productivity by sustainable agricultural and natural resource development[2]. In this light he believes

> We need a new revolution – a doubly green revolution[3], that repeats the success of the old but in a manner that is environmentally friendly and much more equitable. This is going to take the application of modern ecology in such areas as integrated pest management and the development of sustainable agricultural systems. It is also going to need much greater participation in the development process by farmers themselves. But I also believe it is going to need the application of modern biotechnology – to help raise yield ceilings, to produce crops resistant to drought, salinity, pests and diseases, and to produce new crop products of greater nutritional values.

11.2.2 FOOD PRODUCTION AND POPULATION: THE IMPENDING CRISIS

It is estimated that the world population will rise from approximately 6 billion today to a total of 8 billion by 2030 and 11 billion by 2050 (Kendall et al, 1997; Vasil, 1998). Of this increase 95% will occur in the world's poorest nations at a rate of about 1.9% per year. Thus, despite the fact that the overall rate has decreased substantially from a peak of 2% per annum in 1960 to 1.3% today, a substantial absolute increase in the total number of people is expected (Kendall et al, 1997; Dixon, 1998; Vasil, 1998). Global demand for food in the next 30 years is expected to reach 2500 million tonnes, very significantly more than the amount needed today. It is likely that global food production will also increase, albeit at a slower rate of increase of 1.5% per year over the next 30 years (but the reader should recall the problems of distribution mentioned above). In the longer term (certainly by 2050) the two lines (rates of increase of population and food production) are expected to cross and global food production may no longer be sufficient for global food requirements. Indeed, this may occur much sooner than 2050 since there are clear indications that the rate of increase of food

[2] Including the use of genetic modification and other forms of crop biotechnology.
[3] cf. Conway, 1997.

production is declining (e.g. 3% per year in the 1970s; 1.5% per year in the 1990s: Conway and Toeniessen, 1999). This is the impending food crisis. It will affect mostly the less developed countries, which, unless they can increase their own productivity, will need to increase food imports, thereby further exacerbating the problems of poverty and dependence. Even countries such as India, currently self-sufficient in food, will need to become net importers (FAO, 2000).

11.3 FACING THE CRISIS: DIFFERENT POINTS OF VIEW

11.3.1 THE PRIMARY ROLE OF AGRICULTURE IS TO KEEP FOOD PRODUCTION IN PHASE WITH POPULATION

The scientific agricultural scene in most parts of the world is dominated by the view that high-yielding varieties with large inputs and large-scale industrialization of all aspects of agriculture are essential to maintain production at required levels. The Green Revolution provided a much needed fillip to food production and closed a large gap between supply and demand. Since then, supply in many of the countries that adopted the Green Revolution successfully (e.g. India) has managed to keep up with demand, and the only way to keep this advantage is to stay ahead with innovations such as biotechnological applications in agriculture.

However, in many developing countries, food production and population are already out of phase. These countries today are net importers of food. The FAO report (2000) indicates that, on a global level over the next 30 years, developing countries will more than double their imports to feed their people. If this is so, then increasing food production within *developed* countries will become critically important to the net importing, *developing* countries. In general this viewpoint urges the use of biotechnology, in particular GM, in agriculture since it is said to be the only way in the future to produce enough food to feed the starving millions of the world. Nevertheless, as Conway (2000) has emphasised (see above), a solution that just relies on the developed nations is inadequate: there must be a greater correspondence between population and food production in the less developed countries, implying that productivity in the latter needs to be increased.

11.3.2 THE PRIMARY ROLE OF AGRICULTURE IS TO ACHIEVE SUSTAINABLE FOOD PRODUCTION

This point of view posits that the current moves towards even further intensification and industrialization of agriculture are disastrous and we need to stop right now (e.g. see Anderson, 1999). Some of the problems stated to be associated with industrialized agriculture are the following.

- It is non-sustainable because of huge inputs in the form of pesticides, fertilisers and water.
- This in turn causes environmental problems, for which solutions are stated to lie in judicious agronomic practices that minimise input. These types of solution require specialised knowledge such, it is said, as were possessed by previous generations of agriculturists, but which are often lost to present generations. Much of the current 'organic movement' in the UK and USA is based on this type of thinking.
- Many problems that GM technologies, for instance, seek to address are those created by modern agricultural practices (e.g. increases in types and incidence of cotton pests in India due to indiscriminate use of pesticides); without such practices, the 'GM-fix' (as it is perceived) would not be necessary.

According to this viewpoint the first level of solutions should include going back to small-scale, low-input, labour-intensive, organic, and other agricultural practices that may be described as sustainable. In general the purveyors of this viewpoint shun the use of GM technologies since it represents, in their minds, an extension of industrialized agriculture with a potential for unspecifiable future problems. In the UK for example, the Soil Association, which is the certifying organisation for organic agriculture, has outlawed GM technology.

> The first of these two perspectives often maps exactly with the views of those who are pro-GM technology (agricultural scientists, agricultural organisations, most governments, the private sector), while the second maps exactly with the views of anti-GM technology groups (many public interest groups, environmental groups, small farmers' groups). It is important to recognize that the dichotomy presented here represents extreme points at the two ends of a spectrum of views. For example, many scientists who are in principle 'pro-GM', whether from developing or developed countries, advocate that precaution should be exercised in decisions on biotechnological applications, simply because the technology is new and its spread potentially could be far more rapid than that of any other agricultural technology in the past (see also Chapters 8 and 9).

11.3.3   THE ROLE OF AGRICULTURE IN PROVISION OF EMPLOYMENT

Agriculture is important for all nations, but is of prime importance to developing countries, where it is not only a source of food, but also a major source of employment. It is estimated that about 650 million of the poorest people in developing countries live in rural areas where production of food is the main economic activity. Industrialization of agriculture results in unemployment and displacement, both of which have dire consequences for the populations of agricultural societies. Regardless of whether they think they should move toward greater industrialization of agriculture or not, the fact is that developing countries have to consider the importance of employment in the agricultural sector today. Given this, an important role of agriculture is to provide productive employment to people in rural areas and improve their access to food. This clearly needs to be taken into account when considering the role of biotechnology in agricultural development.

11.4   SHOULD BIOTECHNOLOGY BE USED?

11.4.1   INTRODUCTION

This is the question that fuels many discussions on biotechnology. However, it may be too broad a question. Unlike the nuclear bomb, a technology which has undeniably disastrous consequences, the consequences of using 'biotechnology' are not generalisable because of its wide range of techniques and because biological systems are far more complex and variable than any physical system. Therefore, rather than making sweeping statements on whether biotechnology is 'good' or 'bad', or taking absolute positions on its application, we break the question down to more manageable ones. Which biotechnology should be used? When should it be used? Where should it be used? Who should use it?

A general answer to the overall question might be 'if it promises to solve some aspect of the "food problem" or the major problems of agriculture'. However, these are extremely complex problems in which biological, ecological and societal factors are intimately connected and interact in complex ways. Therefore, it would be beneficial to keep in mind that something that intuitively seems to be an ideal technological solution to a problem may have completely unforeseen consequences unless it is evaluated in the context of the problem in all its complexity. Section 11.4.2 deals with this in more detail.

11.4.2   THE GREEN REVOLUTION: A USEFUL MODEL?

The Green Revolution – the introduction into world-wide agriculture of new high-yielding cereal varieties in the 1970s and 1980s – certainly led to increased food production, estimated for the period 1980 to 2000 as 15%, leading many to hail it as a great success. However, maximisation of yield potential relied on large-scale, monoculture-based, mechanised, intensive farming requiring extensive input of agro-chemicals. Is this the way of farming for the future? Can it produce even higher yields to feed a growing world population? Can it overcome the problems that poor farmers face? The answers to these questions are hotly disputed. Despite the success of the Green Revolution in helping to increase global yields it is increasingly coming under some heavy criticism. For example, its performance has been said to be inconsistent and it has been claimed, probably correctly, that it failed in Africa[4]; many countries have not even had the Green Revolution (for mainly socio-economic reasons) and still rely on planting landraces and saving their own seed (see text box) whilst in other less developed countries the chemical-intensive methods are reported to be destroying the environment and the health of farmers[5]. It is clear that there are still millions of malnourished people, who live predominantly in the Third World, who have obviously not benefited. The debate thus continues, but it is clear that the Green Revolution has certain shortcomings that even supporters recognise as needing to be addressed.

According to the UN's Food and Agriculture Organisation (FAO), one in seven people are chronically malnourished, including one in three children. This means that many people in developing countries, if they manage not to die of starvation, lack the necessary levels of proteins, vitamins, minerals and other micronutrients in their diet. Gordon Conway, president of the Rockefeller Foundation in New York, provided some examples that demonstrate the impact this failure is having when he addressed a recent OECD conference on GM food: 'About 100 million children suffer from vitamin A deficiency. They are more likely to develop infections and the severity of the infection is likely to be greater. Each year half a million go blind and some 2 million die as a result (of the deficiency). Iron deficiency is also common. About 400 million women of childbearing age are afflicated by anaemia caused by iron deficiency. As a result they tend to produce stillborn or underweight children and are more likely to die in childbirth.' (Conway, 2000).

A more positive view is put by M.S. Swaminathan, a leading agronomist who is thought of as the father of India's Green Revolution. He argues that not only is there a need to address the very important issues concerning the underlying structural causes of hunger and poverty, but it is also important not to rule out the benefits of the new biotechnologies, which he asserts, based on his experience with the Green Revolution, can play an important role in meeting the need for

[4] Florence Wambugu of the African regional office of the International Service for the Acquisition of Agri-Biotech Applications (ISAAA) in an interview with the UK journal *New Scientist* notes that the Green Revolution failed Africa, because it was alien to them, coming from the West/North, and assumed the adoption of farming practices that were not appropriate for Africa. For example, many African farmers had to be educated in the use of fertilisers.

[5] Greater productivity in terms of producing larger quantities of food to meet rising demands can only be achieved at the devastating and unsustainable expense of cultivating more land, using greater quantities of water (water is predicted to become scarcer in the future due to climate changes affecting patterns of rainfall, urbanisation, soil erosion due to human activity, pollution etc.), herbicides and pesticides. The latter two only add to the feared risks on humans and other living beings. According to the *New Internationalist* magazine, 'The hazards of pesticide use are now widely recognised, although statistics are hard to gather. The World Health Organisation estimates that at least three million people are poisoned by pesticides every year and more than 200,000 die.

food security. Put another way, he argues that the new biotechnologies, including genetic modification, are needed but should only be developed and used according to an integrated vision of environmental and socio-economic sustainability:

> Since there is no option in population-rich and land-hungry countries but to produce more per unit of land, water and labour, there is a need for technologies which can promote and sustain an ever-green revolution rooted in the principles of ecology, economics and social and gender equity. It is obvious that the challenge can be met only by integrating recent advances in molecular genetics and genetic engineering, information and space technologies and ecological wisdom, resulting in appropriate eco-technologies. There should be no relaxation of yield-enhancing research, since there is no other way of meeting global food needs. (Swaminathan, 1997).

This provides a strong argument for the use of crop biotechnology in less developed countries. In the next sections we consider what features of crop biotechnology may be appropriate for application in less developed countries before going on to discuss whether they can be implemented according to Swaminathan's vision.

11.5   WHICH BIOTECHNOLOGY SHOULD BE USED?

11.5.1   INTRODUCTION

Crop biotechnology encompasses a range of modern biological methods that have been applied in agriculture. For our discussion in this chapter the most relevant are (i) micro-propagation, (ii) molecular-marker-aided improvement and (iii) genetic modification (GM) or genetic engineering (GE). Of these, the last is the most contentious.

11.5.2   MICRO-PROPAGATION

This method consists of the cultivation of plant cells or tissues *in vitro* in order to obtain clonal material or to propagate virus-free plants. Both developed and developing countries use these methods in research and in commercial applications. All manipulations and operations are usually within the confines of tissue culture facilities or greenhouses and there is little fear of widespread ecological impact. It may be the least expensive of biotechnologies. Its suitability in any particular context is, therefore, likely to be decided purely on the basis of whether it makes technical and economic sense to do so.

11.5.3   MOLECULAR-MARKER-AIDED IMPROVEMENT

A molecular marker is a particular tract of DNA (coding or non-coding) that is located in a specific part of the genome of an organism. They are useful as markers if they enable the breeder to predict the inheritance of particular traits. Molecular markers may be identified through sequencing or other analyses including randomly amplified polymorphic DNA (RAPD) and restriction fragment length polymorphisms (RFLP). Such markers, once characterised, can be used in a range of applications in plant breeding. The foremost among these uses are in aiding selection of quantitative trait loci (QTLs) and in speeding up the progress of introgression of a desired trait into a suitable genetic background (see also Chapter 8). Other uses of molecular markers may be in basic studies of phylogenetic relationships and in population biology. Further, the study at the molecular level of phylogenetic relationships may itself feed back into marker-aided breeding. For example, in the cereals, which differ greatly in genome size from genus to genus, it has been shown that genes and other markers are at equivalent chromosomal positions in all species (Ahn and Tanksley, 1993; Moore et al, 1995; Bevan, 1999; Goff, 1999). This aids not only the selection of required genetic traits within species but also the transfer of appropriate genes from species to species by GM techniques. The potential value of marker-aided breeding is obvious. It is ironic, in view of the debates about the applications of GM technology in less developed countries (see below and also Chapters 8 and 9), that the development of marker-aided breeding has relied on the use by the plant breeders of GM techniques in the identification, cloning and sequencing of markers.

11.5.4 GENETIC MODIFICATION

Genetic modification involves the isolation, cloning and insertion of a desired gene into a target plant, thus creating transgenic plants. The inserted gene may be obtained from a distantly related organism (e.g. a bacterium) or a closely related organism (e.g. a different species of the same genus). It is different from previous technologies in that it enables the plant breeder to import into a plant genome (e.g. that of an elite crop line) a gene (or genes) from unrelated organisms. The technology and its general ethical implications are discussed in Chapters 1, 8 and 9. Suffice it to say here that it is regarded as much more controversial than other techniques in crop biotechnology and that the controversy arises from several aspects (again as discussed in Chapters 1, 8 and 9). Firstly there are some who have a fundamental, perhaps deontological, objection to moving genes between species. Secondly, particular application of GM to crops, especially food crops, may evoke negative responses in certain situations. Some of these responses, such as those aroused by the concept of inserting a fish 'anti-freeze' gene into plants, may be regarded as 'emotional', but it needs to be remembered that these emotional responses may reflect underlying cultural or religious values. Thirdly, there may be concerns about environmental impacts of crop plants with new combinations of genes. Fourthly, worries have been expressed about application of crop GM technology in relation to issues of corporate power and global justice. The latter is of course very relevant to the use of crop genetic modification in less developed countries, on which most of the remainder of this chapter is focused.

11.6   WHEN SHOULD GENETICALLY MODIFIED CROPS BE USED?

11.6.1   INTRODUCTION

Not all situations are necessarily ideal for the application of genetic modification, and some may be dealt with equally well by other methods, either traditional or modern. However, there may be circumstances under which biotechnological applications have a clear advantage (see also Chapter 8). The major examples of these are to circumvent constraints imposed by lack of suitable genetic variation, to speed up the process of generating new varieties and to reduce the environmental impacts of agriculture (but see Chapter 9).

11.6.2   WHEN GENETIC VARIATION IS A CONSTRAINT

When productivity has been maximised through traditional breeding and no further increase in production is possible then the introduction of a new gene may open up new possibilities. This was the case, for example, when a bacterial gene coding for a protein toxic to insect pests (the insecticidal protein from *Bacillus thuringiensis*, usually referred to as Bt toxin) was inserted into maize and cotton, resulting in plants that themselves act as pesticides, thereby firstly circumventing the relative inefficiency of increasing pest resistance in these species by conventional breeding and secondly reducing the use of insecticide. (See Chapter 8 for a fuller discussion of this application.) Another example in which GM techniques have circumvented the barriers imposed by lack of suitable genetic variation is the development of Golden Rice®. This is discussed more fully in Section 11.11.4.

11.6.3   WHEN TIME IS A CONSTRAINT

GM methods can significantly reduce the time taken to introduce a new gene into a crop. This is because, unlike the several generations of back-crosses required to regain the desired genetic background, the gene of interest is directly inserted into the crop cultivar with the desired genetic background, as described in Chapter 8. In practice, the time saved may not be as great as one might expect. For example, the stated advantage assumes that the gene is inserted into the target cultivar (probably an elite line) directly. However, this is often not so: when the technology is applied in less developed countries, the process should ideally involve the local varieties rather than varieties optimised for conditions in a developed country. There will thus be a longer development phase, in which use of the transgene in the local varieties (which may well not be characterised very thoroughly) is evaluated.

11.6.4 TO REDUCE THE ENVIRONMENTAL IMPACT OF AGRICULTURE

This consideration, already mentioned in passing in Section 11.6.2, may not, in itself, result in increased production, but there is no question that an overall benefit to society results. If a potentially deleterious environmental impact is reduced through the minimisation of pesticide use, then it would also benefit the grower and workers, whose task it otherwise might be to spray noxious chemicals to control pests. Indeed, as mentioned in Section 11.4.2, there is already evidence that use of agro-chemicals, perhaps with fewer safety precautions than are desirable, has affected the health of farmers growing some of the high-yielding Green Revolution crop varieties. Furthermore, since many poorer farmers cannot afford extensive use of agro-chemicals, there may be economic advantages as well as environmental advantages. However, even these applications of GM technology are not without their critics, as discussed in Chapter 9.

11.7 WHERE SHOULD CROP BIOTECHNOLOGY BE USED?

As was discussed in Section 11.2.2, projections for population growth and food production suggest that food production has to increase at an average of 1.5% per year to postpone a major food crisis until after 2030. These projections also suggest that most of this increase will occur in *developed* countries. In other words, even though developed nations have a food surplus today and much of it does not reach the people that need it the most, any hope the world might have of keeping most people away from starvation hinges on the assumption that developed nations will continue to produce this surplus food *and* that the people who need it will be able to get it. Given this, and given that genetic constraints place limits on further increases in food production, it is possible that GM must, necessarily, be a part of the solution for the world's 'food problem'. However, if the implementation of GM only occurs in developed countries, the problems of distribution, discussed in Section 11.2, will remain. Furthermore, even if this problem could be solved (and we have already noted Conway's pessimism about this: Section 11.2), there would be increased dependence of the less developed countries on developed countries, a trend that may well be exacerbated by globalisation; by contrast, the ideal would be the development of sustainable agriculture capable of contributing very significantly to the food security of less developed countries. On this basis, it has been argued that GM technology certainly has a role to play in less-developed countries (e.g. Swaminathan, 1997; Wambugu, 1998; The Nuffield Council on Bioethics, 1999). However, the lessons of the Green Revolution must be learned. The Green Revolution was a success in Asia and Latin America, partly because of the particular farming systems in use and partly because the plant breeders listened to local voices. It did not work in Africa, at least partly because of inadequate background research into African crops/crop varieties and agricultural systems and of a failure to recognise local cultural and social values. Nevertheless, GM crop technology is seen by some as being as applicable to African countries as to other less developed countries (Wambugu, 1998; The Nuffield Council on Bioethics, 1999). Indeed, Florence Wambugu makes the plea that

> Africa . . . is wanting, waiting and hoping that the biotechnology revolution will not pass them by, (as the Green Revolution did), due to a lack of resources and unrealistic controversial arguments from the North, based on imagined risks.

If then we accept that GM technology should be applied in less developed countries, what factors should influence its use?

11.8   USING CROP BIOTECHNOLOGY

11.8.1   THE NEED TO WORK ON APPROPRIATE CROPS

Most of the work so far on GM crops has been on maize, cotton, soya bean and rape and, to a lesser extent, tomatoes, potatoes, wheat and rice. Furthermore, the work has mostly taken place in developed countries. It is important that the range of target species is extended to include crops such as cassava, yams, sorghum, sweet potato, legumes (in addition to soya), millet, bananas and plantain. This is the case whether the work is done on crops for cultivation in developing countries or in developed countries for export to developing countries. In addition to crop species, attention must also be given to what cultivars of particular crops are appropriate for what applications of GM technology. Further, all these considerations must be set against a background of the need to preserve crop biodiversity.

The need to choose a cultivar that would best serve regional needs might seem obvious, but is not always heeded. This can be an issue with traditional methods of breeding too. For instance, members of a tribe in the north-eastern state of Orissa in India (where the national Central Rice Research Institute is located) will not cultivate 'government released' rice cultivars partly because they do not like the taste (Arunachalam, 2001). This issue is even more critical in the GM context. If genes for vitamin A are inserted using GM technology into elite rice varieties such as basmati, then it would defeat the very purpose of using the technology: basmati is an expensive variety that is beyond the reach of the poor; those that can afford basmati have other sources of vitamin A!

11.8.2   THE NEED TO WORK ON APPROPRIATE TRAITS

It seems almost too obvious to state but state it we do, that if crop GM technology is used in food production in less developed countries, the particular applications must address the needs of those countries. In general terms therefore the targets should be to give higher, more stable and sustainable production of food staples, possibly nutritionally enhanced, thereby increasing food security and also supporting the livelihood of poor farmers, small-holders and farm workers.

In accordance with these general aims, the Nuffield Council on Bioethics (1999) developed the following list of suitable trait targets:

• Conversion of inputs of nutrients, water and sunlight into harvestable plant material
• Partition of harvest between edible (or otherwise useful) and other plant material
• Sustainable extraction of soil nutrients and water
• Improved nutritional quality of the crop, including micronutrient and vitamin content (such as the vitamin-A-enhanced Golden Rice®), nitrogen : calorie ratio
• Resistance to/tolerance of pests and of environmental stresses (especially water stress and salinity, but in some instances, waterlogging).

However, it also needs to be recognised that crop GM technology may not always be the way to solve a particular problem. Sometimes it is not a technical 'fix' that is needed but a simple change in agricultural practice. This leads us to discuss specific aspects of the application of crop biotechnology in developing countries.

11.9   APPLICATION WITHIN DEVELOPING COUNTRIES

11.9.1   INTRODUCTION

A major set of questions raised within developing countries often is: who implements GM technology, who benefits the most from it and how can we be sure that it is the best solution for us? These points are discussed in some detail in Sections 11.10 and 11.11, but are raised here to emphasise that, of all concerns, these are often the most troubling to people in these countries. There are technological issues in GM applications that apply anywhere in the world. Here however we point to other features that should be considered specifically in relation to less developed countries, whether the technology is implemented *de novo* in developing countries or is modified from technology already implemented elsewhere.

11.9.2   FARMING CONDITIONS

Farming conditions vary widely between developed and developing countries, and within developing countries. Compared with the situation in developed countries farms in most developing countries are much smaller, and many more people are occupied in agriculture. In India, for example, about 450 million farmers cultivate holdings that average 0.2 ha in area. How might this factor affect the design of a GM project? For instance, in the USA large plots of GM cotton containing the Bt toxin gene are grown with surrounding buffer zones of non-Bt cotton in order to minimise the chances of the insect pests evolving resistance to the insecticide. With extremely small plots and the only available means to manage resistance being a spatial one, what types of resistance management method are likely to be effective in developing countries? It is also important to note that, while average farm size in India is 0.2 ha, size actually varies considerably from farm to farm. This heterogeneity means that uniform criteria cannot be applied across the country.

The incorporation of genes for herbicide resistance is at present (2001) a major feature of GM programmes in the USA and Europe. However, herbicide-tolerant varieties may be of little importance in countries where herbicides are not widely used. In India, for instance, the intensive cultivation of small plots means that the norm is hand weeding; indeed, common 'weeds' such as *Chenopodium* and *Amaranthus* are gathered for use as potherbs, which provide important nutritional supplements to the major grain crop being cultivated.

11.9.3   BIOSAFETY MEASURES

The types and range of health issues relating to GM applications in agriculture are likely to be similar regardless of where the applications occur. An argument could be made that results of similar tranformations in cultivars of the same crop species are likely to be similar. However, given that each transgenic plant line represents an original independent 'event' of insertion into potentially widely different genotypes, the potential health effects will have to be determined on an individual basis. Similarly, while the types of environmental issue to be considered are likely to be similar to those encountered in developed countries, the specific agro-ecological conditions are critically important in assessing the short- and long-term effects of widespread cultivation of transgenic plants. The additional time and cost involved in the necessary regulation and monitoring should be considered at the time when decisions are made to develop a particular GM application. Nevertheless, the Nuffield Council on Bioethics concluded that 'If developing countries stimulate appropriate, regulated, open GM crop research and selective release, they can steer the technology towards activities that are safe, more employment-intensive and better directed towards availability, quality and stability of food for the poor', under the auspices of 'an appropriate regulatory regime'. Widespread application of the biosafety protocol within the

Convention on Biological Diversity would certainly help here but we note that the USA, the country with most commercial power in this area, is not a signatory to the protocol.

11.9.4   EXPERTISE

It is obvious that the effective and appropriate application of crop GM biotechnology requires specialised skills at all stages of development and implementation. Developing countries vary from each other as to the extent and levels of available expertise to effectively establish and implement policy that governs biotechnological applications. Thus, developing countries need to build up such expertise internally: the requirement for training and for indigenous research is vital if the developing countries are to develop 'ownership' of GM technology that is appropriate for their needs. However, this assumes that resources are available to do what is necessary. The capacities and priorities of developing countries may be highly variable in this regard, and blanket statements regarding all developing countries may be of little help in making specific decisions. Nevertheless, these are aims that can be aspired to and which will, it is hoped, avoid the problem of GM technology being regarded as simply an alien import from the developed world, as happened in many African countries in respect of the Green Revolution.

11.10   A REALISTIC VISION?

11.10.1   INTRODUCTION

In earlier sections we quoted the scientists from Asia and from Africa who saw the application of GM technology as vital to the food security of the developing countries. Indeed, Swaminathan (1997) has presented a very positive view of GM technology in harmony with the agricultural systems of developing countries and applied to appropriate crops within those systems. However, as Wambugu (1998) has emphasised, this will rely on the technology actually being available. This raises issues of ownership, benefit and power. The question of who owns the technology and who most benefits from it is the one that underlies the intense reactions that many people have against the use of biotechnology in agriculture, especially in developing countries. The GM Revolution may, as Conway and others argue, be as necessary as was the Green Revolution, but there is a difference in implementation. The Green Revolution was spearheaded by the public sector and by charitable organisations but the GM revolution is being largely fueled by the private (corporate) sector. Does this affect the possibility of applying crop GM technology in less developed countries in the way that Swaminathan envisages?

11.10.2 THE MOVERS AND SHAKERS IN CROP
BIOTECHNOLOGY

Anyone who becomes involved in the debate about GM technology very quickly realises that the main players and drivers are major commercial agro-chemical and/or pharmaceutical companies that have reinvented part of their core business. Indeed, some of the campaigning directed against crop GM technology has been essentially based on the involvement of 'big business' and 'high technology' in food production (e.g. Anderson, 1999) (notwithstanding the fact that food production by non-GM methods is itself a major commercial enterprise involving high technology). Further, through a spate of mergers amongst their own kind and through purchase of seed companies, the number of major players on the world biotechnology stage has been reduced to about six, including Monsanto (USA), DuPont (USA), Novartis (Switzerland), AstraZeneca (UK/Sweden) and Aventis (Germany/France). A World Development Movement report (WDM, 1999) notes that other large chemical companies have also become active in the biotechnology industry, such as 'Dow AgroSciences (USA), the agro-chemical subsidiary of Dow Chemical which owns 69% of US seed company Mycogen, and the German agro-chemical giant BASF'. All this activity, along with the financial 'muscle' they can exercise, has ensured that these companies have a firm grip on the technology and its potential market, and can have a major say in how it will be developed. Their agendas for development and application of GM technology often seem far removed from the vision of application to sustainable food production and food security in developing countries. Furthermore, some regard certain international agencies as working towards the same commercial agenda. For example, a view that has often been expressed recently in India (Sharma, 2001; Shiva, 2001) is that the 2001 UNDP Human Development Report, *Making Technologies Work for Human Development*, is simply another biotechnology-industry-sponsored study that pushes the virtues of modern GM technology from the 'North' to the resource-poor 'South' under the garb of eradicating poverty. Are such worries justified?

11.10.3 THE NEW FRONTIER – REINVENTING THE
MARKETPLACE

Chemical and pharmaceutical companies, with their agro-chemical businesses providing herbicides and pesticides, have grown and prospered on the back of current intensive agricultural practices (including the demand created by the Green Revolution). However, there are indications that the agro-chemical industry *per se* is in decline. Furthermore, the agro-chemical companies accepted in 1998 that the market for their products in the major developed nations was static, and as a result they are eager to target and penetrate markets in less developed countries. It is only natural, in the name of commercial survival, that such companies will seek out new technologies and markets in order to regenerate an important core business. This includes crop GM biotechnology. We have noted earlier that crop GM biotechnology can offer new solutions that overcome some of the weaknesses of Green Revolution farming practices and that offer the possibility of higher crop yields and increased food security in developing countries. How far does this vision tie in with the commercial vision of a major company? Monsanto sees the future for GM resulting in three waves of beneficial products:

> The first consists of GM crops which are resistant to insects and disease, or tolerant of herbicides. These will allow farmers to meet the growing demand for food . . . The second wave . . . will see genetically induced 'quality traits' in food, such as high-fibre maize or high-starch potatoes, some of which will help doctors fight disease. And in the third wave, plants will be used as environmentally friendly 'factories' to produce substances for human consumption.[6]

The market for these GM products is lucrative. According to the International Seed Federation (ISF) the market for GM seeds was about $2 billion in the year 2000 with a total (GM and non-GM) seed market in general around US$23 billion. An IFPRI[7] report records that between 1995 and 1998 the value of the global market in transgenic crops grew from US$75 million to US$1.64 billion. It goes on to say that 'a total of nine countries, five industrial and four developing, grew transgenic crops commercially in 1998. The industrial countries – Australia, Canada, France, Spain, and the United States – contained about 85 percent of the 28 million hectares sown with transgenic crops. Argentina, China, Mexico and South Africa cultivated the remaining 15 percent of land. Argentina devoted the largest area to transgenic crops in the developing world: 4.3 million hectares in 1998; 60 percent of its soybean area was sown with transgenic varieties'. Yet this is just the beginning for this market. If ISF predictions are correct the GM market is set to have a $20 billion share of an expected $30 billion seed market in the year 2010. This prediction thus suggests that approximately 67% of the value of the seed market will be based on GM seeds, controlled by a few multinational companies (compared with a market share of only 9% in 2000).

11.10.4 THE STRATEGIES OF THE GM 'GIANTS'

The new biotechnologies and methods do not come cheap nor do they become commercially viable in the short run. To ensure success, agro-chemical businesses must pursue a number of strategies. They must raise the necessary finance; they also need to quickly (or as quickly as is practicable) turn around any investment in research into commercial and marketable products. Furthermore, they need to identify and broaden existing markets and open up new markets for these GM products. Another important strategy is to convince farmers they

---

[6] According to Dan Verakis, spokesman for the Monsanto Corporation, quoted in *The Observer*, London, UK, 23 August, 1998.
[7] IFPRI: International Food Policy Research Institute, Washington, DC, USA.

need the GM crops in order to create a push for demand. Simultaneously, they need to create a pull from the main markets in the developed world for GM food by convincing consumers of their benefits (although in Europe in the late 20th century this was singularly unsuccessful as various environmental organisations campaigned vigorously against GM crops: see Chapters 2 and 9). Finally, the companies must protect their investments. This is being achieved through a strategy of patenting their developments on the one hand while trying to strengthen international patent laws on the other (This is discussed below in Section 11.11 and also in Chapters 9 and 10). The worry then is that, in the hands of major companies, crop GM biotechnology will be determined by these drivers rather than those needed to help the majority of poor farmers. Take the example of GM crop research in India (Ghosh and Ramanaiah, 2000): of the on-going projects, seven are funded by private seed companies, all of which are to develop pollination control to produce hybrids or are in parental materials that are to be used in hybrid seed production. Of the 17 projects in public research institutes, only one is directed towards the development of pollination control for hybrid seeds. Ultimately it is only publicly (or charitably) funded research institutes that will prioritise the application of GM research to help the poor and needy. We now explore this problem a little further.

11.10.5 PRIORITISING GM SEED DEVELOPMENTS FOR THE DEVELOPED NATIONS' MARKETS – IGNORING THE POOR FARMERS

In order to justify their investment in GM technology companies must concentrate on developments conducive to extracting profitable returns. This is the nature of the capitalist system and it is not our intention here to criticise the system *per se*. It also explains why most of the developments to date are primarily focused on research and products beneficial to developed nations where potential sales are large, patents are well protected and the commercial risks are lower. For example, the first wave of products has included herbicide-tolerant crops such as soya and oil-seed rape. Use of such crops will extend the commercial life of a company's herbicide (e.g. Monsanto's glyphosate – 'Round-up'®), which farmers must use in conjunction with the herbicide-tolerant crop. Further, where GM crops have been grown in less developed countries, they have mostly been either cash crops, such as Bt cotton in China, or crops that are used for cattle feed such as soya and maize (corn)[8].

Research and development on crops beneficial to small farmers is thus ignored, not because of some global conspiracy but because of basic economic factors associated with a concentration of activity in the private commercial sector. There is simply no commercial reason why a company should invest in ventures likely to produce low returns or no returns at all. It is not surprising then that the

---

[8] Some 90–95% of soybean harvests and 60% of traded maize are not consumed by humans but by livestock: see Lappe and Bailey (1998).

GeneWatch report concludes that the majority of GM crop developments are mainly being applied to crops of importance to the developed world and fit 'comfortably into modern foods systems that emphasise food processing, consumer niche markets and production efficiency' (GeneWatch, 1998; see also Rissler and Mellon, 1996). Furthermore, it is not remarkable that there is virtually nothing that is directly relevant to less developed countries. In 1998 GeneWatch noted that 'internationally there are just four 'coherent, coordinated' GM research programmes on Third World crops' (based on a report by Kendall et al, 1997), further adding that 'even these are minimally resourced'(GeneWatch, 1998; see also Puonti-Kaerlas, 1998). Finally they remark that 'the World Panel on Transgenic Crops concluded that technology transfer projects between multinational corporations and less-developed countries were so rare that the examples they cited were *exceptional*. At best, therefore, it seems that application to such countries will be largely incidental, arising from so-called 'spillover innovations' (based on Kendall et al, 1997). However, there is a note of optimism: evidence is now emerging that suggests some of the corporations are working in partnership on projects that can benefit the poor, but have no real commercial value except in terms of public relations (see also Section 11.12).

However, it is also claimed that a major part of this problem here is simply the infancy of the technology. As indicated already, there is still much background research to be done on many of the crops of less developed countries in order to identify and characterise appropriate target species and strains and then to use GM technology to modify those characters and traits listed in 11.8.2. Nevertheless, the current rapid progress being made in genomics will transform plant breeding (both 'conventional', including marker-assisted breeding, and genetic modification) as the functions of more genes are identified. Breeding for complex traits such as drought tolerance, which is controlled by many genes, should then become more common. This is an area of great potential benefit for tropical crops, which are often grown in harsh environments and on poor soils. The problem then becomes one of who pays for the research. In purely commercial terms, the potential customers are too poor to justify the necessary investment. However, even if the large multinational companies did invest in this research, the question remains as to whether the developments will benefit farmers in developing countries or instead lead to further dependency and/or exploitation? This leads to a consideration of one particular facet of the control that may be exerted by the powerful 'players' in the use of this technology, namely intellectual property rights.

11.11 INTELLECTUAL PROPERTY RIGHTS

11.11.1 INTRODUCTION

A major impetus for the development of GM technology is the ability to patent genes and genetic constructs, thereby allowing the owners of the patent to earn royalties for GM crop lines. This topic is dealt with in detail in Chapter 10. The author of that chapter views the practice of patenting genes favourably, albeit with some major reservations. However, whilst recognising that the practice is legal in most developed countries, we have reservations about it. In the absence of specific arrangements to circumvent or waive the rights inherent in owning a patent, the patenting of plant genes can easily become another means of exploitation of the poor by the rich. It may well be, as argued in Chapter 10, that this can be avoided, but a note of caution is certainly appropriate (see also Chapter 9). Furthermore, both governments and commercial organisations in the most powerful countries are seeking to strengthen international law on intellectual property rights via the World Trade Organisation (and specifically via the Trade-Related Aspects of Intellectual Property Rights – TRIPS – chapter of the World Trade Agreement; see also Chapter 10). This leads us to discuss some issues relating to intellectual property rights in crop GM biotechnology and in some instances to suggest means available to circumvent some of the resulting problems. However, as a prelude to that discussion we suggest that if the Japanese *Norin 10* gene in wheat that was bred into the Mexican Sonora varieties in the 1950s (and which was important for the subsequent Green Revolution) had been patented, we might live in a somewhat different world today.

11.11.2   BIOPIRACY

Biopiracy is the exploitation (including the use of intellectual property rights) by organisations usually based in developed countries of biological (including agricultural) resources from less developed countries (see Edwards and Anderson, 1998, for example). It is clearly unethical, by most peoples' standards, for a commercial organisation to obtain crops from less developed countries and then to patent, and restrict the use of genes from such crops (whatever one's view of patenting genes in general).

However, the provisions of the Convention on Biological Diversity allow nation-states autonomy in dealing with their own natural resources (see Chapter 4). This means that a government is entirely at liberty to assign rights to exploit, for example, a gene bank, to an outside organisation.

Nevertheless, there are opportunities for partnership (see Section 11.12.2) here in that governments and/or research institutes in less developed countries may be able to negotiate deals with organisations and companies in developed countries, to the benefit of both (see e.g. Jayaraman, 1998). Such agreements may well stimulate appropriate research on the crops grown by poor farmers without extracting a high price from those farmers for utilising any new varieties that arise from the research. Indeed, according to the Nuffield Council on Bioethics (1999) some such arrangements are already in place. However, even this partnership approach is not without its dangers: Edwards and Anderson (1998) report that the International Centre for Agricultural Research in the Dry

Areas (ICARDA: a transnational research organisation) has signed agreements with research institutes in Australia allowing them to claim rights over seeds developed by ICARDA with the simple proviso that 'they gain approval from the countries of origin'. This proviso is not regarded as a strong enough protection for farmers in those countries and the agreement has evoked widespread opposition in developing countries (Edwards and Anderson, 1998).

11.11.3 'TERMINATOR TECHNOLOGY'

This technology has led to the greatest outcry in developing countries as well as amongst many scientists, citizens and campaigning organisations in developed countries. The technology, still under development, prevents the (clandestine) sowing of a second crop using seeds from the first harvest grown from commercial GM seed. It involves a modification that prevents germination of the next generation of seed whilst not affecting the food quality of the seed. In some versions of the technology the inhibition of germination may be overcome by spraying with a chemical, supplied of course by the same company that sells the seed. Whatever the specific detail, Terminator and related technologies mean that a farmer cannot save seed for the next year's crop, a practice that has been part of agriculture for several thousand years amongst many people across the world (not just in developing countries). In many African countries seeds and knowledge are shared with pride and given away as a great honour. We need to note in passing that the use of F1 hybrid seed also means that the farmer must buy new seed stocks every year and for that reason, hybrid corn (maize) was not widely accepted in many African countries.

Terminator technology is basically designed to lock farmers in to a cycle of buying seed or chemical spray every season, and this is particularly valuable to the owners of the technology in developing countries where patent rights or the law are weak. Thus Willard Phelps, a spokesperson for the USDA, proclaimed that the purpose or goal behind Terminator technology is 'to increase the value of proprietary seed owned by US seed companies and to open up new markets in Second and Third World countries'. As a result he wants the technology to be 'widely licensed and made expeditiously available to many companies'.[9] This goal was reinforced by a Delta Pine Land Company press release claiming that the Terminator technology has 'the prospect of opening up significant worldwide seed markets to the sale of transgenic technology for crops in which seed currently is saved and used in subsequent plantings' and that 'we expect [the new technology] to have global implications, especially in markets or countries where patent laws are weak or non-existent'.[9]

Interestingly, the campaigning against this application, which involved many who are essentially in favour of GM technology, led to the abandonment by at least two companies of Terminator-type technology in commercial crops (see,

[9] RAFI News Release, March/April 1998: The 'terminator technology': new genetic technology aims to prevent farmers from saving seeds.

e.g., Edwards, 1998). Nevertheless, the issues of exploitation and social and global justice were clearly revealed during the debate and will doubtless surface again. Indeed, Terminator-type technology is still under commercial development, albeit by fewer companies than before, and if it is used as aggressively as is implied by Willard Phelps (see above) it will be a major barrier to the use of crop GM technology in poorer countries.

11.11.4   GOLDEN RICE®

We have already mentioned the development by Professor Ingo Potrykus of the vitamin-A-enhanced Golden Rice® (Ye et al, 2000). The work was funded from the public purse in the European Union and Switzerland and by a charitable organisation, the Rockefeller Foundation. It was this development that brought to light some of the barriers that intellectual property rights might place in the way of applying GM technology to agriculture in less developed countries. The number of patents covering the techniques used in its development is about 40 (Kryder et al, 2000; see also Chapter 10). However, many of the rice-growing countries do not recognise the patents or had previously been seen as so unimportant that the patents had not been registered there. Other patents which might prevent freedom to operate in this instance turn out to be too specific to cover this product. In fact only a handful of patents might actually stand in the way of developing this crop further and at the time of writing (2001) it appears that the organisations owning those patents have waived their rights to royalties. In this instance then, intellectual property rights have not stood in the way of the application of GM crop technology in less developed countries. However, there are many such genetic manipulations going on around the world that will inevitably be faced with similar situations. As a consequence, many GM varieties may not be freely available, even though public money funds some of this research.

11.11.5   ENABLING TECHNOLOGY

Several of the patents that might have (but in the end did not) inhibit the development of vitamin-A-enhanced rice essentially cover enabling technology. Therefore, as mentioned above, the problems are likely to recur for many potential applications of GM technology in less developed countries. One way round this is for the development by public or charitably funded organisations of alternative effective enabling technologies that are not subject to commercial considerations. For example, the Center for the Application of Molecular Biology in International Agriculture (CAMBIA) in Australia is developing a range of molecular tools (e.g. marker genes for plant transformation). These will be made freely available to 'empower a wide spectrum of users with the necessary biotechnological tools and freedom to operate'.[10]

11.11.6 APOMIXIS

Apomixis is the formation of seeds that contain viable embryos without the intervention of a fertilisation event, and occurs in many species throughout the plant kingdom. Apomixis can take place by various mechanisms but, in all, a diploid cell in the ovule of a flower behaves like a fertilized egg and goes on to form a normal embryo in a normal seed. This seed, when sown, will grow into a normal plant that will, in turn, form seeds in a similar fashion, because this lineage contains a gene or genes that cause apomixis. The advantage of this is that a particular crop line can be maintained true-breeding for generation after generation (see also Chapter 8). Therefore, a farmer need not go back to buy seed that he or she cannot afford, but instead can take advantage of a high-yielding variety by being able to save the seed and obtain consistent yields. Further, the advances in plant genomics mean that it is now possible to identify the gene or genes that regulate apomixis and eventually to transfer them to crop varieties developed for specific situations in specific countries. However, the potential restrictions imposed by patent rights that we have already noted could equally well apply to the creation of apomictic lines via GM technology. In the face of this, the signatories to the Bellagio Apomixis Declaration (May 1998)[11], including CAMBIA, were concerned that 'current trends towards consolidation of plant biotechnology ownership in a few hands may severely restrict access to affordable apomixis technology, especially for resource-poor farmers'. Thus they argued for 'widespread adoption of the principle of broad and equitable access to plant biotechnologies, especially apomixis technology . . . and . . . the development of novel approaches for technology generation, patenting and licensing that can achieve this goal'. (See also The Nuffield Council on Bioethics, 1999).

11.12 IS THERE A WAY FORWARD?

11.12.1 INTRODUCTION

In earlier sections of this chapter we have noted that there are differences in viewpoint as to whether crop GM technology is appropriate for use in less developed countries and as to whether it has any contribution to make to sustainable food production and food security in those countries. We have also noted the insistence of influential voices in developed countries (e.g. Conway of the Rockefeller Foundation: Conway, 2000) and developing countries (e.g. Swaminathan, 1997, and Wambugu, 1998) that this technology must be incorporated into the agricultural systems of the 'Third World'. However, there are major obstacles in the way. Firstly, the technology, and with it, power, is largely

[10] 'Our institutional ethos is built around an awareness of the need for local involvement in achieving lasting solutions to agricultural problems': see www.cambia.org
[11] See http://billie.harvard.edu/apomixis

in the hands of a small number of commercial organisations in developed countries. As commercial organisations they are involved in agriculture and food production as a business activity. However, an aggressive commercial stance has not helped to transfer the technology to the world's poorer countries. Indeed, there is the very real possibility of serious exploitation of the developing countries by commercial interests in developed countries. Secondly, very little GM research has been carried out on many of the crops grown in less developed countries or on the incorporation of GM technology in an appropriate way into their agricultural systems. The integrated vision of Swaminathan (1997) that we cited in Section 11.4.2 seems a long way off. Is there then a way forward?

11.12.2   THE PARTNERSHIP APPROACH

Both Conway (2000) and the Nuffield Council on Bioethics (1999) have argued for a partnership approach in developing appropriate crop GM technology for the poorer countries. In particular, Conway states

> On the one hand, they [developing countries] could encourage the for-profit sector to develop and market high quality, locally adapted, premium seeds (especially hybrid seed, which farmers can, if they wish, keep for the next season's crop although the yields are likely to be lower[12]) for the commercial and semi-commercial farmers. Protection would be through a modified Plant Variety Rights[13] system. On the other, they would encourage a strong public sector seed system that serves poorer farmers. This would provide an economic incentive for private sector research, innovation, and marketing, and help ensure that the public sector had access to new technologies. Over time, more and more farmers from the semi-commercial sector should be able to buy seeds on a regular basis. And hopefully, many farmers would move from being really poor to the semi-commercial or commercial categories . . . . A key part of such an approach would be the stimulation of public–private partnerships whereby genomic information and technologies are made available to public plant breeders.

This partnership approach has many merits. Developing countries could then see themselves as stakeholders in the technology's development and in partnership with private-sector companies to genetically modify crops consumed by the poor so that they grow better and more abundantly in their environment without making it worse. Indeed, the Nuffield Council on Bioethics (1999) concluded that a 'compelling moral imperative exists to make transgenic crops available to developing countries that want them to combat hunger and poverty. Creative partnerships between developing countries, CGIAR centres, and the private sector could provide the institutional mechanism for sharing the new technologies'.

This has implications for policy making at all levels in order to deliver solutions for the poor that include the benefits of biotechnology. Persley and

---

[12] But we draw attention here to the potential of apomixis in maintaining yields from generation to generation.
[13] See Chapter 10.

Doyle (1999) in their briefing for the biotechnology industry and WTO propose the following:

> The successful application of modern biotechnology to the problems that cause under-nourishment and poverty could be called a 'bio-solution'. The delivery of new bio-solutions to the problems of food security and poverty will require continual policy development and actions at the national, regional, and international levels. These efforts will involve the following five areas: (1) determining the priorities and assessing the relative risk and benefits in consultation with the poor, who are often overlooked while others decide what is best for them; (2) setting policies that benefit the poor and minimise technology-transcending risks that adversely affect the poor; (3) establishing an environment that facilitates the safe use of biotechnology through investment, regulation, intellectual property protection, and good governance; (4) actively linking biotechnology and information technology so that new scientific discoveries worldwide can be assessed and applied to the problems of food insecurity and poverty in a timely manner; and (5) determining what investments governments and the international development community will have to make in human and financial resources in order to ensure that bio-solutions to the problems of food security reach the poor.

Perhaps the first targeted GM crops for developing countries should be those that benefit the poor and their livelihood as a priority. After food security has been achieved then there may be a role for targeting crops suitable for Western demands, where a profit can be made, providing it does not detract from the common good and well-being of all. None of this will be achieved without the will of all stakeholders to shift some of the emphases of GM research, nor without a good deal of effort. However, as noted by the Nuffield Council on Bioethics (1999), 'To forgo such efforts would . . . sacrifice the prospects of major GM crop-based advances in food and agricultural output and employment for the food-poor'. It is therefore noteworthy that the US Congress is considering possible legislation that would authorise the National Science Foundation 'to establish research partnerships for supporting the development of plant research targeted to the needs of the developing world.' (ASPB, 2001).

11.13 AN ETHICAL OVERVIEW

The subject reviewed in this chapter raises several societal and ethical issues that have a bearing on the moral behaviour of individuals, groups, commercial and non-commercial organisations, countries and even groups of countries. In brief the main issues are the following.

- *Use of the earth's resources.* This involves a duty of care for the environment *per se*, including a detached assessment of risk of particular courses of action.
- *Distribution of the earth's resources.* It was noted early in the chapter that the earth produces at present more than enough food to supply the needs of everyone. However, there are still many millions of poor people, mostly living

in less developed countries, who are hungry. The food they require does not reach them.

- One of the factors that leads to this problem is *the inequality in the abilities of different nations to produce food.* Although the Green Revolution enabled several countries to become self-sufficient in food production, it is actually the developed countries that produce most of the excess.

- The development of crop GM technology has the potential to *increase further the inequalities* between the developed and less developed countries.

- *Can a purely utilitarian ethical system* (Chapter 1) *deal with this situation?* If the result of seeking the greatest good is applied to the people of just one nation, then the answer is clearly 'No'. If utilitarianism is to be invoked it must be a utilitarianism that sees the human population of the world as one.

- However, it is probably more helpful here to think in terms of *human rights*, as defined by the United Nations. These include the rights of every human to enjoy an adequate diet.

- Further, the adequate diet should be a *secure provision that is generated in a sustainable way.* Thus, for the human population we speak in terms of food security and for both the human population and the environment we speak in terms of sustainability.

- In common with authors of other chapters, we note that the ethical systems adopted by individuals, by groups and by societies are often influenced by *religion and culture.* This religious and/or cultural element means that one individual or society may well place a very different weight on a particular factor in moral decision-making than another individual or society. Ethical decisions, especially at the societal level, are thus rarely absolute, but require the careful evaluation of the different positions in relation to the problem in hand. A parallel situation has recently occurred in the UK in the decision to allow the cloning of human embryos to generate stem cells (see Chapter 16).

Finally, since this book is called *Bioethics for Scientists*, it is noted that in modern biological and biomedical science, it is impossible for scientists to declare that their work is free from societal values (see Chapter 2). In the field of molecular biology, for example, whether in the context of medicine or of agriculture, the route from 'pure' research to applied research to actual application is seamless. Indeed, much 'pure' research is funded only because society perceives that it may gain some value from that research. Scientists should thus be aware of the wider ethical and societal context in which their work is placed and should make an honest attempt to understand the potential implications of their work. This will mean discussing those implications with those outside the scientific community in order to involve a wider society, at least in terms of the dissemination of information and interchange of views. A little openness will go a long way!

REFERENCES

Ahn, S. and Tanksley, S.D. (1993) Comparative linkage maps of the rice and maize genomes. *Proceedings of the National Academy of Sciences, U.S.A.*, **90**, 7980–7984.

Anderson, L. (1999) *Genetic Engineering, Food and Our Environment*. Green Books, Dartington, UK.

Arunachalam, V. (2001) The science behind tradition. *Current Science*, **80**, 1273–1275.

ASPB (2001) Arntzen, Clutter, testify before Congress on plant biotechnology, genome research. *ASPB News*, **28**, 10–11.

Bevan, M. (1999) The small, the large and the wild: the value of comparison in plant genomics. *Trends in Genetics*, **15**, 211–214.

Conway, G. (1997) *The Doubly Green Revolution.* Penguin, London, UK.

Conway, G. (2000) Crop biotechnology: benefits, risks and ownership. *OECD Conference: the Scientific and Health Aspects of Genetically Modified Foods*, 2000, Edinburgh, UK.

Conway, G. and Toeniessen, G. (1999). Feeding the world in the twenty-first century. *Nature*, **402** (suppl.), C55–C58.

The CornerHouse (1998) *Genetic Engineering and World Hunger*, CornerHouse Briefing **10**.

Dixon, P. (1998) *Futurewise: Six Faces of Global Change.* Harper Collins, London, UK.

Edwards, R. (1998) Devilish seeds: U.S. officials fear a backlash over terminator technology. *New Scientist*, **160**, 21.

Edwards, R. and Anderson, I (1998) Seeds of wrath. *New Scientist*, **157**, 14–15.

FAO (2000) *Agriculture: Towards 2015/30.* http://www.fao.org/news/2000/000704-e.htm

GeneWatch (1998) *Genetic Engineering: Can it Feed the World?* Briefing 3. GeneWatch, Tideswell, UK.

Ghosh, P.K. and Ramanaiah, T.V. (2000) Indian rules, regulations and procedures for handling transgenic plants. *J. Sci. Ind. Res. India*, **59**(2), 114–120.

Goff, S.A. (1999) Rice as a model for cereal genomics. *Current Opinion in Plant Biology*, **2**, 86–89.

Jayaraman, K. (1998) India seeks tighter controls on germplasm. *Nature*, **392**, 537.

Kendall, H.W., Beachy, R., Eisner, T., Gould, F., Herdt, R., Raven, P.H., Gould, T., Schell, J.S. and Swaminathan, M.S. (1997) *The Bioengineering of Crops: Report of the World Bank on Transgenic Crops.* World Bank–CGIAR, Washington, DC, USA.

Kryder, D.R., Kowalski, S.P. and Krattinger, A.F. (2000) *The Intellectual and Technical Property Components of Pro-Vitamin A Rice: a Preliminary Freedom to operate Review*, ISAAA Briefs No. 20, ISAAA, Ithaca, NY, USA.

Lappe, M. and Bailey, B. (1998) *Against the Grain: the Genetic Transformation of Global Agriculture*. Earthscan, London, UK.

Moore, G., Devos, K.M., Wang, Z. and Gale, M.D. (1995) Grasses line up and form a circle. *Current Biology*, **5**, 737–739.

The Nuffield Council on Bioethics (1999) *Genetically Modified Crops: the Ethical and Social Issues,* Nuffield Foundation, London, UK.

Persley, G.J. and Doyle, J.I (1999) *Biotechnology for Developing-Country Agriculture, Problems and Opportunities: an Overview.* Focus 2, Briefing 1. International Food Policy Research Institute, Washington, DC, USA.

Puonti-Kaerlas, J. (1998). Cassava Biotechnology. *Biotechnology and Genetic Engineering Reviews*, **15**, 329–364.

Rissler, J. and Mellon, M. (1996) *The Ecological Risks of Engineered Crops.* MIT Press, Cambridge, MA, USA.

Sen, A.K. (1981) *Poverty and Famine: an Essay on Entitlement and Deprivation.* Clarendon, Oxford, UK.

Sharma, D. (2001) Biotechnology will pass the hungry. *Times of India*, 17 August.

Shiva, V. (2001) UNDP as a biotech salesman. *The Hindu*, 6 August.

Swaminathan, M.S. (1997) ICRISAT in the 21st century: towards sustainable food security. *Environmental Awareness*, **20** (No. 4) Baroda, India.

Vasil, I.K. (1998) Biotechnology and food production for the 21st century: a real-world perspective. *Nature Biotechnology*, **16**, 399–400.

Wambugu, F. (1998) Benefits and risks of genetically modified crops: gathering important insights on research into the benefits and risks of genetically modified crops for man and his environment. *CERES Forum on Food Products from Plant Biotechnology II*, 1998, Berlin, Germany.

WDM (1999) *The battle for international rules on GMOs: the biotech industry versus the world's poor*. WDM campaign briefing, December 1999. See also http://wdm.org.uk/cambriefs/GMOs/battle.htm.

Ye, X., Al-Babili, S., Kloti, A., Zhang, J., Lucca, P., Berger, P. and Potrykus, I. (2000) Engineering the pro-vitamin A (beta-carotene) biosynthetic pathway into (carotenoid-free) rice endosperm. *Science*, **287**, 303–305.

IV Ethical Issues in Biomedical Science

12 Starting Human Life: the New Reproductive Technologies

Linda Baggott la Velle

12.1 INTRODUCTION

The 'bald' facts of life for most people are the following:

1. we have no control over whether or not we become alive;
2. once alive we strive to stay alive;
3. we have a strong and instinctive drive to reproduce, fortified by a strong moral intuition;
4. we all eventually die.

As organisms, human beings are subject to these facts of life, but over the course of human evolution and of the history of civilisation, we have developed the ability to rationalise and to moralise. The philosophical basis of this is discussed in Chapters 7 and 18. Most people believe that human life is a matter of moral seriousness: we readily understand that, as well as ourselves, the basic instinct for life is present in other people, and we appreciate that we have a moral duty to protect that instinct in them. This has been termed the 'presumption in favour of life' (Dunstan and Sellar, 1988), and for humankind it is fundamental to our continued existence. As society has developed, the drive to protect and promote life has become increasingly formalised in people's behaviour, to the extent that from early times it has been enshrined in legislation. However, this raises two fundamental philosophical and ethical questions:

1. When does human life begin?
2. At what point should the 'moral seriousness' of human life require it to be protected by legislation?

Bioethics for Scientists. Edited by John Bryant, Linda Baggott la Velle and John Searle.
© 2002 by John Wiley & Sons Ltd.

12.2   WHAT CONSTITUTES A HUMAN PERSON?

The new reproductive technologies involve the use of what we might coldly call 'human material', i.e. gametes prior to fertilisation and embryos after fertilisation, but what is the moral status of this material? Is the embryo[1] in the dish in the laboratory a human person? Is the foetus[1] in the womb a human person? Many attempts have been made to answer these questions. So, what does constitute a human person? Most obviously a person is a body with intelligence, rationality, self-awareness and the ability to form relationships with other persons, but to what extent do some or all of these attributes have to be present in order to constitute a human person? For example, are individuals in whom these attributes are to a greater or lesser extent impaired by permanent damage to the central nervous system less than human? This question raises so many difficulties as to make us realise that to define a human person primarily in terms of the characteristics determined by the activity of the central nervous system is inadequate (although at the end of life it is increasingly becoming an important consideration: see Chapter 17).

Aristotle believed that the foetus became a human being when it was recognisably a human form, which he claimed was at 40 days for a male and 90 days for a female[2]. Another view is that this happens when the foetus first moves in the uterus. A traditional Christian[3] view is that human life begins at fertilisation and that from that point on it should be fully protected as a person. This view therefore regards all the new reproductive technologies and embryo research as morally wrong because they inevitably involve the *in vitro* creation of embryos and the wastage of many of them. The implications of this prohibition are of course that infertile couples can never have a child and embryo research into the prevention, diagnosis and treatment of genetic disease and the causes of miscarriage cannot proceed. Many argue therefore that such an approach not only prevents benefit being conferred on people but also perpetuates human suffering.

As the new reproductive technologies have developed so new attempts have been made over the last 30 years to find an ethical code which on the one hand respects the 'moral seriousness' of human life from an early stage but on the other allows the benefits of these techniques to be made available to those who need them.

---

[1] Note that we use terms in the following way: *zygote* is the immediate product of fertilisation, from the Greek word *zygosis* meaning coming together. A series of cell divisions, known by embryologists as cleavages, leads to the establishment of the *embryo*. After implantation into the wall of the uterus, the growing organism is called a *foetus*.

[2] Leaving aside the arbitrary nature of this judgement, we now know the 'default' state of the human embryo is female and thus the idea that females develop in the womb more slowly than males is erroneous.

[3] This is the position adopted by the Roman Catholic Church and by certain other groups, including some of the 'pro-life'/anti-abortion groups (not all of whom have a religious basis). It also needs to be stated clearly that many Christians do not take such a 'strong line' on the status of the zygote or of the pre-implantation embryo.

In the United Kingdom this approach was formalised by the Warnock Committee (1985; Warnock, 1998). Its recommendations were subsequently enshrined in the Human Embryology and Fertilisation Act, which came into UK law in 1990. The Warnock Committee examined what was known about the early development of the human embryo, namely the following.

- Fertilisation is a process. It takes about 30 hours from its onset to the complete genetic fusion of the sperm and the egg to form the zygote.
- For up to ten days thereafter, the mass of dividing cells moves down the Fallopian tube and then into the uterus.
- At this stage there are several possible outcomes. Many embryos do not survive; estimates vary between 50 and 80%, implying that in nature there is a significant wastage at this stage. However, if the embryo does survive it may develop after implantation (see next bullet point) into a single foetus with its supporting placenta or it may divide into two with the subsequent development of identical twins.
- At the 30–60 cell stage the embryo, which has now become a hollow ball of cells, begins to attach itself to the wall of the uterus. Even at this stage, the fates of the cells of the embryo are not determined. It is only as implantation is properly established that embryo and placenta start to differentiate and then the embryo begins to exhibit polarity (i.e. to exhibit a front and a back, a left and a right and an up and a down). For the mother, this stage is very significant because implantation leads to the missing of the menstrual period and some women claim that they 'feel' pregnant at this point.
- At 14 days the process of differentiation has led to the development of the primitive streak, from which the central nervous system will develop.

In summary, between the onset of fertilisation and the next 14 days of development, the survival of the embryo is by no means certain. Survival depends on the successful embedding or implantation into the wall of the uterus and many embryos fail to do this. During these 14 days the cells are pluripotent, that is, having the potential to develop into any kind of cell, either as part of the embryo itself, or of the surrounding membranes, including the placenta, which act as a support system only, and are discarded at birth. Finally, while these dividing cells contain the potential of a nervous system there is no evidence in this early stage of nervous tissue, which is an essential part of being human.[4]

It was thus concluded that fertility and embryo research were justifiable up to 14 days. However, the Warnock Committee did not say that *anything* could be done with these early embryos (referring specifically to those created by *in vitro* fertilisation), but that activity should be restricted broadly to the diagnosis,

---

[4] It is for these reasons that most of the people who are concerned about these issues, including many adherents of the Christian and other faiths, hold that the pre-implantation embryo is not a person although it has the potential to become a person.

prevention and treatment of human disease[5] (see Section 12.8). The Warnock Committee Report is thus a classical example of recognising the complexity of the issues, of bringing together ethical considerations with scientific data and formulating recommendations which are both principled and pragmatic.

12.3   THE BIOLOGICAL DRIVE TO REPRODUCE

Some non-living artefacts such as the motor car display various characteristics of life such as movement, nutrition (fuel supply), excretion (exhaust system), respiration (carburettor) and even, in the most up-to-date models, sensitivity and response (detection of light intensity, and rainfall on the windscreen), but growth and reproduction are – as yet – not seen in manufactured articles such as cars, so it can be argued that reproduction is the single most distinguishing characteristic of life. The genetic sorting that occurs during sexual reproduction is also a means by which natural selection, that driving force of evolution, can occur, and so is responsible for the diversity of plant and animal species on Earth. People, as organisms, are just as subject to these forces as any other with which we share the planet, and indeed the urge to reproduce is felt at the most powerful and basic biological level at various times in our lives. Since fertility is so much a part of normal life, it follows that infertility can be regarded as a pathological state. Most people have a basic instinct to have children, and when they learn that this may not be possible, they may experience a form of sorrow similar to the grief of bereavement. During the course of their lives, most people can expect that others close to them will die, but that in time the sadness that results from it will diminish. Infertility, however, can be a life-long affliction: infertile people have said that the fertile seem to take their inheritance for granted, and the loss experienced by the involuntarily childless is a form of grief which is difficult to come to terms with because there is no focus for it – the grief is for the baby who never was. This bereavement may be because of what has been termed 'genetic death' (Snowden and Snowden, 1993), which is to say the person's genetic inheritance will not be passed on to the next generation; there is little doubt that those whom it affects suffer very badly. Nowadays however, issues such as involuntary childlessness are more openly and frequently discussed, and with the increasing scientific and medical understanding of the causes, considerably more can be done to help the subfertile. When a couple who decide to have a child discover that they are unable to conceive, for them it is a personal tragedy. If they are advised that assisted reproduction may offer them a chance of parenthood, they then have to decide whether it is physically, emotionally, morally and financially acceptable to them. Infertility treatment is expen-

---

[5] It was thus interesting that Mary Warnock, as a member of the upper house in the UK parliament, voted against the use of *in vitro* embryos as sources of cloned stem cells. Her reasoning was that although personhood is not ascribed to early embryos, nevertheless they should not be commodified.

sive, both financially and emotionally, invasive of the most intimate areas of people's lives, and for many deeply unnatural, but such is the desperation of some childless people that they are willing to undergo many weeks, months and even years of treatment. The conditions under which they should receive help constitute one of the big ethical problems of contemporary society. Those who are opposed to infertility treatment question the 'right' of everyone to have children. An argument often put forward is that expensive medical resources should not go into infertility treatment when there are many children already born who need parents. These resources, they contend, could be better expended in promoting improved parenting, or in other forms of medical care. Their argument is based on the view that bringing up children is more important than 'incubating' a baby. A more reactionary opposition to infertility treatment is based on the view that infertility helps to ameliorate the problem of world over-population. However, none of these views take into account the consequences for the childless individuals whose family line will end – those facing 'genetic death'.

But is it everyone's birthright to reproduce? Natural selection is no respecter of people's emotional longings, and if there is a good biological reason for being unable to have children, then it could be argued that this is 'nature's way' of strengthening the species to ensure its survival. However, as will be discussed later in this chapter, humans have overcome many problems of infertility with increasingly sophisticated and invasive techniques, and it can equally be argued that to help someone to reproduce who might otherwise be facing 'genetic death' is not a problem, because if the rate of technological advance is plotted into the next generation any defective genes that would otherwise have died out will probably be easily dealt with by genetic testing/diagnosis or even, in the future, by genetic modification techniques (see Chapter 14).

Of course, not everybody does want to have children; some have them by 'accident', whilst others decide consciously to avoid having a family, but for the majority of people having children is a natural and essential part of their lives. This drive appears to be independent of sexual orientation, as has been seen in the recent (2001) case of a British male homosexual couple resorting to assisted reproduction and surrogacy (in the USA) in order to have children carrying some of their own genes. Further, fertility clinics regularly have requests from lesbians for donor insemination. It is not surprising therefore, that all aspects of human reproduction have over the course of history been the subjects of intense moral and ethical debate. As medical and scientific advances have enabled more and more control over reproduction, and further means of promoting it and controlling it have been discovered, the debate has intensified. Recently, however, technical advances have moved faster than the moral discussion with the result that people have expressed their emotions, outrage and anger with increasing ardour about these issues. This chapter considers aspects of the moral and ethical issues around the new reproductive technologies – the scientific and medical techniques that enable infertile or subfertile people to have children.

Because of the technicalities involved, some detailed explanation is necessary of the biological structure and function of the reproductive process and what can go wrong with it and the medical interventions that can be made.

12.4 FERTILITY, SUBFERTILITY AND INFERTILITY

A popularly held view, encouraged by media coverage, is that global over-population is responsible for many problems in the world. It may therefore be surprising to learn that human fertility is relatively poor. Compared for example with rodents, reproduction in people is very inefficient, and the fact that there are so many of us on this planet is due to improved survival rather than reproductive success. Even in couples who have already conceived (i.e. are of proven fertility) and are having normal unprotected sexual intercourse, the average monthly chance of getting pregnant is only 20–25%, and just by chance about 10% of these fertile couples will fail to conceive during their first year of trying. It is only after about this period of time, during which their family doctor will have carefully explained to them how to maximise their chance of conceiving, that they are referred to a specialist fertility clinic.

The basic requirements for conception are shown in Figures 12.1 and 12.2. Failure of any of these will result in infertility, but some are much more likely to fail than others. The pie chart in Figure 12.3 shows the relative frequency of the various common causes of infertility.

Frustratingly, the cause of a couple's infertility frequently remains unexplained. Even after extensive investigation of both partners, no apparent problem may be discovered. Obviously this means that no specific treatment can be prescribed. However for couples with unexplained infertility of less than 3 years, most are within the normal range of fertility as described above, and are likely to

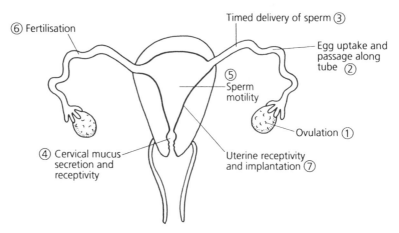

Figure 12.1. Conditions needed for conception – female. Reproduced with permission, from Baggott, 1997

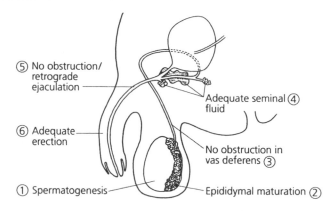

Figure 12.2. Conditions needed for conception – male. Reproduced with permission from Baggott, 1997

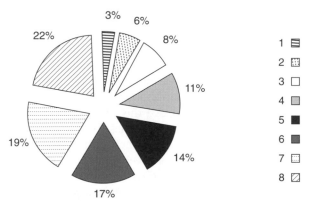

Figure 12.3. Frequency of common causes of infertility: 1, ovulatory failure; 2, tubal damage; 3, endometriosis; 4, cervical mucus defect/disorder; 5, sperm defect/disorder; 6, other male factor; 7, coital failure; 8, unexplained. The total comes to more than 100% because some couples have more than a single cause (Hull et al, 1985)

conceive naturally within a further 2 years. If they have been infertile for more than 3 years, it is unlikely that the woman will get pregnant by natural means, and the best hope for them to have a child that is genetically theirs is to have assisted reproduction treatment.

The *complete* inability to conceive a child – infertility – is very rare. This would only be the case for example if the woman had completely blocked Fallopian tubes, or premature menopause, or if the man had a complete lack of sperm. Absolute infertility in both partners who are of reproductive age means that there are no treatment options open to them, and their only means of having a family would be to adopt or foster children.[6] The anxiety of wondering whether

---

[6] Although it has been suggested that cloning, using genetic material from one parent inserted into an

they will be acceptable as foster parents or adopters can be as stressful as the treatment for infertility. If complete infertility is diagnosed in either partner, both, either separately or together, will be offered counselling. This is one of the most important aspects of the work of fertility centres. To help them to come to terms with their problems, people need information, and to talk to someone both informed and sympathetic about the implications of childlessness and/or fertility treatment. Most young couples trying to start a family do not expect that they may be unable to do so; infertility comes as a shock, and of course it not only affects the infertile person, but his or her partner as well. It has been seen that the worries and anger that the infertile may experience are often focused around their treatment (Jansen, 1996). This means that to be effective in the long term (whatever the outcome of the treatment) and to help prevent other problems such as their relationship breaking down, the counselling of the couple must be closely linked to the treatment. Many counsellors are now specialising in infertility counselling, which means that patients can have the benefit of talking to someone who is not only independent of the medical and scientific staff who are involved in their treatment, but also has the time and expertise to listen to and help them to deal with their anxieties. They may choose to opt for adoption or fostering. There are however, relatively few healthy babies available for adoption in the UK or USA nowadays, mainly because of improved contraception, greater acceptance of single parenthood and an increased frequency of termination of pregnancy (abortion). All this has led to an increase in recent years in the number of couples adopting babies from comparatively poor and over-populated or war-torn countries elsewhere in the world. The procedures involved in adopting children from less developed countries are often far from straightforward, and many people question the morality of removing children from their ethnic and cultural roots even if it can be argued that they could have a materially and emotionally better upbringing. In the countries where there are few babies for adoption, babies have even been offered for sale, thus turning babies into a commercial commodity. For example, in a notorious case early in 2001, a British couple purchased twins from the USA after answering an advertisement placed on the Internet.

An alternative to adoption is fostering. Many children who require fostering have a physical, mental and/or emotional handicap, and caring for them may require a rather different commitment from 'normal' parenting. However, although fostering is often a short-term arrangement, and parting can be very painful for both the child and the foster-parents, it has enormous rewards and many people have reaped great happiness and fulfilment from it. Nevertheless, some infertile people may decide that neither adoption nor fostering is an option for them, and they must then get on with their lives without having children of their own.

enucleated oocyte from a donor, might in the future be a possible way for such couples to have a baby that was genetically related to one of them (but see Chapter 16).

12.5 OPTIONS FOR THE SUBFERTILE

12.5.1 INTRODUCTION

More options are open to the couple if only one partner is infertile, and even more if the specialist decides that the 'infertility' is in fact subfertility, which is therefore treatable. Studies in many Western countries have shown that as many as one in six couples seek specialist help at some time in their reproductive lives because of difficulty in getting pregnant, and that a similar number may be unable to have a baby, but do not seek help (Thompson et al, 1985). This means that infertility/subfertility is a fairly common problem. In times past, this was very much a taboo subject: childless women were said to be barren, and were often much pitied or even reviled. It is noteworthy that in those days, infertility was almost invariably assumed to be the woman's problem, possibly because male infertility was often disguised by infidelity. It has been claimed that up to 20% of people in Western culture are not fathered by who they think they are! (Ridley, 1995).

12.5.2 THE REGULATION OF MEDICAL TREATMENT FOR SUBFERTILITY

Much medical and scientific research is focused upon causes of and treatments for subfertility, and it was partly due to public concern about these that following the report of the Warnock Committee (Warnock, 1985) the Human Fertilisation and Embryology Act was passed in the UK in 1990. A Government 'watchdog' called the Human Fertilisation and Embryology Authority (HFEA) was established to make provision to regulate and monitor treatment centres, and to ensure that research using human embryos is carried out in a responsible way. The HFEA does this by means of a licensing system, which covers any fertilisation treatment involving the use of donated eggs or sperm or embryos created outside the body, the storage of gametes and embryos and research on human embryos. Licensed centres in the UK that offer assisted conception treatment must conform to a code of practice issued by the HFEA. An important aspect of this code of practice is that account should be taken of the welfare of the child that might result from fertility treatment. If the centre believes that for any reason the people seeking treatment are not in a position to bring up the child in reasonable circumstances, they are obliged under the Act to withhold treatment (Morgan and Lee, 1991). Of course, this is highly judgmental, and in this circumstance it is the assisted conception team, usually led by a consultant gynaecologist, which decides whether that patient may be helped to conceive in that centre. Many centres for example refuse treatment to single or lesbian women on the grounds that the child has the right to a father. The issue of homosexuality and parenthood is one that is regularly in the news, and, with the new possibilities afforded by reproductive technologies, likely to appear with increasing frequency. Other reasons for refusal of treatment may include health factors and criminal record. The issue of who receives treatment is highly controversial, not least because there are no clearly drawn lines. As part of their professional ethical code, doctors give precise and careful consideration to these matters, and are obliged as a profession to accept responsibility for their decisions: the buck stops with them.

12.6   TREATMENT OF HUMAN SUBFERTILITY

12.6.1   INTRODUCTION

Problems of subfertility affect both sexes approximately equally. Little help was available until the pioneering work of Margaret Jackson, a family doctor in Exeter, UK, who used artificial insemination by donor to help some childless couples. This remained the only real technique involving the use of gametes outside the body until in the UK gynaecologist Patrick Steptoe and embryologist Robert Edwards brought about the fertilisation of human eggs by human sperm in the laboratory (Edwards et al, 1980). Subsequently they successfully transferred the resulting embryos back into the uterus of the woman, resulting in the birth of the first 'test tube' baby, Louise Brown, in 1978. Since then this treatment for subfertility has become available to more and more couples. Many of the procedures of reproductive technology are surrounded with considerable moral and ethical controversy, and the final part of this chapter is devoted to a discussion of the main issues.

12.6.2   TREATMENTS FOR SPERM PROBLEMS

Relatively speaking, less is known about the causes and fewer remedies are available for problems of male subfertility. However the study of male reproduction (andrology) is now the focus of an increasing amount of research. Problems of gamete production account for much male subfertility, and various options for the treatment of poor sperm quality are available. Although a few cases of azoospermia (total lack of sperm in the semen) have been successfully treated with administration of hormones, very little could be done until recently about either azoospermia or asthenospermia (poor sperm quality). These couples were usually advised to consider donor insemination. Now, however, various techniques of micromanipulation of gametes have been developed (see Figure 12.5). Although technically difficult and very costly, using these techniques, babies have been born to couples in which the man's sperm is either of very poor quality or absent from the ejaculate. These procedures and their attendant ethical problems will be discussed more fully below. Oligospermia (low sperm count) can often be helped by taking general measures to improve health, such as losing weight, stopping smoking, reducing alcohol intake and avoiding stress. Assisted reproduction remains the best hope for a man with such a diagnosis, and the various techniques, described in the following sections of this chapter, are aimed at overcoming the natural barriers presented to the sperm in its journey to fertilise an oocyte, thus bringing the egg and sperm closer together.

Assisted reproduction technologies

Assisted reproduction technologies is a general term covering those procedures that help a couple to conceive by manipulating their own, or donor, gametes *outside* the body. The two main methods are artificial insemination, and *in vitro* fertilisation (IVF) and its variations.

Artificial insemination

Artificial insemination is the term used for the technique of introducing sperm to a woman's body by a syringe and catheter. AIH (artificial insemination by husband's – or partner's – sperm) involves a sample of carefully prepared sperm cells being deposited at and around the top of the woman's vagina or cervix. Intrauterine insemination (IUI) is a variation on this, and is a procedure in which the prepared sperm cells are put directly into the woman's uterus, via the vagina and cervix. The chance of successful conception is maximised by careful monitoring of the time of ovulation (egg release from the ovary). The men for whom artificial insemination is a suitable treatment for subfertility include those who have a problem with normal sexual intercourse, perhaps caused by impotence, failure to ejaculate properly, or spinal injury. Artificial insemination is less successful in men with oligospermia. IUI can also help couples in which the woman has identified subfertility, such as producing cervical mucus that is hostile to the sperm. These methods are also employed in cases of unexplained infertility, in which they are occasionally successful, but most couples are carefully counselled, and their hopes should not be raised too high.

Artificial insemination techniques can, of course, be used with donor sperm. The term used for this is donor insemination (DI). The method of depositing the sperm in the woman's reproductive tract is exactly the same as for AIH, but the sample used originates from an especially recruited, fertile donor. Donor sperm may also be used in other fertility treatments, such as IVF (*in vitro* fertilisation) or GIFT (gamete intrafallopian transfer). The couples for whom it offers hope of having a baby, which is genetically the woman's, include those where

- the man has oligospermia or azoospermia, or asthenospermia
- the man has had a vasectomy, which is unreversed
- the man has had surgery, radiotherapy or injury to his reproductive tract
- the couple have incompatible blood groups
- the man is a carrier of a serious genetic disease.

Some clinics will also treat women who have no male partner, but wish to have a baby without having had a heterosexual relationship. Decisions on these ethically sensitive cases are always made after very careful consideration of the welfare of the child, as is required by the HFEA's code of practice.

Treatment of the couple by DI is the culmination of much commitment by many people. There are three main steps: recruitment and screening of donors; testing, freezing and preparing the sperm and finally insemination. Relatively few DI centres have their own donor clinic. This is because there must be a large population of potential donors to make it worthwhile. Towns and cities with a large student population often have donor clinics, but recruitment can also be from other sectors of the population, such as police, fire and ambulance services and also from men of proven fertility who have decided to undergo vasectomy. Those centres that do not recruit their own donors buy in frozen samples from larger centres. Potential donors are first interviewed to find their family and sexual history, so that possible chromosomal and serious sexually transmitted diseases may be identified. For example someone who had had a close relative die of Huntington's disease, or who admitted to a promiscuous lifestyle, would probably be dissuaded from sperm donation at an early stage. Travelling expenses only are paid to donors, so they must be very committed, because they may have to donate as often as twice per week for many months. They must be aged between 18 and 55, and be fit and healthy. Each donor's semen sample is then analysed in the laboratory, to assess its volume, density (sperm count), motility, morphology and presence of antibodies, and also to screen for infections such as hepatitis, syphilis and HIV. To be acceptable for storage, three samples, three months apart must have at least

- sperm count greater than 60 million cm^{-3} (40 million cm^{-3} is required for fertilisation)
- ejaculate volume greater than $2\,cm^3$
- 50% progressive motility of sperm (swimming forwards)
- freeze–thaw survival greater than 40%
- 50% normal sperm morphology
- no antisperm antibodies (which would affect the sperm function in the woman's body)
- negative for viral or bacterial infections.

Not surprisingly, up to 95% of potential donors, for one reason or another, fall away from the programme. There is something of a shortage of good quality sperm samples, particularly among some ethnic groups. Careful record is made of the donor's physical characteristics, and when selection of a donor is made prior to insemination, choice will be made on the basis of a match with the social father-to-be, normally the woman's husband or partner. The criteria for matching donors and recipients are usually ethnicity, body build, hair and eye colour, height, blood group and religion. If a donor's sperm meets all the criteria listed above, subsequent samples from him are prepared in the laboratory for freezing in liquid nitrogen at −196 °C, where they can be stored indefinitely in special tubes called 'straws'. This process is known as *cryopreservation*. After six months of donation and storage, the man is screened again for infections such as HIV. If this proves negative, his samples can be released for use in treatment. A specialist fertility nurse, who will have monitored the patient's menstrual cycle until ovulation, most often performs the DI. At this time, the nurse selects a donor straw, and allows it to thaw at room temperature. Having updated the storage records, and checked the thawed sperm cells are alive, the nurse performs the insemination procedure as described above. Success rates for DI average at around 10% for intracervical, and up to 20% for IUI per cycle of treatment. Couples are usually advised that they should think of their treatment as a six- or eight-month course, rather than a series of individual treatments. It has been estimated that 50–60% of women undergoing a series of six treatment cycles become pregnant.

The donor, although the genetic father of any child born as a result of DI treatment, has no legal obligation to the child, and has the right to remain anonymous. This is an important function in the UK of the Human Fertilisation and Embryology Authority, which keeps a record of all sperm (and oocyte donors), but maintains confidentiality of identification. If someone over 16 who knows or suspects that they were conceived as a result of DI (or oocyte donation) inquires of the HFEA about their genetic parentage, the Authority can tell them whether they were born as a result of a donated gamete, and also whether they are related to anyone they may wish to marry. The actual identity of the donor is never disclosed. It is interesting to note, however, that at present the legal limit for pregnancies from any one donor is ten, and as soon as that donor reaches this figure, his sperm cannot be used in further DI treatment, and must be discarded. Obviously the chance of one person born as a result of DI treatment finding out from the HFEA that the person whom they wish to marry was also born as a result of gametes from the same donor is extremely remote. A recent initiative in the USA involves a very different, open approach to the identity of sperm donors. This is emerging in response to the need of children born as a result of DI to know the identity of their biological father. It poses an interesting ethical dilemma, and one that is set to be debated for some time to come. Among some people there remains a real fear about cultural acceptance of a child conceived by DI. There is also the matter of the harmfulness of keeping secrets that will not go away, perhaps leading to the possibility of the child rejecting the social parents when it finds out the truth of its biological parents. Often it is the infertile husband or partner for whom maintaining secrecy is the most important (Jansen, 1996). There seems to be a cultural pattern here. In the West, many countries have legislation similar to the HFE Act (1990) in the UK, requiring anonymity of donors. However, in Indian Hindu society, where extended families are the norm, a brother's sperm donation is the favoured option.

12.6.3   FEMALE INFERTILITY AND OOCYTE (EGG) DONATION

For a woman who is unable to produce oocytes at all, but who has a functional uterus, donated oocytes offer the only hope of becoming pregnant with a child that is genetically that of her partner. The donated oocytes are placed in a specially prepared dish containing culture medium with the partner's sperm, and replaced in the usual way for a routine IVF treatment cycle (see below). For two important reasons, oocyte donation cannot be regarded as directly comparable to sperm donation. Firstly, oocytes are in much shorter supply, as a donor must go through the procedure of ovarian stimulation and oocyte retrieval (described below), and may yield at best only a few oocytes. Understandably, very few women are able to make this commitment, which, compared with the method used by men to produce sperm samples for donation, is inconvenient to say the very least! Secondly, cryopreservation of oocytes is less certain than freezing of sperm because the delicate meiotic spindle apparatus can easily be irreversibly damaged by the freezing procedure and the oocyte is unable to divide after it is thawed. This means that complicated management of recipients and donors is necessary if suitable oocytes are to become available to women requiring this type of fertility treatment. Consequently, donated oocytes are in extremely short supply. They are needed not only for treatment of infertile women, but also for research on such projects as development of improved protocols for oocyte cryopreservation, development of a contraceptive vaccine and development of micromanipulation techniques, as well as for basic developmental biology research.

Potential donors must be between 18 and 35 years old if the oocytes are to be used in the treatment of others. They must also be offered counselling, and given full information about the implications of their donation. In common with sperm donors, an oocyte donor is required to give full details of her medical history, and of any inherited diseases in her family. She will also be tested for such infections as hepatitis B and HIV. Any child born as the result of treatment with her donated oocytes, although genetically her offspring, will not be legally hers, and as with sperm donors, the HFEA maintains a register of oocyte donors, but does not disclose identifying information.

Scientists and doctors have responded to the problem of oocyte shortage by seeking new sources of eggs. Research suggested that ovarian grafts from live donors or from cadavers and ovarian tissue from aborted female foetuses could be used to treat some infertile women. In 1994 the UK's Human Fertilisation and Embryology Authority issued a consultation document on this issue, and wide-ranging public debate followed.[7] The main issues were the following. Should ways of increasing the number of eggs available for treatment be sought? Should tissues and eggs from cadavers and aborted foetuses be used in treatment

---

[7] For example, the three editors of this book, together with others making up the Exeter Ethics Group, participated in the consultation: Human Fertilisation and Embryology Authority: Public Consultation document: Donated ovarian tissue in embryo research and assisted conception. A response (May, 1994).

or research? Who should give consent to use the tissue or eggs, when should consent be given, and in what form? The Polkinghorne Report (1989) gave guidance on the research use of foetuses and foetal tissue. It was stated that tissue from a therapeutically (legally) aborted foetus could be used 'provided any decision concerning its use is separated from the decision to induce abortion'. It is the principle of separation that is important in this context, because it means that the abortion cannot be performed in order to provide tissues for other use, and the mother should not be influenced in her decision to have an abortion for this reason. The report also suggested that a mother who had a miscarriage or a termination of pregnancy because the foetus had died should not be approached to give her consent for the use of the foetal tissues. During the debate on the HFEA's consultation paper (1994), it became clear that for many people in the UK there was an intuitive revulsion (which may have reflected some strongly held ethical views) towards the use of foetal tissue in this way. How the child might feel if it were to learn that its biological mother had been an aborted foetus can only be speculated upon. Many people believed that the law, based on moral principles, should ensure that no one should ever have to face this particular knowledge about their genetic heritage. As a result of consultation it was decided that the use of oocytes from cadavers or aborted foetuses should not be permitted in fertility treatment (HFEA, 1994).

One of the most important issues surrounding assisted reproduction that involves the use of donated gametes – sperm or eggs – is that of consideration for the welfare of the child born as a result of treatment of this kind. The HFEA will only grant a treatment licence to a centre that satisfactorily counsels its patients. Among the questions explored with the patients is whether, when and how to tell the child of the circumstances of his or her conception. Before DI became a relatively common procedure, it was very rare for parents to tell a child that he or she was conceived as a result of DI. However, since the HFE Act came into force in 1990 the legal status of the child is protected, and patients are often advised that it is in the child's best interest to know about this aspect of his or her origin. When single women or lesbian couples request DI treatment, the centre is obliged to consider the matter of the child's need for a father before offering treatment. Many centres refuse to treat such women on these grounds. This, together with the matter of consent, was the central issue in the case of Diane Blood, a British woman who wanted to use her deceased husband's sperm to have a baby by assisted reproduction. The husband, Stephen, had died suddenly of meningitis before he could give his consent for the use in this way of his sperm (or indeed for its collection), and the HFEA, acting strictly within the law, refused to grant Mrs Blood's request. However, the decision was overturned in the High Court, and, following treatment in Belgium, Mrs Blood eventually gave birth to her son in 1999.

In some rare cases gametes of a matching ethnic group are unavailable to a couple seeking treatment. This situation parallels the difficult social and/or moral problems that may occur when children of a different ethnic background are fostered or adopted by a family. Again, before offering treatment, the centre must give careful consideration to the circumstances surrounding any particular case of this kind. The HFEA's Code of Practice requires centres to take into account each couple's preference in relation to the general physical characteristics of the donor that can be matched according to good clinical practice. It would however not be regarded as good clinical practice if a couple were allowed to be treated with gametes of a different ethnic origin than the man or woman.

It has recently become possible, using donated eggs, to treat postmenopausal women to enable them to have children. It has been argued that men can father children at a very advanced age, and so medical science should enable women to become mothers later in life as well. Some people see this as a great advantage for women who want to build their careers before having children. Most women go through the menopause in their early 50s, but the range is 45–55. At around 38–40 a woman's fertility declines rapidly, and the chance of miscarriage rises, whether she has conceived naturally or by assisted reproduction technology. It is widely held that it is not fair on a child to have a much older mother, who may die while the child is still young or indeed who may find it difficult because of age to be an active parent right through the child's pre-adult life. However, the HFEA takes the view that each case should be considered individually, and that it is not necessary or advisable to fix an upper age limit for the treatment of infertility.

The next set of causes of subfertility involves problems of gamete interaction. Any blockage of the male reproductive tract that prevents the maturation and subsequent passage of sperm through the excurrent ducts, epididymis, vas deferens or sperm ducts will compromise fertility. This may be caused by disease, or may have been carried out surgically, as for example in vasectomy; more rarely, congenital abnormalities may result in blockage. These ducts are extremely narrow, but developing techniques of microsurgery can sometimes re-establish patency of these tubes with the result that sperm can again be present in the ejaculate. If this procedure fails, the only recourse is to a relatively new method called microepididymal sperm aspiration (MESA), in which a very few immature sperm are withdrawn, under anaesthetic, directly from the epididymal duct using a very fine needle. They are then used in assisted reproduction procedures. MESA is not generally available, as there are only a few practitioners able to do it. The failure rate is very high. If the entire epididymis is blocked, it is possible to remove sperm still undergoing spermatogenesis from the testis and inject them directly into oocytes. Developmental work on this procedure is still on-going, but the number of live births resulting from it is increasing.

Tubal obstruction in the female tract is a common cause of infertility, and can also be remedied by microsurgery. Providing the blockage is not too complicated, this technique is relatively successful. In an alternative to tubal surgery the Fallopian tubes are by-passed completely using assisted reproduction methods.

12.6.4   *IN VITRO* FERTILISATION (IVF)

In vitro means 'in glass' and it is either in a test tube, or, more usually, in a type of petri dish that the gametes of the man and the woman are actually mixed in the laboratory in such conditions that fertilisation *in vitro* may occur. Following fertilisation and early development, up to three embryos may, under the Human Fertilisation and Embryology Act (1990), be replaced in the woman's uterus. The success rate as measured by the number of live births per treatment cycle is increasing, but must still be regarded as poor: the UK national average ran at 11.1% per cycle of treatment in 1990, improving to 18.2% in 2000 (HFEA, 2000) (see Figure 12.4). Many fertility centres have their own websites on which they publish their success rates. Some clinics in the USA claim nearly 40% live births per treatment cycle. It should be borne in mind that this high rate is probably due in part to careful selection of patients – for example younger women, those who have had previous children and those with simple reasons for subfertility have a far greater chance of success with IVF.

Couples with a range of reasons for subfertility can be helped with IVF. Among these are

- women with blocked Fallopian tubes,
- men with oligospermia that is not so severe that there is no realistic chance of fertilisation,
- couples in whom the sperm and cervical mucus are incompatible,
- women with endometriosis, but whose ovaries are still functional,
- couples who have more than one cause of subfertility,
- couples who have unexplained infertility,
- women who have complete ovarian failure, and who are receiving donated eggs fertilised by their partner's sperm.

IVF is a complex and very demanding treatment for all who are involved in it, and ethical issues arise at each stage. The main steps are the following.

1. *Stimulation of ovulation* during the treatment cycle (sometimes called 'superovulation'). The best chance of getting pregnant from IVF occurs when more than one embryo is replaced in the uterus. A regime of hormone administration is needed to mature more than one Graafian follicle in each cycle. Commonly, the natural functioning of the woman's pituitary gland is 'down-regulated' by using drugs that block the release of the hormones that control the egg maturation function of the ovaries. Further drugs are given to stimulate the production of many eggs. Ultrasound scanning is used to monitor the development of the eggs. When they are mature, further drugs control their release from the ovary. There is some risk that this regime may induce a potentially fatal condition known as ovarian hyperstimulation syndrome. Because of this, and also because it may be the couple's preference, some treatment centres use 'natural cycle IVF', in which no superovulatory drugs are employed. As the

woman almost always produces a single egg, her chance of success is significantly reduced.

2. *Semen analysis and preparation.* On the morning of the woman's egg collection, the man produces a semen sample, having already had a full semen analysis, to ensure that there are no indications that his sperm would not be able to fertilise an egg. The sperm to be used for the IVF are carefully prepared, and approximately 100 000 motile sperm are selected for the *in vitro* insemination. Sometimes, perhaps due to stress and anxiety, the man is unable to produce a useable sample, and in spite of the preparation of his partner, the treatment cycle may have to be abandoned if the sample cannot be produced in time.

3. *Oocyte collection.* There are two main methods for this: ultrasound guidance and laparoscopy. The former is more usual, and can be done with the woman sedated, in which case she comes in for the oocyte collection as an out-patient. Light intensity in the theatre and embryology laboratory is kept to a minimum to protect the oocytes from ultraviolet light, which is thought to be harmful to them. A fine needle is passed through her bladder or vagina and guided by ultrasound to the ovary. Each follicle is visualised, and in turn is punctured, the oocyte and its surrounding fluid being gently sucked (aspirated) into a tube containing culture medium. This tube is immediately handed to the embryologist who examines the contents under a microscope, recording any oocytes collected. The surgeon then proceeds to the next follicle, and the process is repeated until all the follicles on both ovaries are aspirated. After the oocyte collection is complete, the woman is allowed to rest until she has sufficiently recovered to go home. Use of a laparoscope (an instrument allowing the surgeon to look directly into the abdomen) for oocyte collection gives a very clear view of the follicles. A general anaesthetic is needed for this method, so recovery is not so rapid. The collecting needle is inserted through the woman's abdominal wall separately, and the procedure for follicle aspiration is the same as before. The oocytes are placed in separate labelled tubes, then taken in a heated test tube block back to the laboratory.

4. *Insemination.* The oocytes are graded according to their maturity, and each is put into a separate droplet of fresh culture medium, which is held in steady conditions in an incubator. The 100 000 sperm are added to each droplet and returned to the incubator until the following day.

5. *Fertilisation and embryo culture.* The oocytes are inspected 12–18 hours after insemination to see whether they have been fertilised. Re-insemination of apparently unfertilised oocytes sometimes brings about fertilisation, but any abnormal looking oocytes are discarded. Some treatment centres transfer the fertilised oocytes at this stage to the woman, but others return them to the incubator for a further 24 hours, during which time the first embryonic cleavage divisions occur. Following this, the embryos are graded according to quality. The best three are selected to be transferred back to the woman, and any others that are suitable may be frozen for possible use in subsequent cycles. These 'extra' embryos can also be donated to another couple who are unable to create

their own, or, alternatively, given to research. This raises again the issue of the ethical status of the early embryo that was discussed earlier in the chapter. For those who believe that personhood is established by fertilisation and syngamy the concept of spare embryos and their use in destructive procedures is abhorrent. However, for most people the creation of spare embryos is regarded as a necessary part of the *in vitro* procedure.

6. *Embryo transfer.* The woman lies comfortably on the bed, often with her partner nearby. As soon as the woman is ready, the embryos are drawn up into a fine flexible catheter with a small amount of medium. The embryos are gently flushed into the uterus via the cervix. The woman lies quietly for an hour or so, and then goes home. Embryo transfer is quite painless, and she is able to go about her usual life straight away. Some treatment centres give additional hormones at this time in order to maximise the chance of implantation. Sensitive pregnancy tests can be given 14 days after embryo transfer. The chance of multiple pregnancy is quite high, for example 27% of all pregnancies achieved by IVF in 1989 were multiple pregnancies.

As IVF is such a complicated procedure, reasons for failure can occur at any stage:

- the ovaries may fail to produce follicles (15%)
- oocyte collection may be impossible because of the inaccessibility of the follicles (5%)
- the oocytes may not be fertilised (20–25%)
- the embryos may fail to develop normally (20%)
- the embryos may fail to implant, and a menstrual period follows. This is by far the greatest cause of failure of IVF, and the one that is least well explained. There is evidence that this happens relatively frequently in normal fertile women after fertilisation has occurred following sexual intercourse (as was discussed in Section 12.2).

As mentioned immediately above, the most common cause of IVF failure is failure of the embryo to implant in the lining of the uterus. Implantation requires normal function of the endometrium, or lining of the womb. Endometriosis is one of the common causes of this type of subfertility. This condition can be treated either medically, in which case drugs are administered to suppress menstruation, or surgically when the affected areas of tissues are removed. The former treatment protocol usually takes at least 6–9 months, in which conception cannot occur, and there is no guarantee that the endometriosis will not return as soon as normal hormonal function is restored. The outlook with the surgical approach is comparable. Mild cases of endometriosis can be treated such that pregnancy is possible, but the chance of this is far less with increasing severity of the disease. Some women with congenital abnormalities of the uterus can have corrective surgery. Often however, there is no apparent explanation for failure of an embryo to implant, but medical research, probing at the microscopic and even molecular level of structure of the reproductive tract, indicates that abnormalities of the membranes of the endometrial cells may be responsible.

There are a number of variations on the basic IVF procedure described above. One commonly used technique is GIFT (gamete intrafallopian transfer), which involves oocyte collection and immediate replacement into the Fallopian tube together with 2–300 000 sperm prepared in the same way as for IVF. It is less successful than IVF, but is a simpler and cheaper procedure, requiring less complex laboratory facilities. ZIFT (zygote intrafallopian transfer) involves transferring oocytes to the Fallopian tube following a period of incubation with sperm. The transfer is done before any cell division is seen, but it does require laboratory incubation in order to achieve syngamy. However, because the fertility team cannot be sure that the oocyte is in fact a zygote (fertilised oocyte), this method is not often used.

Following pioneering work done in Belgium (Silber et al, 1995), and subsequently in the UK by Simon Fishel's team in Nottingham (Fitzpatrick, 1999), micromanipulation techniques in which the natural barriers to the incorporation of the genetic material of sperm into the oocyte itself are overcome, have become available for treatment of subfertility (see Figure 12.5). The outer coats of the oocyte present considerable obstacles to sperm that have poor motility or morphology. The cumulus cells surrounding the oocyte when it leaves the ovary can easily be dissected away by hand and the protein coat, called the zona pellucida, can be breached to allow more easily the entry of sperm. This can be done by forcing a very fine glass needle through it (partial zona dissection), or directing a jet of acidic medium at it (zona drilling: Figure 12.5a). Two additional procedures go a stage further. They are both extremely difficult technically, but have the potential to overcome virtually any fertility problem involving the gametes. Subzonal insemination (SUZI: Figure 12.5b), a technique pioneered in the UK (Fishel et al, 1990), involves introduction of a single sperm cell between the zona pellucida and the oocyte surface membrane. The sperm still has to bind with the oocyte membrane to gain entry to the interior of the oocyte, but the barriers presented by the outer coats are overcome. As a method of assisted reproduction it is likely to be superseded by intracytoplasmic sperm injection (ICSI: Figure 12.5c), in which mature, immotile or very immature sperm cells alike can be induced to fertilise an oocyte by being injected via a micropipette directly into the cytoplasm of the oocyte. The success rate of these methods is relatively low, although they are improving, and many researchers and clinicians working in the treatment of subfertility believe that these are important developments for the future.

12.6.5   THE COST OF FERTILITY TREATMENT

There are very few clinics in the UK at which patients can have IVF paid for by the National Health Service. This means that unless they live in certain health

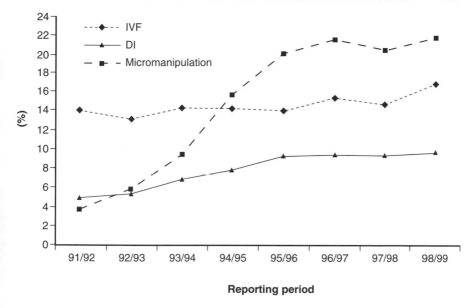

Figure 12.4. Live birth rates per treatment cycle for licensed treatments 1991–1999 (HFEA Annual Report, 2000)

authority areas and can wait for a considerable period for an appointment, if they want or need IVF treatment, they must go to a private clinic. This can be very expensive. There is no certainty of, or even strong likelihood of, a positive outcome from IVF treatment. The success rates of even the best centres rarely exceed an average of 15% live births per cycle of treatment. Most centres charge a fee per cycle, and a course of treatment can easily run into thousands of pounds. It could be argued that if people who are healthy apart from being unable to conceive want this labour intensive, 'hi-tech' treatment, then they should be prepared to pay for it. On the other hand it could be said that everyone who wants a baby but is unfortunate enough to be unable to have one ought to be treated free. In fact, it is a measure of the desperation of the involuntarily childless that they are prepared to incur large debts in order to pay for the treatment. It is also a very sad fact that the rate of marital breakdown of couples who have undergone failed IVF treatment is said to be greater than the national average. Those couples who experience repeated failures of IVF treatment may also experience an additional burden of grief over and above that caused by their childlessness.

In countries such as the UK that have state-funded healthcare the issue of whom should pay for the cost of infertility treatment is based, at least in part, on the type of pathological condition it is considered to be. It might be likened to (a) an ordinary disorder such as appendicitis, which is deserving of NHS treatment, (b) a dysfunction of a particularly tragic and compelling kind, such as spina

(a) Partial zona drilling (PZD)

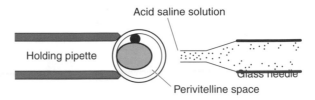

(b) Subzonal insemination (SUZI)

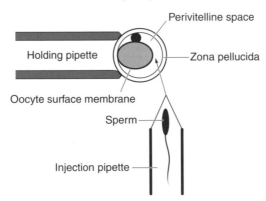

(c) Intracytoplasmic sperm injection (ICSI)

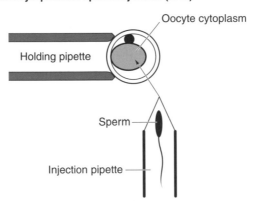

Figure 12.5. Diagram of oocyte showing a) partial zona drilling (PZD), b) Subzonal insemination (SUZI) and c) intracytoplasmic sperm injection (ICSI). Reproduced with permission, from Baggott, 1997

bifida, or (c) a complaint of questionable status, such as baldness, which a person might want treated for relatively trivial reasons. The question of what constitutes human nature, already discussed in this chapter, lends evidence to the argument for infertility falling into the (b) category. Most people seem to have a natural desire to have children, and intrinsic to this desire seems to be having children in the ordinary way, i.e. sexual intercourse, followed by pregnancy and labour of the genetic mother, then birth and nurturing of the children within the family. Infertile couples are unable to have children in this way, and are therefore unable to experience full satisfaction of this natural desire. By definition, no form of infertility treatment can satisfy a couple in all aspects of this desire: it is a damage-limiting approach. This has led to some people forming the opinion that infertility treatment comes into the (c) category listed above. However, most couples who have received infertility treatment would say that it is better, sometimes very much better, than nothing.

12.7 RELIGIOUS VIEWS ABOUT FERTILITY TREATMENT

Some of the existing views about assisted reproduction based on religious beliefs centre around what is deemed to be natural or unnatural and whether that which is considered to be unnatural is therefore immoral. This is by no means the straightforward question it might appear to be. It could be argued that medical treatment for example is unnatural, but few would argue that it is immoral. The philosopher David Hume (1711–1776) said 'I am surely morally permitted to get out of the way of a falling rock which might otherwise kill me, despite the fact that the motion of the rock is the outcome of natural law'. Such arguments have been developed by many ethicists and it is generally accepted today that naturalness versus unnaturalness is very far from being a reliable criterion in reaching ethical decisions as shown, for example, by Reiss and Straughan (1996) and in a specifically Christian context by Bruce et al (2001). However, the statement that some procedure is not natural, and by implication therefore is not acceptable, is still made often enough to be considered here.

What is natural or unnatural with respect to human nature is very difficult to disentangle. It is part of human physical, psychological and emotional behaviour to form pairs, reproduce by sexual intercourse and raise their own children: this is human nature. To try to 'outwit' this nature by intervening in reproductive processes with for example abortion, contraception, assisted conception and masturbation could be said, on the basis of an ethical framework based on natural law, to be at best futile, and at worst dangerous and potentially destructive. However, 'natural' facts about species are actually facts only about most of their members. Common and familiar exceptions to this view of normal human nature exist: humans are sighted, but some people are born blind, humans are mammals, but not all women are able to breast feed, humans have a drive to reproduce, but some people are voluntarily childless, and others kill their own children.

Furthermore, in the case of assisted reproduction the argument that it is unnatural bites both ways. If most humans have a natural desire to form reproductive pairs and raise their own young, then the attempt to thwart this desire is as potentially calamitous or futile as the attempt to do anything else unnatural. Hence it can be argued that fertility treatment is acting in accordance with nature, rather than against it.

However, one of the most influential organisations that oppose the technologies of assisted reproduction, namely the Roman Catholic Church, does so in effect on the basis of natural (i.e. God's) law. It is embedded in Roman Catholic doctrine that the natural purpose of sperm is to fertilise, within the female body. It should be remembered that this doctrine was developed before any understanding of the existence of male and female gametes. The man was said to provide the 'seed' and the woman was simply the 'garden' in which the seed was incubated. Spilling the seed, whether by masturbation or by *coitus interruptus*, was and is, despite a proper understanding of the role of both male and female in reproduction, still regarded as sinful, as is the frustration of the sperm's natural function by contraception.[8] This has clear implications for the production of sperm by masturbation and its subsequent manipulation, whether for donor insemination or for *in vitro* fertilisation. It needs also to be mentioned here that in some African cultures masturbation is believed to compromise potency, and thus men refrain from it. This has implications for donor insemination programmes, because men of this ethnic origin rarely volunteer to be sperm donors.

Finally it must be stressed that, in contrast to the Roman Catholic Church, other Christian denominations do not object to either contraception or to assisted reproductive technology. Neither does the Jewish faith oppose these interventions; indeed in the UK at the time of writing, one of the leading medical practitioners in the field of human fertility, Lord Robert Winston, is a practising Jew.

12.8   SPARE EMBRYOS AND EMBRYO RESEARCH

The processes of superovulation and IVF often produce more embryos than are required for a treatment cycle, or even a series of treatment cycles. As has been mentioned earlier, for those who believe that these early embryos are people, this is a very contentious issue. The couple undergoing treatment sometimes gives their consent to these 'spare' embryos being used in research. In the UK the Human Fertilisation and Embryology Act (1990) states that any research pro-

---

[8] The main basis for this is to be found in the story of Onan, Genesis chapter 38, verses 1–10 in the Old Testament of the Bible. Ironically (remembering that the Old Testament is actually a Jewish book), current Jewish doctrine does not forbid either the manipulation of gametes *in vitro* or contraception.

ject involving the use of live human embryos must be licensed by the HFEA. The licences are only issued after thorough and careful scrutiny of the research proposal. The type of research that can be undertaken is strictly limited and very closely monitored by the HFEA as it proceeds. Licences are only given to projects if they are considered to

* promote advances in the treatment of infertility,
* increase knowledge about the causes of congenital disease and to develop methods for detecting genetically abnormal embryos before implantation,
* increase knowledge about the cause of miscarriage,
* develop more effective methods of contraception.

Experimental procedures on human embryos are only permitted under the law for the first 14 days after the mixing of the gametes – in other words before the appearance of the primitive streak. This has led to the coining (not by clinicians, but by biologists) of the term 'pre-embryo', perhaps in an effort to defuse any possible ethical objections of the type hinted at above ('pre-embryo' may sound rather less human than 'embryo'). Under the UK's Human Fertilisation and Embryology Act (1990), certain types of research on human embryos are prohibited. These include

* replacing a human embryo in an animal,
* nucleus substitution – this procedure comprises removal of the nucleus of an embryonic cell (blastomere) and replacing it with the nucleus taken from the cell of another person or embryo,
* altering the genetic structure of any cell while it forms part of an embryo,
* cloning of human embryos for the purposes of infertility treatment.

Other than in these circumstances, it is only after this period of 14 days from fertilisation that legally protected human life begins. The embryo that has been experimented upon must not be maintained either *in vitro*, frozen, or replaced into a woman after that time; it must be destroyed.

The mixing of human gametes with those of another species is also prohibited unless a licence is granted. A diagnostic test known as the hamster egg penetration (HEP) assay is sometimes used to see how well sperm can penetrate an egg. The hamster egg is unique in that when its outer coat is removed chemically it will permit binding and penetration of a wide variety of sperm from other mammalian and even non-mammalian species. In the HEP assay, about 40 hamster eggs are harvested from a freshly killed, superovulated hamster. The eggs are incubated with the test sperm, and after 3–4 hours the eggs are fixed and stained, and the percentage of eggs with sperm heads inside them is calculated. Although this procedure gives some measure of the fertilising ability of the sperm, it is by no means routinely performed in centres offering assisted conception treatment. However, it is a useful research tool under certain circumstances, but requires a licence under the Act. For some people, the use of animals in this way is in itself a serious ethical issue (see Chapter 19).

12.9   CURRENT AND FUTURE DEVELOPMENTS

12.9.1   ETHICAL CONSIDERATIONS

Such is the sensitivity of feeling about human embryo research that the development of any new technique for the treatment of infertility must be strictly regulated. Research licences are only granted when it can be demonstrated that the researchers will add to the body of knowledge, and in doing so gain technical competence. Only then can they be allowed to use a new technique in the treatment of patients. It is with this careful and painstaking approach coupled with on-going public debate that big issues, such as what has been termed by the media 'designer babies', can be approached responsibly. There is no doubt that research will continue to strive to improve on nature: indeed, the elimination of undesirable genes may be one of the consequences of the Human Genome Project, but this should always be tempered with the question 'what is undesirable?' For further discussion of these major issues, see Chapters 13 and 14.

Although there is little doubt that elimination of certain genes, such as those for Duchenne muscular dystrophy, cystic fibrosis or haemophilia, from the human population is desirable, anxieties surrounding these ideas do raise the spectre of genetic and ethnic 'cleansing'. Lessons learned from past and recent history must inform each step on the middle ground in the implementation of advances in reproductive technology.

12.9.2   SEX SELECTION

Techniques are being developed that may be able to separate sperm bearing an X chromosome from those bearing a Y chromosome. This means that, using assisted conception procedures such as IVF, it may become possible much more reliably to choose the sex of the baby. This has become known as sex selection. As yet this has not reproducibly been achieved with human sperm, in spite of the claims of some practitioners. Success has, however, been claimed with bull sperm, which have differently shaped heads compared with human sperm. This enables the bull sperm, when placed in an electromagnetic field, to be separated fairly reliably according to electrical charge, which appears to correspond to the chromosomal content of the nucleus.

There are two main ethical problems surrounding sex selection. The first concerns the medical reasons for doing it, and the second concerns the social reasons. In cases where a woman risks having a child with a life-threatening, sex-linked disease, choosing the sex of the child may ensure that it is born healthy. Most people, given a choice, would rather have a healthy child than one with a severe disability, and most people, given the same choice for themselves, would rather be healthy than disabled. However, this is not to diminish the value or contribution to society made by these disadvantaged people. On the other hand, in a family for example with one or two children of one sex, the couple may want to have another child of the opposite sex. Alternatively, some cultures may attach higher status to one sex, and wish to ensure that this is the gender of the child they have. Such social reasons for influencing the proportion of girls and boys born are not considered by the HFEA to be desirable. The arguments put forward in favour of sex selection were

- that there is no evidence that it may have an adverse effect on society,
- that couples wanting a child of a different sex may carry on having more children than they really want, or can support,
- that sex selection is an important medical advance, which it would be wrong to deny people who wanted it.

In formulating the policy on this issue, these arguments in favour of sex selection were felt to be outweighed by those against it.

- Sex selection for social reasons is rarely in favour of girls. For example, inheritance of titles or wealth are often only legally possible in the UK through the male line, and in some ethnic populations there may be a considerable financial implication of having girls when it is time for them to marry.
- Some people think that it is important for the first born child in a family to be a boy. Position in the family is known to have an effect on a child's psychological development, and the effect of having a majority of first-borns as boys may reinforce adverse sexual stereotypes, perceptions of gender status and patterns of sex discrimination already present in society.

It will be readily seen that these are culturally based arguments against sex selection for social reasons. In the absence of this cultural backdrop, for example if society were matriarchal instead of patriarchal, or if social order were neither matriarchal nor patriarchal, but equal, these arguments would not exist, and there would be no need to deny sex selection. The arguments against sex selection for social reasons are therefore based on the admission that gender has value, and it was seen that this is not an appropriate attitude to encourage, and the practice on these grounds is not permitted under the HFE Act (1990). Now, sex selection is permitted for sound medical reasons, but only where post-fertilisation techniques are used.

In a specific case, the Masterton family, from Scotland, had four boys and then a girl, after the birth of whom Mrs Masterton was sterilised. The daughter was killed in a tragic accident, and the Mastertons applied to one of five clinics licensed by the HFEA to undergo IVF treatment with sex selection to try to have another daughter. They were advised by the HFEA that permission could only be given on medical grounds, usually only given for families with a history of inherited, gender-specific diseases. The Mastertons' case, it decided, was made on social grounds and so turned down their application for treatment in the UK. The couple contested this decision in court under the European Convention on Human Rights, and finally went to Italy for the treatment. However, the only successfully fertilised egg resulted in a male embryo, and the Mastertons decided not to proceed with embryo transfer. Although they rejected the embryo, they gave permission for it to be stored for donation to a childless couple. They told their sons that they might have a brother born in another family. The moral objections to the use of embryo sex selection in this case were based on the precedent it set for 'design' of babies. The pro-life charity 'Life' held that this was an example of 'simple eugenics', and expressed its opposition to manipulation of gender numbers in society. A spokesman for the Roman Catholic Church also condemned it not only because of the sex selection, but also on the basis that male embryos would be destroyed in the process (in addition to that church's objection to IVF in general), although the Roman Catholic authorities and 'Life' sympathised with the tragic case. However, a Scottish Episcopal bishop thought that this was an exceptional case, and that the HFEA should consider relaxing its policy in such circumstances[9].

12.9.3   GENETIC SELECTION

In the recent case in the USA, baby Adam Nash was especially conceived *in vitro* after precise genetic selection to be a bone marrow donor for his sister, who suffered from the rare and fatal blood disorder Fanconi's anaemia, inherited from the Nash parents (see also Chapter 15, where Peter Turnpenny and John Bryant discuss this case in the context of genetic enhancement and 'designer babies'). The case involved a high level of clinical and scientific expertise, and not a little good fortune, as, of 12 embryos produced from the IVF procedure, only one was found to match the criteria (i.e. absence of Fanconi's anaemia coupled with an immunological match to the unwell sibling). After the embryo transfer, in spite of the high probability of failure with the implantation of only one embryo, baby Adam was delivered in good health. Stem cells were removed from his umbilical cord, and transplanted into his sister in the hope that they will reconstruct healthy bone marrow in her. It can be argued that this practice has some ethical currency in that the sister's life might be saved at no 'cost' to the brother. However, that the baby was 'designed' for the purpose, and did not give his informed consent to donating his stem cells, may to some be unacceptable.

12.10   SURROGACY

Some women who have functional ovaries but are unable to become pregnant have undergone oocyte retrieval and IVF with their husband's sperm, but make an arrangement with another woman to have the embryo(s) transferred to her uterus, with the hope of pregnancy and a live birth. This is surrogacy, and a special provision has to be made in court to allow the genetic parents to become the legal parents, because in law the woman who bears and gives birth to a child

[9] *Daily Telegraph*, London, UK, 5 March 2001.

is the legal mother, regardless of the child's genetic origin. In the UK the court will only grant the parental order if it is satisfied that no money has been involved in the surrogacy. In other words, the surrogate mother cannot be paid for carrying the other couple's child. This issue again brings into question the concept of 'having' children. It is the perception of some, probably most, people that pregnancy, labour and birth are essential features of this process, but clearly, as in the case of surrogacy, it is possible using assisted conception techniques that a couple can achieve the birth of their genetic offspring without the genetic mother going through pregnancy and birth. It is interesting to speculate from the ethical point of view exactly who has had the baby after the full procedure: drawing up the legal contract between the commissioning couple and the surrogate mother, the gamete collection from the commissioning couple, the IVF by the embryologists and andrologists, the embryo transfer by the medical team and the surrogate mother's pregnancy and birth.

12.11   CONCLUSION

The increase in pace of the revolution in reproductive technology and medicine has outstripped the moral debate, with the result that almost weekly there is a new sensation in the media. However, this is not to say that people are slow to come to terms with the importance or even the complexity of new techniques or possibilities for treatment. The ethics surrounding the issues of human reproduction are at the same time fundamental and extremely complex: this makes it difficult, if not impossible, to be completely right or wrong about them. The controversies give rise to real dilemmas because of technical difficulties, uncertain outcomes and variance of principles.

Critical reading and frequent discussion help to inform a person's views, and this is very important because although laws exist on such matters as fertility treatment this legislation should always at least take account of the views of society as a whole, although it can also be argued that public opinion can be a very dangerous arbiter of the law. This is not to say that the relationship between social morality and the legislature is static: indeed, one informs the other. Engaging in debate and voicing opinion is particularly important when new guidelines are being drawn up. The HFEA regularly puts out public consultation documents on such issues as sex selection and the use of donated ovarian tissue in embryo research and assisted conception, and the questions raised by these matters are discussed by treatment centres, church groups, political gatherings, academic departments and so on. Above all it is most important for everyone to keep trying to come to terms with these difficult moral problems, by thinking carefully about their significance not just for oneself, but also with empathy for those more directly affected.

12.12   SUMMARY OF THE KEY ETHICAL ISSUES

1. Artificial reproductive technologies separate sexual intercourse from procreation.
2. Donor insemination breaks the link between the genetic parent and the nurturing parent.
3. In any treatment involving the donation of gametes, there are issues about the confidentiality of the genetic parent(s).
4. The issue of the moral status of the human embryo and the question of when life begins are at the heart of any procedure involving the creation of embryos outside the body, and their subsequent destination.
5. In the last few decades, the attitude of many people to what they are owed by the world has changed. Some now feel that they have the right to have what they want, including children. This may be in spite of the fact that they suffer the pathology of infertility.
6. The adequacy of funding for medical treatment and the fairness of its distribution raises the issue of priorities.
7. There may be circumstances in which the harm caused by artificial reproductive technologies outweighs the benefits.[10]

REFERENCES

Baggott, L.M. (1997) *Human Reproduction.* Cambridge University Press, Cambridge, UK.

Bruce, D., Horrocks, D., Bryant, J.A., Burnside, J., Carling, R., Carruthers, P. Hall, A.M. and May, A. (2001) *Modifying Creation? GM Crops and Foods: A Christian Perspective.* Paternoster, Carlisle, UK.

Dunstan, G.R. and Sellar, M.J. (eds) (1988) *The Status of the Human Embryo: Perspectives from Moral Tradition.* Exeter University Press, Exeter, UK.

Edwards, R.G., Steptoe, P.C. and Purdy, J.M. (1980) Establishing full term human pregnancies using cleaving embryos grown *in vitro. British Journal of Obstetrics and Gynaecology,* **87**, 737–756.

Fishel, S., Antinori, S., Jackson, P., Johnson, J., Lisi, F., Chiariello, F. and Versaci, C. (1990). Twin birth after subzonal insemination. *Lancet,* **335**, 722–723.

Fitzpatrick, M. (1999) New technique treats male infertility. *Journal of the American Medical Association,* **282**(15), http://jama.ama-assn.org/issues/v282n15/ffull/jmn1020-4.html

Human Fertilisation and Embryology Authority (HFEA) (1994) *Report on Donated Ovarian Tissue in Embryo Research and Assisted Conception.*

Human Fertilisation and Embryology Authority (HFEA) (2000) *Ninth Annual Report and Accounts of the HFEA.*

Hull, M.G.R., Glazener, C.M.A., Kelly, N.J., Conway, D.I., Foster, P.A., Hinton, R.A., Coulson, C., Lambert, P.A., Watt, E.M. and Desai, K.M. (1985) Population study of causes, treatment and outcome of infertility. *British Medical Journal,* **291**, 1693–1697.

---

[10] For example, there has been a case in which a woman sued a clinic following the birth after IVF of triplets rather than the twins she specified.

Jansen, R. (1996) *Overcoming Infertility: a Compassionate Resource for Getting Pregnant.* Freeman, New York, USA.

Morgan, D. and Lee, R.G. (1991) *Blackstone's Guide to the Human Fertilisation and Embryology Act 1990.* Blackstone, London, UK.

Reiss, M.J. and Straughan, R. (1996) *Improving Nature? The Science and Ethics of Genetic Engineering.* Cambridge University Press, Cambridge, UK.

Ridley, M. (1995) *The Red Queen: Sex and the Evolution of Human Nature.* Penguin, London, UK.

Silber, S.J., Nagy, Z., Lui, J., Tournaye, H., Lissens, W., Ferec, C., Liebaers, I., Devroey, P. and van Steirteghem, A.C. (1995) The use of epididymal and testicular spermatozoa for intracytoplasmic sperm injection: the genetic implications for male infertility. *Human Reproduction*, **10**, 2031–2043.

Snowden, R. and Snowden, E. (1993) *The Gift of a Child.* Exeter University Press, Exeter, UK.

Thompson, W, Joyce, D.N. and Newton, J.R (eds) (1985) *In Vitro Fertilisation and Donor Insemination*, Proceedings of the 12th Study Group of the Royal College of Obstetricians and Gynaecologists.

Warnock, M. (1985) *A Question of Life: The Warnock Report of the Committee of Enquiry into Human Fertilisation and Embryology.* Blackwell, Oxford, UK.

Warnock, M. (1998) *An Intelligent Person's Guide to Ethics.* Duckworth. London, UK.

13 Genetic Information: Use and Abuse

Bartha Maria Knoppers

13.1 INTRODUCTION

Privacy as a human right and confidentiality within the physician–patient relationship are well accepted principles. Recent developments in, human genetic research have, however, brought some unique issues to the fore. The first is that of maintaining the confidentiality of medical records (to say nothing of the research records themselves) while accepting access by researchers; the second is that of access by third parties such as insurers and employers; the third is that of familial disclosure, and finally, the fourth is the emerging trend towards the anonymising of DNA samples and information so as to avoid any possible abuse. Taking these issues in turn, we shall argue that the first should be clarified through legislation, the second through a moratorium for insurers and an amendment to discrimination laws for employment, the third through ethical codes of conduct and that the last approach, while ethically and legally expedient, needs to be rethought. Indeed, automatic anonymisation to ensure confidentiality is both ethically and scientifically short-sighted. We would contend that only the 'normalisation' and integration of genetic information as protected medical information can avoid abuses worse than the ones currently imagined.

13.2 CONFIDENTIALITY OF RECORDS

Most legislative or ethical codes protecting medical records foresee specific exceptions for research purposes, although genetic research is not mentioned. At the international level, only the *Universal Declaration on the Human Genome and Human Rights* (UNESCO, 1997) and the World Health Organisation's *Proposed International Guidelines on Ethical Issues in Medical Genetics and Genetic Services* of that same year specifically address the issue of access to genetic information in medical records for research purposes. UNESCO's position is that

Bioethics for Scientists. Edited by John Bryant, Linda Baggott la Velle and John Searle.
© 2002 by John Wiley & Sons Ltd.

'. . . genetic data associated with an identifiable person and stored or processed for the purposes of research or any other purpose must be held confidential in the conditions foreseen by law' (art. 7).

One international directive that specifically addresses the issue is the Human Genome Organisation's (HUGO) Ethics Committee *Statement on the Principled Conduct of Genetic Research* (1996), which states 'That recognition of privacy and protection against unauthorized access be ensured by the confidentiality of genetic information. Coding such information, procedures for controlled access, and policies for the transfer and conservation of samples and information should be developed and put into place before sampling'. This was followed by the 1998 *Statement on DNA Sampling: Control and Access*, which went even further in allowing researchers access both to genetic information and to 'left-over' tissues removed during medical care provided that patients were notified and did not object.

At the regional level, the 1997 *Convention on Biomedicine* of the Council of Europe limits predictive or carrier testing to 'scientific research linked to health purposes' (art. 12) and mandates confidentiality generally:

1. Everyone has the right to respect for private life in relation to information about his or her health;
2. Everyone is entitled to know any information collected about his or her health. However, the wishes of individuals not to be so informed shall be observed;
3. In exceptional cases, restrictions may be placed by law on the exercise of the rights contained in paragraph 2 in the interests of the patient (art. 10).

Different countries however have taken explicit positions on access to genetic information for research. The Advisory Committee on Genetic Testing of the United Kingdom in its *Advice to Research Ethics Committees* (1998) permitted access but distinguished between anonymised and coded information, the latter requiring an explicit consent from the patient before access for research would be provided (see Section 13.5). This position typifies that of most countries.

One interesting further addition to this common national approach is that of the USA's National Action Plan on Breast Cancer (1999). This group maintained that the 'privacy protections for experimental research data in which health care is not delivered should exceed the protections established for medical records. Rules for third-party access to medical records should not be uniformly applied to experimental research data' (rec.1). 'Neither should. . . researchers. . . place individually identifiable experimental research data not utilized for health care in the medical record' (rec. 2). This position is notable in that it addresses research records *per se* and mandates increased protection. While it is common practice for researchers to keep such data separate, few countries have specific statutory rules on genetic research records and on the need for different approaches.

Even if the information in the medical record is medically relevant, the patient must explicitly consent to such inclusion. Indeed, according to the United Kingdom Advisory Committee, '. . . concern over disclosure is particularly acute where the research involves late onset disorders, but remains a significant issue in all genetic testing' (art. 2.1). The Québec Network of Applied Genetic Medicine in Canada also introduced a new element in its May 2000 *Statement of Principles: Human Genomic Research*, by extending the obligation of professional confidentiality to all members of the research team (prin. III, 1).

Finally, no discussion of the confidentiality of and access to genetic research results would be complete without mention of increasing surveillance of how this information is shared among researchers. In particular, international collaboration has come under scrutiny. For example, the Swedish Medical Research Council in its *Research Ethics Guidelines for Using Biobanks* (1999) maintains that only coded material can be sent abroad. Publication of the results of genetic research should also avoid any possible identification of individuals.

It is evident from the above that two approaches are possible here. The first is to tighten the legislation and deontological rules governing the protection of medical information and explicitly include all research records within such increased protection; the second would be to adopt legislation specific to research records. This issue is all the more important when we consider possible access by employers and insurers.

13.3   INSURERS AND EMPLOYERS

Again, it bears mentioning that the absence of an explicit text on access by insurers and employers to genetic data does not mean that such access is either possible or denied. Indeed, the prohibition against 'discrimination based on genetic characteristics' or the protection of personal data generally, or of medical information specifically, can determine access by such third parties.

At the international level, the 1997 World Health Organisation's *Proposed International Guidelines* go further in not only specifically prohibiting access to genetic information (Table 4 in the document) but in the case of banked samples state '. . . insurance companies, employers, schools, government agencies and other institutional third parties that may be able to coerce consent should not be allowed access, even with the individual's consent' (Table 10 in the document). This recognises the vulnerability of applicants for insurance and employment, who will almost always automatically 'consent'. The HUGO Ethics Committee attacks the problem another way by stating in its 1998 *Statement on DNA Sampling* 'Unless authorized by law, there should be no disclosure to institutional third parties. . .'.

At the regional level, as early as 1992 the Council of Europe's *Recommendation on Genetic Testing and Screening for Health Care Purposes* maintained that 'Insurers should not have the right to require genetic testing or to enquire about the results of previously performed tests, as a pre-condition for the conclusion or modification of an insurance contract' (prin. 7). It should be noted however that such limitations do not prevent employers from requesting medical information when relevant to particular aptitudes required for a job or from screening employees for the purposes of job safety or environment. Insurers too have always used family history or health questionnaires to determine premiums.

At the level of individual countries, four positions have been taken with regard to life and disability insurance in countries that already have universal health insurance coverage: (i) legislative prohibition (e.g. Belgium, 1992), (ii) voluntary moratorium by the industry (e.g. France, 1999), (iii) a proportional approach (e.g. The Netherlands, 1999) and a hybrid between (ii) and (iii) (as in the UK, 2000–01). The proportional approach means that access to genetic information or tests will only be required when the amount of insurance sought exceeds income or a certain specified amount. No such clear position has emerged concerning employment. The situation is quite different in the United States, where lack of universal health insurance and the fact that most employers are also insurers means that at present (early 2001) state laws address only the issue of health insurance by prohibiting access to genetic information or tests for the purpose of obtaining such insurance.

Obviously, neither life insurers nor employers have the state's role to provide social security. Nevertheless, much can be said for advocating a moratorium on requesting access to genetic test results to ensure that patients continue to see their physicians and participate in research. Such a moratorium would also provide the industry with the necessary time to ensure that their actuarial tables are accurate, considering the complexity of genetic information. Another possible reform in the area of employment is to extend the definition of prohibited discrimination to include perception of handicap, since most cases of employment discrimination occur when the 'genetically at-risk' are perceived and treated as already ill.

13.4 FAMILY MEMBERS

Genetic information is not only personal or social, as we have just seen in discussing its economic impact on access to insurance and employment, but it is also necessarily familial. This raises unique issues since medical secrecy has always been considered a legal and professional obligation between a physician and an individual.

It is interesting to note the relative degree of uniformity at the international level for access by family members 'when serious burden can be avoided' (Table 7 in the World Health Organisation's *Proposed International Guidelines*, 1997). The HUGO Ethics Committee in its 1998 *Statement on DNA Sampling* is more explicit as to the precise conditions when such exception could be made: 'Special considerations should be made for access by immediate relatives. Where there is a high risk of having or transmitting a serious disorder and prevention or treatment is available, immediate relatives should have access to stored DNA for the purpose of learning their own status. These exceptional circumstances should be made generally known at both the institutional level and in the research relationship'.

Surprisingly, regional statements are relatively absent on this issue though, in 1997, the Group of Advisors on Ethics of the European Commission in an *Opinion on the Ethical Aspects of Prenatal Diagnosis* suggested that '. . . the woman or couple should be strongly recommended to allow the release. . .' of genetic data relevant to other family members (art. 2.8). In 1998, the American Society of Human Genetics Social Issues SubCommittee issued a *Statement on the Professional Disclosure of Familial Genetic Information.* The Society was careful to not advocate a legal duty but rather considered such disclosure to be 'ethically permissible'. It did so after an extensive review of international and national positions. It also maintained that 'At a minimum, health care professionals should be *obliged* to inform patients about the implications of their genetic test results and about the potential risks to their family members'. Moreover, after the death of a person, the province of Québec has gone so far as to legislate a right of access by blood relatives in the case of medical need for familial or genetic conditions.

We recommend that the exceptional circumstances described above should be adopted in ethical codes of conduct so that the considerations underlying this exception to medical secrecy are clear for health professionals.

13.5   CODING AND ANONYMISATION

The advent of genetic testing and research has, as we have seen, created new issues and challenges. Concerns about the possible uses and abuses led to the proposition in the early 1990s to protect genetic information through coding. The advent of DNA banking however, together with the international collaboration common to genetic research, led to the elaboration of different, more stringent standards. Indeed, ethics review committees and researchers themselves began to anonymise samples. Anonymisation involves not only irrevocably removing all identifiers but also ensuring that the samples are not identifiable in any way. Some clinical and demographic data however can still accompany the sample. However, the ethical and legal safety offered by anonymisation carries scientific risks. Once anonymised, the clinical data accompanying the sample cannot be updated since the source, i.e. the person, cannot be found. This makes their scientific usefulness short-lived unless used as controls.

The HUGO Ethics Committee was the first to warn against this trend in its *Statement on DNA Sampling* (1998). It is also evident that research with anonymised samples means that relevant findings can never be given back to the participant. These limitations have led Estonia (in contrast to Iceland) to advocate using coded samples in their national DNA bank so as to be able one day to give medical information back to its citizens.

This issue remains to be addressed by most countries. Further distinctions also need to be made concerning the use of archived samples remaining after medical care. Here too, increasing restrictions are denying access in the absence of an explicit consent to research. While, obviously, patients requiring care are not signing up for research and while the old practice of using such abandoned material without consent no longer corresponds to modern values and concerns, perhaps a system of notification upon admission to hospital together with the possibility of objection ('opting-out'), and the anonymisation of these leftover samples would be a better approach. We recommend therefore that serious consideration be given before automatically requiring that research samples be anonymised before secondary use and that the use for research of leftover samples from medical cases be subject to notification.

13.6   CONCLUSION

There is no doubt that, like information about mental illness 50 years ago, genetic information is still considered stigmatising. The rapid increase of knowledge of genetic risk factors in all common disease may serve to temper this fact. Such cultural 'normalisation' may also serve to encourage family members to share information with other at-risk relatives. The confidence to share information, to be tested or to participate in research or to reveal one's status to employers or insurers, however, will be largely dependent on the protection afforded such medical information or research records. Furthermore, the current trend towards genetic exceptionalism rather than integration is skewing the control of and access to DNA samples, whether obtained specifically for research or left over after medical care. While we should proceed with caution, it is not certain that anonymisation in DNA banking necessarily corresponds to the will of participants. If asked, such participants may well agree to coding both for their own health needs and for those of future generations. Imagined abuses should not overshadow the reality of present and future therapeutic uses.

Finally, we may define a number of general over-arching principles that should be considered in the use of genetic information. These are

- confidentiality and the right to privacy (but see below: responsibility)
- no discrimination, for example by employers, on the basis of genetic information
- reasonableness; for example, insurers have a legitimate interest in medical history in order to assess risk – the question is are they being reasonable or so unreasonable that they are being entirely motivated by profit rather than by the legitimate interests of the client?
- responsibility – for example to disclose to family members a genetic diagnosis

of which ignorance would cause harm or prevention from gaining access to benefit or therapy.

Reference to these principles will contribute significantly to the prevention of the abuse and the promotion of the appropriate use of genetic information.

REFERENCES

Advisory Committee on Genetic Testing (ACGT) (1998) *Advice to Research Ethics Committees*, UK Department of Health, London, UK.
http://www.doh.gov.uk/pub/docs/doh/recrev3.pdf

American Society of Human Genetics (ASHG) (1998) Professional disclosure of familial genetic information. *American Journal of Human Genetics*, **62**, 474–483.

Belgian Parliament (June 1992) Loi sur le contrat d'assurance terrestre. *Moniteur Belge,* 20 August, 18 283.

Council of Europe (1992) Recommendation No. R (92)3 of the Committee of Ministers to Member States on Genetic Testing and Screening for Health Care Purposes. *International Digest of Health Legislation*, **43**, 284–289.

Council of Europe (1997) Convention for the Protection of Human Rights and Dignity of the Human Being With Regard to the Application of Biology and Medicine: Convention on Human Rights and Biomedicine. *International Digest of Health Legislation*, **48**, 99–105.

Fédération Française des Sociétés d'Assurance (FFSA) (1999) *Les Assureurs Français Renouvellent leur Engagement de ne pas Utiliser les Tests Génétiques*, press release.

Group of Advisers to the European Commission on the Ethical Implications of Biotechnology (GAEIB) (1997) *Ethical Aspects of the 5th Research Framework Programme*.
http://europa.eu.int/comm/sg/biotech/en/ biotec13.htm

Human Genome Organisation (HUGO) (1996) Statement on the principled conduct of genetic research. *Genome Digest*, **3**, 2–3.

Human Genome Organisation (HUGO) (1998) Statement on DNA sampling: control and access. *Genome Digest*, **6**, 8–9.

National Action Plan on Breast Cancer (1999) Recommendation to protect privacy in genetics research. *Science*, **285**, 1359–1361.

Netherlands Parliament (1999) Medical checks law. *International Digest of Health Legislation*, **50**, 68–70.

Network of Applied Genetic Medicine (2000) Statement of principles: human genomic research. In *La Recherche en Génétique Humaine: Cadre Éthique*. Network of Applied Genetic Medicine, pp. 2–17.
http://www.rmga.qc.ca/doc/principes_en_2000.html

Swedish Medical Research Council (1999) *Research Ethics Guidelines for Using Biobanks, Especially Projects Involving Genome Research*.
http://194.52.62.221/SinglePage/SinglePage.asp?ItemID=670

UNESCO (1997) *Universal Declaration on the Human Genome and Human Rights*.
(http://www.unesco.org/ibc/uk/genome/project/index.html

United Kingdom Parliament (2000–01) *Genetics and Insurance* (House of Commons Select Committee on Science and Technology). HMSO, London, UK.

World Health Organisation (WHO) (1997) *Proposed International Guidelines on Ethical Issues in Medical Genetics and Genetic Services*. WHO/HGN/GL/ETH/98.1.
http://www.who.int/ncd/hgn/hgnethic.htm

14 Human Genetics and Genetic Enhancement

Peter Turnpenny and John Bryant

14.1 INTRODUCTION

In this chapter we discuss the possibilities for, and the ethical and social implications of, genetic modification and enhancement of humans. The discussion needs to be set in the contexts of firstly the development of techniques for *in vitro* handling of gametes and embryos, and for genetic manipulation of non-human mammals, and secondly, increases in knowledge of human genetics, including the information provided by the human genome project (HGP). It is also very relevant to consider the impact of past and current developments in human genetics on medical practice and patients' needs. The early sections of this chapter introduce this technical background.

14.2 GENETIC MODIFICATION

The advent of recombinant DNA techniques and of new reproductive technologies, especially those associated with *in vitro* fertilisation (Chapter 12), provided a powerful combination of tools that raised the possibility of genetic modification of mammals. The ethical issues raised by the genetic modification of mammals (and other animals) are discussed extensively in Chapter 6. In this section we confine ourselves to the technical aspects (but return to ethical issues when discussing humans: Sections 14.4.1 and 14.7).

The theoretical basis is straightforward. Foreign DNA is introduced into oocytes (unfertilised ova) prior to *in vitro* fertilisation or into newly fertilised ova. The genetically modified embryo is then introduced into the uterus in the 'normal' fashion and, provided a pregnancy is established successfully, a *transgenic* (genetically modified) mammal will eventually be born. The progress from this theoretical framework to practical reality requires answers to several technical questions. Firstly, can DNA vectors be developed that will deliver 'foreign' genes to the target cell without the possibility of the vector itself having deleteri-

Bioethics for Scientists. Edited by John Bryant, Linda Baggott la Velle and John Searle.
© 2002 by John Wiley & Sons Ltd.

ous effects? Secondly, will the incorporation of exogenous DNA into the host cell genome affect either the fertilised ovum's ability to establish a pregnancy or the development of the foetus? Thirdly, will the foreign gene be appropriately expressed in the mammal that is born? Fourthly, will the foreign gene be transmitted to the progeny?

Using mice as the initial experimental animals, these questions were at least partially answered in the early 1980s (see, e.g., Brinster and Palmiter, 1984; Jaenisch, 1988). Injection of a plasmid, i.e. a small circular piece of DNA, containing the gene of interest into the nucleus ('germinal vesicle') of an unfertilised oocyte (followed by fertilisation), or into the pronucleus of a fertilised ovum, leads to the integration of the foreign DNA into the nuclear genome of the recipient cell. Use of a plasmid as vector avoids the possibility of side effects (such as tumour formation) that had been observed when vectors based on certain viruses were used. The establishment of pregnancy with embryos from *in vitro* fertilised ova had been achieved in the 1970s – indeed, the first *human* 'test-tube baby' was born in 1978. (This is discussed further in Chapter 12). The success rate for transgenic mouse embryos is lower, however, than with unmodified embryos. Presumably the 'trauma' of injecting the DNA into the egg is sufficient to induce abnormal early development in a significant proportion of the embryos, leading in turn to failure at the stage of implantation. Thus, in early experiments, the success rate for birth of mice (and indeed other mammals) from embryos that had been genetically manipulated was only about 25% of that achieved with non-manipulated embryos (Church, 1987). Although this figure has been improved on since then it is still clear that the success rate is lower with transgenic embryos.

So it is possible to obtain the birth of mammals, mice in this discussion, from transgenic embryos. The remaining technical questions relate to the function of those genes. The exogenous genes are indeed expressed and the pattern of expression is appropriate for the promoter (the tract of DNA adjacent to a gene that contains the gene's 'on–off' switch) that is spliced to the gene. Thus, in pioneering experiments (reviewed by Jaenisch, 1988) on the expression of the rat growth hormone gene in mice, the gene encoding growth hormone was spliced to the promoter for the metallothionein gene (metallothioneins are small proteins produced in response to heavy metals). Exposure of the transgenic mice to the heavy metal inducer of the metallothionein gene led to the rat growth hormone gene being switched on, with dramatic effects. Experiments in which the transgene was placed under the control of developmentally regulated promoters (i.e. promoters that switch genes on at particular times during development of the animal) also gave expression at the appropriate developmental stage in the appropriate cells. Indeed, this has been the basis of many of the applications of transgenic mammals (see below). However, the *level* of expression of the exogenous gene varied considerably from animal to animal. As with plants (Chapter 8), this is generally ascribed to position effects, i.e. the variation in the level of expression is associated with variation in the position at which the transgene integrates into the host genome. Finally, the exogenous genes were indeed shown to be transmitted to the germline and thus subsequently to progeny, although, again, the extent to which the gene was expressed in progeny was variable (possibly because of the phenomenon of gene silencing). These variations in expression, both in the original transgenic mice (the 'T0' generation) and in the progeny (T1), have implications in relation to the possible application of these techniques to humans (Section 14.7). However, these problems have not prevented the use of transgenic animals in research and in biotechnology. For example, mice can be modified with mutant genes that cause monogenic diseases such as cystic fibrosis or Alzheimer's or Huntington's diseases, or with oncogenes that can be activated to induce tumour formation. Indeed, in the UK alone, tens of thousands of transgenic mice have been used in biomedical research.

Larger mammals have also been targets for genetic modification. For example, there are transgenic sheep that produce pharmaceutical proteins such as blood-clotting factor 9 in their milk. Indeed, it was from a transgenic sheep that Dolly was cloned (Chapter 16). With Dolly, the motivation for such cloning was to reproduce the genetic makeup of the nuclear donor sheep without the risk of the variation in expression in the progeny (and incidentally, avoiding the need to sex-select the progeny), thus replicating a commercially valuable animal. Further discussion of this lies outside the scope of this chapter but these examples serve to illustrate that genetic modification techniques first developed for mice have now been successfully applied to other mammals, including, most recently a primate, namely the rhesus monkey (see commentary by Vogel, 2001). The birth at the Oregon Primate Research Center of ANDi, the rhesus monkey carrying a foreign gene, has, it is claimed (see e.g. Highfield, 2001) brought the genetic modification of humans closer. However, in a technical sense this is not true because firstly, the *in vitro* fertilisation and embryo implantation techniques used have been much more fully optimised for humans than for monkeys and secondly, the genetic modification procedure was one that could have already been applied to humans. Nevertheless, those who subscribe to the 'slippery slope' argument believe that in genetically modifying a primate, the psychological barrier to genetically modifying a human has been breached or at least dented (see also Sections 14.5 and 14.7).

14.3 GENETIC SELECTION

In mammalian embryos, the definition of cell lineages occurs relatively late. This means that the cells generated during the early cell divisions (termed 'cleavages' by embryologists) are non-determinate. Further, experimental manipulations have demonstrated that it is possible to remove one cell from an eight-cell embryo without disturbing subsequent development (Figure 14.1) Thus, if there is a need to ascertain whether or not an embryo possesses a particular gene, one

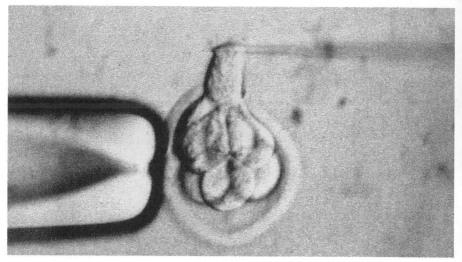

Figure 14.1. Removal of one cell from an eight-cell embryo in order to carry out pre-implantation genetic diagnosis. Photograph reproduced by permission of Dr A.G. Schmutzler.

cell may be removed for testing before placing the embryo in the uterus (or *re*placing it into the uterus if it has been obtained by embryo rescue after a normal *in vivo* fertilisation). Such testing may be used to detect the presence of the foreign gene after injection of DNA into a fertilised ovum, or to conduct genetic testing on the embryo which is appropriate for the situation. This is discussed more fully in Section 14.6.3.

14.4   HUMAN GENETICS AND THE HUMAN GENOME PROJECT

14.4.1   SOME HISTORY

One of the frustrating aspects of reporting science to a wider public is the frequency with which one achievement or advance is confused with another. Thus, we often hear it stated that Watson and Crick discovered DNA. In fact, DNA was discovered in the 19th century and its key role as the genetic material was established in the 1940s. It was precisely because of this role that Watson and Crick were attracted to work on it and their major achievement was to develop a model – a model that has stood the test of time – for its structure (Watson and Crick, 1953). We see a similar confusion when it comes to human genetics. A slightly premature announcement in the year 2000 that the 'first draft' of the human genome sequence was complete (Butler and Smaglik, 2000; Macilwain, 2000; Marshall, 2000a)[1] led many commentators to suggest that

---

[1] The announcement was made somewhat prematurely because a commercial company, Celera

there had not been any work on human genes before the establishment of the Human Genome Project. The truth is rather different and is set out here because of its relevance to the wider applications of science and to the ethical implications of those applications. Long before the establishment of genetics as a science in the early years of the 20th century there was awareness (as documented by Steve Jones, 1996) that certain traits seemed to be transmitted from generation to generation or to be associated with particular families. Indeed, it was a mistaken extrapolation from this awareness that led to the establishment in the 1880s of the eugenics movement in the UK and then in other countries. However, the re-discovery in 1900 of the work of Mendel resulted in a clearer understanding of the nature of inheritance and indeed, led to the Eugenics Society of the UK making a study of the inheritance of haemophilia in the royal families of Europe. This work, published in 1911, provided clear evidence that the gene is sex linked (although, as noted by Jones, there was some inkling as long ago as AD 200 that uncontrolled bleeding may be a familial condition: a Jewish rabbi exempted from circumcision the cousins of a boy who had bled to death after being circumcised).

The first genetic trait to be characterised at the molecular level was sickle-cell anaemia, in which the mutation affects the amino acid sequence of the blood protein haemoglobin. In these days of rapid DNA sequencing it is interesting to note that the amino acid difference between normal and sickle cell haemoglobin (a single amino acid change from glutamic acid to valine) was identified as long ago as 1957 by laboriously sequencing the purified proteins. Further, the genetic code, i.e. the relationship between the sequence of bases in DNA and the sequence of amino acids in protein, was unravelled in some very elegant biochemical experiments in the 1960s. Thus it was possible to predict the base change in DNA that led to the amino acid change in the protein. However, direct confirmation of that prediction had to wait until DNA sequencing methods became available after the advent of recombinant DNA techniques.

As already mentioned, an early impetus for study of human genetics was eugenics, the theory that human society could be improved by encouraging the reproduction of the 'stronger' elements in society and discouraging the reproduction of the 'weaker' elements. From the end of the 19th century through to the Second World War, eugenics movements and societies were well established in several countries, including the UK and the USA. In several countries, eugenics policies were adopted by governments, leading to programmes of compulsory sterilisation of those perceived to be 'feeble-minded' or 'degenerate' (see Dyck, 1997; Wikler, 1999). The supposed scientific basis of this was a social application of Darwinism, aimed at ensuring that the 'fittest' survived. Indeed, the founder of the eugenics movement, Francis Galton, was a cousin of Charles

---

Genomics, which had invested in 300 DNA sequencing machines, threatened to 'scoop' the publicly and charity funded Human Genome Project (Macilwain, 2000; Marshall, 2000a). This tension between the commercial and non-commercial research programmes is still not fully resolved (Marshall, 2000b; Smaglik, 2000). See also Chapters 13 and 15.

Darwin. Ethically, the policies were based on a type of utilitarianism in which the outcome was seen as improving the health (in its widest sense) of society at large. However, as is pointed out by Reiss in Chapter 1 of this book, a utilitarianism which simply aims at increasing the overall level of happiness, pleasure or well-being in society can lead to the legitimate rights of individuals being ignored or over-ridden. This was certainly true in the eugenics movements in which programmes such as compulsory sterilisation (see above) negated the autonomy and right to self-determination of individual humans. Further, those making the decisions as to who should be sterilised had obviously already classified themselves as being amongst the stronger and more desirable elements in society, a view which was doubtless reinforced by the power that was exercised over those who were the subjects of their decisions.

Eugenics fell out of favour, largely as a result of the extreme eugenic practices in Nazi Germany, and indeed the word eugenics, which was originally coined to mean 'good inheritance', is now a pejorative term. However, it has been suggested by some authors (see below) that certain applications of our new knowledge of human genetics may in fact be eugenic in that they seek to eliminate (as far as is possible) certain genetically based disease states from human society. We discuss this further in Sections 14.6 and 14.7 but readers who wish for more in-depth treatments of the topic are referred to Dyck (1997), Kuhse (1999) and Wikler (1999).

Since the middle of the 20th century one of the main motivations in the study of human genetics has been to understand the genetic basis of disease. About 4500 diseases, some relatively common but most of them unusual if not extremely rare, are known to be caused by mutations in single genes, whilst approximately 20 000 chromosomal aberrations have been reported to date (C. Scott[2], personal communication, 2001), a figure which is likely to rise as more micro-abnormalities will be discovered with emerging technology. Further, as we now know, mutations and polymorphisms within certain genes are involved in determining *predisposition* to certain diseases (see Section 14.6). Study of genetic diseases in humans is of course limited by our inability to undertake mating experiments between people possessing particular alleles (although apparently some such experiments were enforced in Nazi Germany). Thus, for example, prior to the advent of DNA analysis, the assessment of genetic risk for a couple wishing to have children depended on the pattern of inheritance of the condition in their family and knowledge of the population frequency of particular gene mutations, which could be calculated by epidemiological studies and statistical methods. There were just a few conditions where the estimation of genetic risk could be modified (not necessarily accurately) by other means, e.g. creatine kinase testing in X-linked Duchenne muscular dystrophy, hexosaminidase assay in autosomal recessive Tay-Sach's disease, and haemoglobin electrophoresis in the (autosomal recessive) thalassaemias. The development of recombinant DNA

---

[2] The Oxford Chromosome Abnormality Database, funded by the South East National Health Service Executive.

technology in the 1970s was highly significant and began to impact clinical medicine by the mid-1980s. The ability to use microbial cells to 'grow' genes, combined with techniques to dissect DNA and detect particular sequences, meant that for the first time it was possible to directly investigate at least some of the genetic changes that lead to the development of a disease. Thus, the genes involved in several heritable diseases, including cystic fibrosis, had been identified, cloned and characterised by the end of the 1980s (see also Section 14.6) prior to the establishment of the HGP. This in turn meant that diagnostic and detection techniques based on actual gene sequences were becoming available.

14.4.2 THE HUMAN GENOME PROJECT

In 1988 a consortium of scientists in the USA persuaded Congress that the time was right for the establishment of a coordinated programme to sequence the entire human genome (for a fascinating account of the history of the project, the reader is referred to Shapiro, 1991). The programme was set to run from 1990 to 2005 and, in the USA, $3 000 000 000 was initially allocated to the project. This sum included 5% allocated to study the ethical and social implications of the project. The decision to set up the HGP was not without it critics. In the USA some members of the public (and presumably some members of Congress: the decision was not unanimous) believed that the money could have been spent in other ways, raising questions about allocation of limited resources (in fact, an almost constant theme in the realm of publicly funded activities). Even amongst the science community there were some who, at least at that time, believed the establishment of the project to be a major mistake (e.g. Rechsteiner, 1991).

However, the project went ahead and, in many ways, the criticisms have been answered. The spin-offs for other areas of science have been dramatic. The extent of assured funding for the research led to significant investment in the development of technology to facilitate the research and much of this has been equally applicable in, for example, plant molecular biology. Focusing on the HGP itself, about two-thirds of the work has been undertaken in the USA with most of the rest being split between the UK, Germany, France, Japan and Canada. The results have been spectacular. By the middle of the year 2000, some four and a half years before the formal closure of the project, the sequences of the majority of the genes of an 'average' human were known (Butler and Smaglik, 2000; Macilwain, 2000; Marshall, 2000a), as was the identity of many disease-causing mutations. This acceleration in acquisition of knowledge about human genetics has almost immediate applications in medicine, as we describe in Sections 14.6 and 14.7. However, before moving on to those applications it is necessary to consider, in the light of increased knowledge of human genes, the possible use of genetic modification techniques with humans since this has some bearing on possible future choices in human genetics.

14.5   GENETIC MODIFICATION: APPLICATION TO HUMANS

In Section 14.2 we showed that various forms of genetic manipulation of mammals are not only feasible but are actually in use. At this point it is appropriate to remind ourselves that many of the techniques developed for mice will be equally applicable to humans. Indeed, as is shown in Chapter 12, the human was the second species for which *in vitro* fertilisation was successfully achieved. Further, the development of a variety of techniques for delivery of sperm to the unfertilised ovum, e.g. 'ICSI' – intracytoplasmic sperm injection, in order to optimise fertilisation (Chapter 12), means that there is extensive experience in the manipulation of ova. Section 14.4 illustrates that our knowledge of human genetics has increased significantly as a result of the application of molecular techniques to the study of human genes (including the work of the HGP). In particular, this research has facilitated the identification of individual genes. For many of these the function is not at present clear, although knowledge of gene function is growing apace. Nevertheless, for a significant proportion of genes their function has been clearly defined. The research has also led to extensive experience of the manipulation of human genes. The combination of knowledge and experience in human embryology and in human genetics thus sets the scene for the possible genetic manipulation of humans. There are three general approaches to this:

- gene detection/genetic diagnosis (with the concomitant possibility of rejection of particular genotypes),
- removal or replacement of genes,
- addition of genes (either to confer a specific genetic trait or to over-ride the effects of a mutant gene, as in gene therapy).

We now turn to discuss these aspects in relation to their applications, both actual and possible.

14.6   GENETIC DIAGNOSIS IN CLINICAL PRACTICE

14.6.1   INTRODUCTION

As described above, the advent of recombinant DNA techniques led to a previously undreamed of capability to identify, isolate and characterise genes, including human genes. This gave a new impetus to human genetic research and even before the formal establishment of the HGP many human genes had been analysed. For many geneticists, a particular milestone was the isolation in 1989 of the large and complex gene that encodes the cystic fibrosis transmembrane regulator (CTFR). Since then, the isolation, sequencing and further characterisation of human genes has accelerated dramatically. Indeed, as already mentioned, the 'first draft' of the human genome was published in 2000. Progress is set to continue well into the 21st century as the fruits of the HGP are realised together with technological advances, such as DNA micro-arrays, that will permit ever more rapid analysis. In an astonishingly short period of time human DNA analysis will have progressed from crude and laborious fragment sizing using a limited range of restriction endonucleases to rapid, far ranging, and relatively inexpensive methods of direct sequencing with the potential to identify a myriad of polymorphisms and mutations.

Two main difficulties will accompany this huge surge in descriptive knowledge of our molecular genetic anatomy. Firstly, the interpretation and health implications of many sequence variations may not be clear for a long time and, secondly, the applications of the technology will generate socio-ethical dilemmas due, at least in part, to the current paucity of beneficial interventions. The unfortunate reality is that treatment in genetic medicine lags far behind our ability to diagnose uncommon inherited disorders and to identify susceptibility risk factors for common conditions.

Despite this, it should not be assumed that making genetic diagnoses has no value. Francis Collins, the second director of the HGP (the first was James Watson) describes the relief of a woman shown not be carrying the BRCA1 (breast cancer pre-disposition) mutation present in other members of her family (Collins, 1999). However, positive diagnoses may also be beneficial. Although little may change for the affected individual, particularly in cases of moderate or severe mental retardation, there are often positive results for the patient and/or family through empowerment conferred by the knowledge and sense of control that is brought to the situation. The resolution of diagnostic uncertainty, and the fact that subsequent investigations are focused on a known rather than unknown problem, means that patients and parents are usually very relieved. Many parents attest to the importance of being able to explain their child's condition when others ask what is wrong (Skirton, 2000), and in very practical terms having a diagnosis often expedites support through the education and social benefit systems. New insights into the complications and possible prognostic scenarios may accompany a diagnosis, there may be genuinely useful medical interventions that can be tried, and comfort and solidarity may be available through the excellent work of the relevant patient support group. These aspects of genetic science in the clinical setting may seem very subliminal compared with the headline issues and major breakthroughs that take the world by storm. However, they are significant in our understanding of 'enhancement' and tell us a lot about what is, or may be, important for many people, namely a sense of control (Berkenstadt et al, 1999), a degree of certainty (Skirton, 1999), and the need for 'cognitive closure' (Webster and Kruglanski, 1994).

Parents may also push for closure in relation to resolving the genetic 'carrier' status of their child(ren) with respect to cystic fibrosis, balanced chromosomal translocation, or any condition/gene that is present in the family but unlikely to have implications for the child until adult life or leading up to reproductive decisions. The genetic testing of children has been dealt with thoroughly elsewhere (Report of a Working Party of the Clinical Genetics Society (UK) 1994; Clarke, 1997) but is an important issue in the enhancement debate. Few would have any objections to testing children if their adverse genetic status could be managed, and therefore enhanced, by practical and beneficial medical interventions or surveillance. If this does not apply, however, whose quality of life is enhanced by the testing? The child has been denied the opportunity to make a fully informed decision about testing as an adolescent or adult and may feel their right to autonomy has been abrogated. The evidence that testing in childhood for carrier status is harmful is not, however, strong or convincing (Fryer, 2000) and the whole debate is likely to move on as families become better informed and seek more control over their information. At present, most parents understand the issues about respecting the child's autonomy when this is fully explained, though a substantial amount of childhood genetic testing is being undertaken, mainly due to parental pressure (Proctor et al, 1999). Furthermore, it would appear from an international survey that geneticists in Asia, Latin America, and southern/eastern Europe are more likely to accede to parents' wishes for their children to be tested compared with geneticists in northern/western Europe and the USA (Wertz, 1998). Prenatal testing, of course, may diagnose carrier status, in which case the situation is known throughout childhood, but this represents a special circumstance. The special situation for adopted children (Turnpenny, 1995), and those conceived by gamete donation, is often difficult and complex.

14.6.2   PRENATAL SCREENING AND TESTING

Inevitably, a recurring issue in the moral and ethical debate is termination of pregnancy, and in this context the use of screening programmes as well as specific prenatal and pre-implantation genetic testing and diagnosis. As with molecular genetics, what is possible today could only have been dreamed of a generation ago. The discovery that alphafetoprotein (AFP) was greatly raised in the amniotic fluid of neural tube defect (NTD) pregnancies (Brock and Sutcliffe, 1972) led the way for the widespread introduction of antenatal screening programmes in the 1980s. The technique of amniocentesis was first introduced for prenatal diagnosis in the 1950s for rhesus iso-immunisation (Bevis, 1953), and subsequently it became a route for treating this condition in the foetus by intra-peritoneal blood transfusion (Liley, 1961, 1963). In the 1960s amniocentesis was used to analyse sex chromatin (Riis and Fuchs, 1960; Serr and Margolis, 1964) for severe X-linked recessive disorders, as well as some inborn errors of metabolism (Jeffcoate et al, 1965; Nadler, 1968), and the first prenatal diagnosis of Down syndrome was made (Valenti et al, 1968). The procedure extracts a small quantity of amniotic fluid from around the foetus, usually at 16 weeks gestation, and can be used for biochemical analysis of the fluid itself (e.g., AFP), foetal karyotyping (chromosomal analysis) or quantitative enzyme testing (for inborn errors of metabolism) on cultured amniocytes. Most of the analyses performed on amniocytes can be undertaken on chorionic villus (CV) tissue obtained by biopsy of the edge of the developing placenta from 11 weeks gestation, a technique pioneered during the 1970s (Hahnemann, 1974) and developed during the 1980s (Ward et al, 1983). If the sample is adequate analysis can usually be direct without the need for culturing cells, and so a result is achieved perhaps eight weeks earlier than would have been the case by amniocentesis. For this reason CV biopsy is often preferred when there are indications for planned prenatal testing. Occasionally it is desired to look directly at foetal blood and this requires the skilled procedure of sampling from the umbilical cord from about 18 weeks gestation; even less often biopsy of foetal tissue such as skin, liver or muscle might be undertaken for very specific tests. Each of these invasive procedures carries a small risk of provoking miscarriage.

None of these procedures would be performed today without the aid of ultrasonography, the most widely used of all antenatal screening modalities. Indeed, ultrasound examination has all but taken over from AFP in screening for NTD. Whilst very sensitive in being able to detect structural abnormalities in the foetus, it is not necessarily very specific in matching anomalies with a precise diagnosis. This may pose serious dilemmas for clinicians and extreme anxiety for parents, some of whom would prefer, with hindsight, not to have known. Accurate measurement of foetal dimensions, for instance, can identify short limbs and predict that the child will have a form of dwarfism but not necessarily distinguish which of the many different types affects the unborn baby. As different forms of dwarfism are accompanied by different medical complications and prognosis, which is exactly the sort of information parents seek, counselling is often, at best, somewhat vague.

Since the mid–late 1980s maternal serum screening of AFP and human chorionic gonadotrophin (hCG) in early pregnancy, combined with maternal age, and sometimes other serum markers and foetal measurements on ultrasound (Wald et al, 1997, 1999), has given rise to an entire industry aimed at prenatal detection of Down syndrome. Many original research articles on every conceivable aspect of this subject have been published, which, *ipso facto*, could be interpreted as a statement from a section of the medical research community that Down syndrome individuals are of less worth and value than 'normal' people. The alternative position is one where the status of all foetuses is something very much less than fully human, combined with the right of the pregnant woman/couple to make an autonomous decision about the fate of her foetus. Most health authorities in the UK fund Down syndrome screening programmes, which are therefore available to any pregnant woman. Antenatal screening programmes are also available in most countries of western Europe (Ireland, Norway and Sweden are exceptions), North America and Australia and expanding in Asia and South America (W.J. Huttly[3], personal communication, 2001).

Prenatal screening, testing and diagnosis in all its forms has greatly extended

[3] The Wolfson Institute, London, UK.

the element of 'choice' about whether or not to continue an established pregnancy, the principle having been given basic legality in the UK by the Abortion Act of 1967. Previously, choice was available only at the stage of whether or not to risk conception, or whether to seek an illegal termination of pregnancy, although as we have seen, 'medical' termination of pregnancy was becoming an option for X-linked recessive disorders. In some countries and societies the limits of choice remain as they were in the UK prior to 1967. One result of 'choice' has been a dramatic decline in the population of newborn babies available for adoption in the last 20 years as abortion numbers have increased, combined with general acceptance of the status of single parenthood.

The conditions currently targeted by publicly funded antenatal screening programmes are primarily NTDs (spina bifida and anencephaly) and Down syndrome. These qualify for expensive screening by virtue of their frequency and severity. Acceptance of these programmes is not, however, universal. Objections are frequently based on the status of the foetus (see Chapter 12) but may also centre on what constitutes a 'severe' condition. Furthermore, not everyone is convinced by arguments based on the cost-effectiveness of screening and there is unease in many quarters about public health 'search and destroy' policies whose concerns are primarily financial. In addition, there may be only limited explanation and counselling given in relation to the screening tests for Down syndrome, so that women are not, paradoxically, provided with full, informed choice (Al-Jader et al, 2000). Indeed, the information about testing may be presented in such a way that it is perceived as routine rather than voluntary (Al-Jader et al, 2000), thus being more difficult to opt out rather than opt in. Nevertheless, a majority of unsuspecting parents who find themselves faced with a choice of whether or not to terminate a Down syndrome foetus will do so (Wald et al, 1997; Al-Jader et al, 2000). Not to do so is perceived as what could be termed 'disenhancement' to their future quality of life. This presupposes that their plans and aspirations *for themselves* (and their 'perfect' child) will be life enhancing to an extent that cannot be matched by caring for, and developing a relationship with, a Down syndrome child. But very few of these (potential) parents have ever had any real contact or experience of children with Down syndrome, so the decision to terminate is often based on a very limited appreciation of the condition. Some parents testify to the positive enhancement that a Down syndrome child has brought to family life, so much so that they choose not to have screening for Down syndrome in subsequent pregnancies.

It is not unusual, of course, that foetal karyotyping for Down syndrome, apart from revealing the sex of the infant (parents can choose whether or not to have this disclosed), identifies some other chromosomal abnormality such as Turner syndrome (females with only one 'X' – karyotype 45, X) or Klinefelter syndrome (males with an extra 'X' – 47, XXY). Parents are often unprepared for findings of this kind and are presented with a choice that was not part of their planned approach to prenatal testing. This requires urgent counselling in order for parents to make an informed decision about the pregnancy but the quality of the information may vary substantially (Abramsky et al, 2001).

The vast majority of rare inherited and congenital disorders are not catered for in routine prenatal screening, though those including structural defects might be detected by ultrasound. Many (but not all) conditions can be tested for when the foetus is known to be at risk because of a positive family history where a precise diagnosis has been made, often when a couple have already had an affected child. In these cases planned prenatal diagnosis is a choice based on very intimate knowledge of the condition in question, balanced with the effect that a second affected child would have on the entire family dynamics. For many couples the decision is far from easy. They may already have had to come to terms with a sense of guilt at being parents of one affected child, perhaps now loved and absorbed into family life, but face the prospect of condemning a subsequent affected child *in utero*, and by inference casting a judgement on their first. In their case the decision about prenatal testing is more complex and calculated, though again primarily concerns the avoidance of future disenhancement.

Experience in clinical practice suggests that the decision to undergo prenatal testing and termination of pregnancy, as a generalisation, becomes easier as the severity of the medical condition worsens. Inborn errors of metabolism leading to neuro-degeneration and early demise are frequently viewed as being devoid of any real hope because they usually breed true and the outcome is inevitable and predictable. These decisions may be more difficult in the future if realistic treatment options become available.

14.6.3   PRE-IMPLANTATION GENETIC DIAGNOSIS

There are many couples, and not just those with objections on grounds of religious faith, for whom termination of pregnancy is not acceptable as a way of exercising choice in the face of genetic risk. This may be due to their views on the status of the foetus, the physical and emotional trauma of undergoing termination, or a combination of these. Some take steps not to have further children but others are attracted to pre-implantation genetic diagnosis (PGD). In this technique eggs cells from the mother are harvested for fertilisation by her partner's sperm but the procedure requires hormonal super-ovulation and surgical recovery. *In vitro* fertilisation (IVF) techniques (Chapter 12) then produce embryos which are subjected to biopsy of a single cell at the eight-cell stage, about 48 hours after fertilisation (see Figure 14.1). To succeed, DNA analysis must be possible on the genetic material from this single cell. Even then, if pregnancy is successfully established after implantation of embryos (usually two) selected for low risk, centres providing this service offer the patient-clients conventional prenatal testing in order to be sure that no mistake has occurred. The hurdles to be overcome are considerable because of limited availability of the expertise, a success rate in the region of only 20% per cycle of treatment, and lack of public health service funding, which means that couples often have to raise much of their own finance. This raises ethical issues regarding equality of access to these services, which is part of a wider debate about the rationing of resources and the ethical principle of 'justice' in the delivery of health care. In the UK treatment of this kind is tightly regulated through the Human Fertilisation and Embryology Authority (HFEA), which is dealt with elsewhere (Chapter 12). For individuals and couples who hold the view that full human status begins when a sperm fertilises an egg, PGD is not acceptable. However, PGD is acceptable to many couples who cannot put themselves through the emotional trauma of conventional prenatal testing and possible termination of an established pregnancy. This raises very interesting issues regarding the way these couples perceive the status of the embryo as compared with the foetus. Technically, 'embryonic' life is that period of early development during which the vast majority of body structures are formed, which for humans is the first eight post-ovulatory weeks (approximately 10 weeks *gestation*); the foetal period extends from eight weeks to birth (O'Rahilly and Müller, 1992). To many couples the acceptability of testing the very early embryo reflects, in a very experiential way, the reasoning behind the Human Fertilisation and Embryology Act (1990) in the UK, which allows the use of (and research on) human embryos up to 14 days (see Chapter 12). However, it seems unlikely that the views of couples requesting PGD have been directly influenced by these regulations.

Currently, PGD in the UK is only granted a licence when it is undertaken for clear medical reasons, i.e. in situations of known genetic risk. This includes X-linked disease where it is not possible to undertake specific genetic analysis to determine which of the male embryos are at high or low risk. In these cases only female embryos are implanted. It is not currently permissible to use PGD for sex selection for purely social or family reasons. This was challenged in the autumn of 2000 by a Scottish couple with four sons who lost their only daughter in an accident (Cramb, 2000). The mother was sterilised after their daughter's birth and sought PGD to implant only female embryos – but this is not allowed within the UK regulations. They were treated in Italy but gave up the only embryo arising from the treatment because it was male (Wormersley, 2001). The motivation to use the technology in this way could be regarded as a form of 'commodification', or 'instrumentation' – reproductive treatment undertaken to meet very specific aspirations and needs of the parents. The same would be true of reproductive cloning if, for example, a sample of tissue from the deceased child had been available in culture. The concept of the newly created child as a commodity or instrument is an extension of the normal desire of couples to become parents to a level where having the child is conditional on certain characteristics or qualities which the parents require; in the case cited the condition is female sex.

The situation where X-linked genetic risk is less easily defined is not clear. An example would be a couple who had two sons with some form of communication or autistic disorder, though without a specific genetic diagnosis. Epidemiological evidence suggests boys are more commonly affected than girls, prompting the couple to request sex selection on these grounds. The difficulty here is knowing whether the population ratio can be meaningfully applied to an individual family because there are many different underlying causes of communication/autistic disorder, most of them not yet elucidated, and there can be no certainty that X-linked inheritance is operating.

In the USA, where reproductive technologies are unregulated, the case of Adam Nash (Hall and Davis, 2000), who was born in August 2000, has brought a new dimension to embryo selection for health reasons. His sister Molly, six years old when Adam was born, has the condition 'Fanconi's anaemia', and her bone marrow is failing. Inherited as an autosomal recessive condition, both parents are healthy carriers of the faulty gene, with a one in four risk of having an affected child in each pregnancy. With the support of their reproductive physicians the parents opted for PGD, not just to avoid Fanconi's anaemia but also to positively select for a tissue match for Molly by 'HLA testing'. From 12 fertilised eggs only one proved to fulfil the dual requirements of not having Fanconi's *and* being a perfect tissue match. Implantation of this embryo was successful and stem cells harvested from the umbilical cord at birth were subsequently infused into Molly in the hope that her bone marrow will be repopulated by normal tissue. It is the first time that PGD has been specifically carried out for the purpose of providing medical help for another person, in this case a sibling, and can also be considered as a form of 'commodification' in the embryo selection and genetic enhancement debate. Objections can be raised on the grounds that the child has no choice about being a tissue-matched donor, or 'instrument', for the sick sibling, and therefore the child's feelings were not seriously taken into account. We do not know whether the child will, in the future, be proud to have helped the sibling in this way or perhaps experience some sort of identity crisis at having been an instrument for the purpose of the sibling's health. Will they be closer siblings because of this treatment, or will they have a very unnatural relationship? If the sick sibling were not to respond to treatment and die anyway, how would such an outcome affect the donor sibling's psyche and self-identity? On the other hand, some might argue, none of us actually 'chooses' to be brought into the world, many of us are unplanned 'accidents', and this undermines arguments about utilitarianism and identity. In the UK, the HFEA relaxed its guidelines in late 2001 to grant a licence for a case similar to that of the Nash family.

An important issue in the Nash family case, but common to all IVF, whether for PGD or infertility treatment, is the creation of many surplus embryos that will never be used. Whether this is acceptable clearly depends on one's views on the status of the early embryo but the ethical debate must also now address the potential creation of human embryos purely for stem cell research – so-called 'therapeutic cloning'. This has recently been approved in principle by a large majority vote in both houses of the UK Parliament (December 2000 and January 2001), in stark contrast to a European Parliament resolution in September 2000. These important issues are covered more fully elsewhere (Chapters 12 and 16).

14.6.4   FUTURE CHOICE

The likely development of relatively rapid and less expensive techniques for analysing hundreds of thousands of DNA sequences inevitably raises the question of how this might eventually be applied to prenatal screening and testing, as well as medical examination for employment and insurance. Concern over the prenatal issues prompted two public consultation exercises held in the UK during 2000, one from the HFEA dealing with PGD, and the other from the Advisory Committee on Genetic Testing (now absorbed into the Human Genetics Commission) dealing with more conventional prenatal screening and testing. Key questions will be the extent to which extended screening will be regarded as cost-effective by public health and the demand shown by the general public. However, there is also the issue of whether extended prenatal screening will be perceived as a form of eugenics and the way forward is likely to be very cautious. Significantly, the HFEA public consultation on the use of PGD showed support in cases of serious genetic disorders but concern over wider implications of the technique. Any additions to the range of conditions screened will need to be deemed 'serious' in the perception of both the public and relevant medical specialists. If such screening depends on invasive prenatal procedures to make a definitive diagnosis, thus carrying a risk to the pregnancy, it is unlikely to be attractive to pregnant women/couples. For these reasons it seems likely for the foreseeable future that screening and diagnosis for rare conditions will be restricted to those families directly affected, unless non-invasive tests are developed which can modify the risk (as in maternal serum screening for Down syndrome). There is, in fact, great interest in a technique that may have the potential to dramatically alter the way screening is offered, namely the ability to harvest foetal cells from the maternal circulation. Foetal cells differ from maternal cells in size and other characteristics but of course there are relatively few of them. If techniques to separate them efficiently are perfected screening may be possible without the risks associated with invasive procedures.

If extended screening at the embryo stage in PGD, and the foetus in established pregnancy, becomes feasible for a range of severe mendelian disorders it will also be technically possible for mutations and polymorphisms that confer risk, or predisposition, for important medical conditions. Well known examples are mutations in BRCA1 and BRCA2 for breast and breast–ovarian cancer, and the equivalent mutations in DNA mismatch repair genes for hereditary colorectal cancer. In both these familial forms of common cancers the mutations confer a high risk of developing the disease, though it is not inevitable. The penetrance is therefore not 100% but somewhat lower, perhaps 60–85%. Then there are risk factors with far less predictive value, such as the ApoE4 allele in Alzheimer's disease. Many such risk factors are likely to be identified in the next wave of human genome research with so much emphasis on single nucleotide polymorphisms (SNPs), various of which might be found to have predictive value in, for example, schizophrenia, manic depression, autism, diabetes, and various cancers.

Will there be requests for tests of this kind if they are technically possible? At present this is difficult to say. Even with a serious, late onset degenerative condition demonstrating full penetrance, such as Huntington's disease, experience gathered through the UK Predictive Testing Consortium indicates that prenatal tests are relatively uncommon, with 162 having been performed over the complete six year period 1994–99 (S.A. Simpson, personal communication, 2001). A small number of resolute couples are taking the PGD route. Some choose gamete donation but most couples elect for the time-honoured options of having children regardless of the risk, or not having any (or more) children once they know the risk. For some of those who start a family regardless of the risk, one factor which often influences their decision, particularly in the late onset disorders, is the hope that treatment for the disease will be found in time to benefit the next generation, and therefore enhance life for their children.

This experience in clinical practice is supported by evidence from research. Data from the *British Social Attitudes* survey of 1998 (Stratford et al, 1999), including a wide range of questions on genetic research, suggest that people understand and draw a distinction between conditions which are 'seriously' handicapping, whether mentally or physically, and those that might lead to premature death in adult life or give rise to differences of 'appearance', such as extreme short stature. This is reflected in responses to questions focused on medical conditions that might justify abortion. In clinical practice, of course, experience relates overwhelmingly to individuals, couples and families personally affected by specific problems, whereas a public survey asks questions of people who, for the most part, respond in accordance with their innate feelings and/or belief systems. This is important in trying to evaluate the potential uptake of prenatal genetic testing, for perception of risk is not merely a function of the arithmetical chance but more so of the severity or 'burden' of disease, which itself includes crucial elements of chronicity in the prognosis, and therefore the time available for reasonable quality of life.

Future choice in genetic testing at all stages of life may therefore, in theory at least, be confusingly extensive, but the use of such tests to enhance individual or family life seems likely to be conducted within the framework of cautious restraint that characterises current practice, applicable mainly to those personally affected. The *British Social Attitudes* project plans to repeat its survey at intervals in order to provide a measure of changing attitudes with time.

14.7 MANIPULATION OF GAMETES AND EMBRYOS FOR HEALTH AND ENHANCEMENT

14.7.1 GENE THERAPY

The term *gene therapy* refers to any procedure or technique used to treat a medical condition by genetic modification of cells or tissues in the patient. The concept encompasses many possible strategies but current approaches generally involve the transfer of genetic material, i.e. genes, into the patient's cells. Gene therapy, or genetic manipulation in any form, is currently only permissible on *somatic* tissues and not *germline* tissues. Manipulation of the early embryo should be included with the latter since this can reasonably be expected to affect the germline tissues in later development. The reasons for this legal position are easily appreciated when there is so much uncertainty about the safety and side effects of manipulating human genomic DNA at such a fundamental stage. For example, the position effects to which we referred in Section 14.2 mean that at present it is not possible to predict the level at which the inserted gene will be expressed in an individual. This problem will not be overcome until it is routinely possible to insert genes into specified locations within the genome. Limited success for targeting specific genomic locations has been achieved for mice (Thompson et al, 1989) and more recently for sheep (McCreath et al, 2000) but to date for no other mammal. However, even if gene targeting is achieved there may still be variation between individual transgenic animals. Safety issues and concerns about the long term consequences and hazards, possibly persisting over many generations, are additional major concerns about germline, as opposed to somatic, gene therapy. We simply do not know enough about the potentially damaging and irreversible perpetuation of genomic changes in succeeding lineages. At present, therefore, there is substantial consensus that humankind is not yet ready for this. Some take the view that human genetic engineering at the level of the gamete or early embryo will never be acceptable. This is largely because of fears that it represents a new techno-eugenics that will lead ever increasingly to attitudes of genetic determinism in relation to disability and behaviour, and discrimination at all levels in a society increasingly commercially driven. In addition, it may be seen as a step too far in tampering with our own biological nature.

In this whole debate about the future applications of genetic science to humans there are many 'What if. . .?' scenarios. One of these, undoubtedly, is 'What if gene therapy and genetic manipulation of the germline can be shown to be safe, practical and effective?' Should this transpire there are likely to be significant voices advocating use of the techniques to treat, or more accurately, *prevent*, disease, especially where a family has already been affected. In the absence of regulation, if it *can* happen, it almost certainly *will* happen, as the case of Adam Nash demonstrates. Once a precedent is set, other cases generally follow. On one level it can be argued that germline gene therapy represents a *higher* ethical standard than embryo selection or termination of pregnancy, especially if it reduces suffering and the side effects of conventional therapies. However, it seems unlikely that an excess production of embryos with some form of selection can be avoided. This objection would be circumvented if genetic manipulation of gametes could be successfully achieved because most people do not ascribe to sperm and ova the status of 'life' that they ascribe the zygote (i.e. the newly fertilised ovum).

There are counter ethical arguments to genetic manipulation of either embryos or gametes. Firstly, there is still concern about the irreversible effects of manipulation for the succeeding lineage, and many take the view that this is always going to be unacceptable for humankind. This is very much seen as 'unnatural' and 'playing God', building on the notion that it is inappropriate for humans to be 'tampering' with natural forces, controlling their own destiny, and putting faith in genetic determinism. Others see this as an over-sentimental attitude, arguing that humankind has been modifying (and indeed polluting) the environment and the biosphere over aeons in far more significant ways. We have also clearly exercised a great deal of choice in the selection of mating partners, often operating within societies which are heavily stratified from the social and economic viewpoints. Is this, in itself, not a form of biological selection, and if so, is it natural or unnatural? There is a school of thought which very much sees the 'natural' versus 'unnatural' debate primarily as a function of how we view the world 'now', i.e. the 'unnaturalness' of much that developed societies take for granted is not recognised. Further, the division between natural and unnatural does not provide us with clear-cut ethical guidelines (Reiss and Straughan, 1996). Rather, our ethical assessment of a particular intervention may depend on whether actions or judgements are viewed 'intrinsically/deontologically' (see Chapter 1) or 'consequentially'. The 'intrinsic' viewpoint regards an action or judgement as either right or wrong in itself, and by implication cannot identify any circumstances where it is justified to change position. The 'consequential' viewpoint allows for flexibility because rightness or wrongness is judged by the effects or consequences that an action has. Even recent history tells us that much that was considered 'wrong' in the past is now at least tolerated, if not fully accepted. However, the intrinsic (deontological) and consequential approaches need not necessarily be seen as opposite ways of making ethical judgements. Rather, they may each be seen as the context within which the other operates. Thus there may be intrinsic ethical arguments about not pursuing genetic technologies, but not to do so may be to neglect an opportunity, and therefore 'fail', to prevent and relieve human suffering. However, to pursue them for reasons other than the prevention or relief of human suffering would be wrong. Thus the two approaches may act as constraints, one upon the other.

A second major argument against the genetic manipulation of embryos centres around the acceptance of handicap and disability in society and therefore parallels the concerns relating to prenatal Down syndrome screening, as discussed already. If we strive for a society in which genetically determined handicaps and disabilities are reduced, are we making a statement that the handicapped and disabled have no value, or, if they have value, that it is very much inferior to those who are able in mind and body (Shakespeare, 1998; Kuhse, 1999)? In relation to termination of pregnancy for handicap and disability, some would see this as an echo of the eugenics movement of the 19th and early 20th centuries, as discussed in Section 14.4.1. Failure to take account of the views of the handicapped and disabled exposes serious insensitivity and failure to be inclusive, at best, and deficiencies in our understanding of what it means to be human, at worst. The worth of human civilisation may justifiably be measured by its level of care for the sick and underprivileged, who will always be there. However, an underlying tenet in the art and science of medicine is to treat and prevent disease, and relieve suffering. On this basis it may be fully justified to include genetic manipulation as a form of therapy to prevent handicap and disability. In relation to the world's major health problems of infectious epidemics and famines, humankind has not, hitherto, exercised any restraint in trying to abolish these on the basis that they are somehow 'good for us'. In fact, as we now appreciate, a certain level of infectious disease probably is good for us *as a species* – the 'herd immunity' effect – but no individual or family would choose to be sufferers. Extending this reasoning to genetic disease, humanity benefits from the unique contribution of the handicapped and disabled because they are individuals with equal value to the able-bodied and society learns that caring for this group is part of what it means for all human beings to coexist with mutual respect and inter-dependency. However, for a family already severely affected by a serious genetic condition, one understands, and has sympathy with, the choice they may make to prevent the condition appearing again.

These arguments are likely to continue to be held in balance in the whole debate. If genetic manipulation of embryos and/or gametes becomes feasible, safe and effective, however, it is likely that it will be applied, and probably limited to, those couples and families who have already been affected by a serious genetic disease, which is very close to current medical genetic practice as it relates to prenatal diagnosis.

14.7.2   GENETIC ENHANCEMENT

Genetic manipulation of embryos and gametes as it relates to 'enhancement' takes the whole debate to another level, and conjures up images of a brave new world of 'designer babies'. If it becomes possible to modify stature, hair or skin colour, athletic or musical ability, for example, by genetic engineering, will there be a demand for it? The answer may well be 'yes', but this does not make it generally acceptable. There is a powerful argument that it represents a form of eugenics, or at the very least, once again, 'commodification' – parents making their children into instruments of their own pleasure, meeting their own aspirations rather than fully acknowledging them as individuals in their own right. One difficulty, however, is how we perceive the distinction between treatment and enhancement. This is well seen in relation to short stature and the publicity surrounding a young adolescent girl undergoing limb lengthening surgery in the UK, at the expense of the National Health Service, in order to give her enough height to fulfil her ambition of becoming an air hostess (Alderson, 2000). There are similar controversies in relation to breast enlargement or reduction surgery, sex change operations for gender identity disorder, and pure cosmetic surgery. One person's enhancement is another person's treatment and there are grey areas for which there is unlikely ever to be a consensus. Self-image, peer pressure, and the cultural norms of a society are all important factors in determining individual perception of enhancement versus treatment. However, in the UK the 'Clothier Report' (*Report of the Committee on the Ethics of Gene Therapy*, 1992), has rejected genetic manipulation for 'cosmetic' reasons. It can be argued, of course, that many, if not most, parents already practice a form of commodification by bringing up their children a certain way, with their own values or within their own belief system, or giving them privileged opportunities through private education or expensive extra-curricular activities. Whilst this touches on many social issues which are not within the scope of this chapter, most people draw a distinction between this type of enhancement and premeditated, biologically engineered enhancement which leaves much less to the imagination and is clearly aimed at some ulterior motive of self-gratification and/or competitive advantage.

Genetic enhancement through manipulation of embryos or gametes strikes at the very heart of what it means to have one's own identity through natural laws of chance, the acceptance of what 'nature' (or for many people, 'God') allocates to an individual, as if in a personalised ticket. This, it seems, is a powerful undercurrent in the understanding of who we are as individuals and as a species. It is supported by evidence from the *British Social Attitudes* survey (Stratford et al, 1999). This highlighted support for genetic research and gene manipulation for the detection, prevention, and treatment of disease but there were deep reservations for the application of this technology for enhancement. The responses in relation to modifying antisocial behaviours such as violence and aggression were less clear, thus highlighting a grey area of the kind that will always court controversy.

It remains to be seen how public opinion will evolve as the technology moves on. We should not assume that human genome research is leading steadily and inexorably towards 'designer babies' and there is no suggestion at present that this is acceptable either to scientists or the public. As one writer put it in relation to publication of the first draft of the human genome, 'This is the key to life but not the secret of a perfect one' (Daniels, 2000). One could also paraphrase a well known biblical quotation: *Man shall not live by DNA alone.* There is increasing promise of successful forms of treatment for disease that do not require genetic manipulation of embryos and gametes, and so the framework of ethical issues may develop a different focus. It is important that the ethical debate remains vigorous, keeping pace with scientific progress and seeking broad views through public consultation.

14.8 CONCLUDING REMARKS

Human genetic research has already brought enormous change to clinical medicine, and on a daily basis individuals, couples and families are benefiting from existing applications in genetic testing and experiencing enhancement in different ways. Genetic manipulation poses new and exciting possibilities as well as huge challenges and ethical dilemmas. However, it is likely that the focus for the foreseeable future will continue to be on patient-clients for whom it will be directly applicable. It is also increasingly probable that human genome research will influence public health policies as more is learned about disease susceptibility in different groups and populations. The challenges will be to translate and communicate anticipated benefits and health enhancements to the public, as well as protecting them from abuse and misuse of the technology (see also Chapter 13).

REFERENCES

Abramsky, L., Hall, S., Levitan, J. and Marteau, T.M. (2001) What parents are told after pre-natal diagnosis of a sex-chromosome abnormality: interview and questionnaire study. *British Medical Journal*, **322**, 463–466.

Al-Jader, L.N., Parry-Langdon, N. and Smith, R.J. (2000) Survey of attitudes of pregnant women towards Down syndrome screening. *Prenatal Diagnosis*, **20**, 23–29.

Alderson, A. (2000) *The Sunday Telegraph (London)*, 19 November 2000, 4–5.

Berkenstadt, M., Shiloh, S., Barkai, G., Katznelson, M.B.-M. and Goldman, B. (1999) Perceived personal control (PPC): a new concept in measuring outcome of genetic counseling. *American Journal of Medical Genetics*, **82**, 53–59.

Bevis, B.C.A. (1953) Composition of liquor amnii in haemolytic disease of the newborn. *Journal of Obstetrics and Gynaecology of the British Commonwealth*, **60**, 244–251.

Brinster, R.L. and Palmiter, R.D. (1984) Transgenic mice containing growth-hormone fusion genes. *Philosophical Transactions of the Royal Society, London, Series B*, **307**, 309–312.

Brock, D.J.H. and Sutcliffe, R.G. (1972) Alphafetoprotein in the antenatal diagnosis of anencephaly and spina bifida. *Lancet*, **ii**, 197–199.

Butler, D. and Smaglik, P. (2000) Draft data leave geneticists with a mountain still to climb. *Nature*, **405**, 914–915.

Church, R.B. (1987) Embryo manipulation and gene transfer in domestic animals. *Trends in Biotechnology*, **5**, 13–19.

Clarke, A. (ed.) (1997) *The Genetic Testing of Children*, Bios Scientific Publishers, Oxford, UK.

Collins, F. (1999) The human genome project: tool of atheistic reductionism or embodiment of Christian mandate to heal? *Science and Christian Belief*, **11**, 99–111.

Cramb, A. (2000) *The Daily Telegraph (London)*, 5 October 2000, 3.

Daniels, A. (2000) *The Daily Telegraph (London)*, 25 June 2000, 35.

Dyck, A.J. (1997) Eugenics in historical and ethical perspective. In *Genetic Ethics: do the Ends Justify the Genes?* Kilner, J.F., Pentz, R.D. and Young, F.E. (eds), Erdmans, Grand Rapids, USA and Paternoster, Carlisle, UK, pp. 25–39.

Fryer, A. (2000) Inappropriate genetic testing of children. *Archives of Disease in Childhood*, **83**, 283–285.

Hahnemann, N. (1974) Early prenatal diagnosis: a study of biopsy techniques and cell culturing from extraembryonic membranes. *Clinical Genetics*, **6**, 294–306.

Hall, C. and Davis, S. (2000) *The Daily Telegraph (London)*, 4 October 2000, 1.

Highfield, R. (2001) *The Daily Telegraph (London)*, 12 January 2001, 1, 6.

Jaenisch, R. (1988) Transgenic animals. *Science*, **240**, 1468–1474.

Jeffcoate, T.N.A., Fliegner, J.R.N., Russel, S.N., Davis, J.C. and Wade, A.P. (1965)

Diagnosis of adrenogenital syndrome before birth. *Lancet*, **ii**, 553–555.

Jones, S. (1996) *In the Blood: God, Genes and Destiny*. Harper Collins, London, UK.

Kuhse, H. (1999) Preventing genetic impairments: does it discriminate against people with disabilities? In *Genetic Information: Acquisition, Access and Control*, Thompson, A.K. and Chadwick, R.F. (eds), Kluwer–Plenum, New York, USA, pp. 17–30.

Liley, A.W. (1961) Liquor amnii analysis in management of pregnancy complicated by rhesus sensitization. *American Journal of Obstetrics and Gynecology*, **82**, 1359–1370.

Liley, A.W. (1963) Intrauterine transfusion of foetus in haemolytic disease. *British Medical Journal*, **2**, 1107–1109.

Macilwain, C. (2000) World leaders heap praise on human genome landmark. *Nature*, **405**, 983–984.

Marshall, E. (2000a) Human genome: rival sequencers celebrate a milestone together. *Science*, **288**, 2294–2295.

Marshall, E. (2000b) Human genome: storm erupts over terms for publishing Celera's sequence. *Science*, **290**, 2042–2043.

McCreath, K.J., Howcroft, J., Campbell, K.H.S., Colman, A., Schnieke, A.E. and King, A.J. (2000) Production of gene-targeted sheep by nuclear transfer from cultured somatic cells. *Nature*, **405**, 1066–1069.

Nadler, H. (1968) Antenatal detection of hereditary disorders. *Pediatrics*, **42**, 912–918.

O'Rahilly, R. and Müller, F. (1992) *Human Embryology and Teratology*. Wiley-Liss, New York, USA.

Proctor, A.M., Clarke, A. and Harper, P.S. (1999) Survey of genetic testing in childhood. *Journal of Medical Genetics*, **36** (*Suppl. 1*), S73.

Rechsteiner, M.C. (1991) The human genome project: misguided science policy. *Trends in Biochemical Sciences*, **16**, 455.

Reiss, M.J. and Straughan, R. (1996) *Improving Nature? The Science and Ethics of Genetic Engineering*. Cambridge University Press, Cambridge, UK.

Report of the Committee on the Ethics of Gene Therapy. (1992) HMSO, London, UK.

Report of a Working Party of the Clinical Genetics Society (UK) (1994) The genetic testing of children. *Journal of Medical Genetics*, **31**, 785–787.

Riis, P. and Fuchs, F. (1960) Antenatal detection of foetal sex in prevention of hereditary disease. *Lancet*, **ii**, 180–182.

Serr, D.M. and Margolis, E. (1964) Diagnosis of fetal sex in a sex linked hereditary disorder. *American Journal of Obstetrics and Gynecology*, **88**, 230–232.

Shakespeare, T. (1998) Choices and rights: eugenics, genetics and disability equality. *Disability and Society*, **13**, 665–682.

Shapiro, R. (1991) *The Human Blueprint: the Race to Unlock the Secrets of our Genetic Script*. St Martin's, New York, USA.

Skirton, H. (1999) Genetic nurses and counsellors – preparation for practice with families at risk of cancer. *Disease Markers*, **15**, 145–147.

Skirton, H. (2000) A longitudinal study of genetic counselling for families – needs, expectations and outcomes. *PhD Thesis*. University of Exeter, UK.

Smaglik, P. (2000) Forces for collaboration falter with human genome in sight. *Nature*, **408**, 758.

Stratford, N., Marteau, T. and Bobrow, M. (1999) Tailoring genes. In *British Social Attitudes 16th Report*. Jowell, R., Curtice, J., Park, A. and Thomson, K. (eds), Ashgate, Aldershot, UK, pp. 157–178.

Thompson, S., Clarke, A.R., Pow, A.M., Hooper, M.L. and Melton, D.W. (1989) Germline transmission and expression of a corrected HRPT gene produced by gene targeting in embryonic stem cells. *Cell*, **56**, 313–321.

Turnpenny, P.D. (ed). (1995) *Secrets in the Genes: Adoption, Inheritance, and Genetic Disease*. British Agencies for Adoption and Fostering, London, UK.

Valenti, C., Schutta, E.J. and Kehaty, T. (1968) Prenatal diagnosis of Down syndrome. *Lancet*, **ii**, 220.

Vogel, G. (2001) Infant monkey carries jellyfish gene. *Science*, **291**, 226.

Wald, N.J., Huttly, W.J. and Hennessy, C.F. (1999) Down's syndrome screening in the UK in 1998. *Lancet*, **352**, 336–337.

Wald, N.J., Kennard, A., Hackshaw, A. and McGuire, A. (1997) Antenatal screening for Down's syndrome. *Journal of Medical Screening*, **4**, 181–246.

Ward, R.H.T., Modell, B., Petrou, M., Karagozlu F. and Douratsos, E. (1983) Method of sampling chorionic villi in first trimester of pregnancy under guidance of real time ultrasound. *British Medical Journal*, **286**, 1542–1544.

Watson, J.D. and Crick, F.II.C. (1953) Molecular structure of nucleic acids: a structure for deoxynucleic acids. *Nature*, **171**, 737–738.

Webster, D.M. and Kruglanski, A.W. (1994) Individual differences in need for cognitive closure. *Journal of Personality and Social Psychology*, **67**, 1049–1062.

Wertz, D.C. (1998). International perspectives. In *The Genetic Testing of Children*. Clarke, A. (ed), Bios, Oxford, UK, pp. 271–287.

Wikler, D. (1999) Can we learn from eugenics. In *Genetic Information: Acquisition, Access and Control*, Thompson, A.K. and Chadwick, R.F.(eds), Kluwer–Plenum, New York, USA, pp. 1–16.

Wormersley, T. (2001) *The Daily Telegraph*, 5 March, 7.

FURTHER READING

In addition to the above references we suggest for further reading
Marteau, T. and Richards, M. (eds) (1996) *The Troubled Helix*. Cambridge University Press, Cambridge, UK.

15 Patenting Human Genes: Ethical and Policy Issues

Audrey R. Chapman

In February 2001 two rival teams utilizing different methodologies published working drafts of the human genomic sequence. The Human Genome Project has been a massive project, on a scale unparalleled in the history of biology, with very significant scientific and human health implications. It will help us to understand the biological substructure of human nature. The sequencing of the human genome will also provide scientists with tools to identify the genes whose defects or mutations are assumed to be the cause of genetically based diseases and to develop new therapies and approaches for a wide range of diseases and problems (see Chapter 14).

However, we have a long way to go to be able to apply this knowledge. Although scientists have a very clear idea of the biochemical mechanisms involved in the function of genes (i.e. in providing the 'instructions' for synthesis of proteins) they are still very far from understanding the complex interplay of genes and the regulatory cascades and networks that must operate in the development and indeed in the daily life of multi-cellular organisms. One of the startling discoveries of the Human Genome Project is that we apparently have only 26 000–30 000 genes, many fewer than the 100 000 that were anticipated. This means that human genes (and, likely, mammalian genes in general) and gene control mechanisms are probably far more complex than those of other organisms; indeed it is likely that as yet unknown regulatory sequences exist, perhaps amongst the non-coding DNA, sometimes referred to in the past as 'junk DNA'. The unexpectedly low number of human genes compared with the actual genetic complexity of humans suggests that specific human genes carry out a far greater range of functions than those in other organisms; the model of 'one gene codes for one protein' may have to be discarded. Thus the sequencing of the human genome represents, not an ending, but the beginning of a new approach to biology.

Clearly the new genetic knowledge we are gaining will be significant to the human future. How genetic data will be used and whom they will benefit are critical issues. This chapter argues that the resolution of these issues will depend

Bioethics for Scientists. Edited by John Bryant, Linda Baggott la Velle and John Searle.
© 2002 by John Wiley & Sons Ltd.

at least in part on seemingly arcane decisions about the patenting of genetic information. It thus asks whether it is ethically or scientifically appropriate to grant patents or other forms of intellectual property rights for genetic information.

15.1   CURRENT SITUATION

Two consortia with major differences in approach to patenting have now sequenced the human genome. One of the teams, the International Human Genome Sequencing Consortium, is a predominantly public effort headed by the US National Center for Human Genome Research, which has also received significant support from the Wellcome Trust in the UK, one of the largest private medical foundations in the world. This team is committed to free and unlimited access to its data in order to promote the greatest public benefit. By doing so, it is giving expression to a central principle of the United Nations' Universal Declaration on the Human Genome and Human Rights: 'The human genome underlies the fundamental unity of all members of the human family, as well as the recognition of their inherent dignity and diversity. In a symbolic sense, it is the heritage of humanity.'[1]

The rival team is a private initiative by Celera Genomics, a US based corporation in the vanguard of efforts to use the human genome sequence to make money for private investors. Celera had a major advantage over the public team because it was able to make use of its data without having to reciprocate. Currently, Celera is restricting access to its database and charging fees for most uses of this information. It is also staking out patent claims on those genetic sequences that have the most promise for developing diagnostics and treatments.

Moreover, Celera is not alone. Stock market analysts estimate that there are more than 100 'drug-platform' biotechnology companies attempting to exploit genomic information by developing drugs. The bioinformatics market will soon exceed $1 billion per year.[2] The highly competitive effort to patent genes has been variously compared with the 19th century partition of Africa by the colonial powers and the California gold rush (Marshall, 1997).

Given the issues at stake, there are four significant questions that should be considered. The first is whether human genetic information can fulfill the technical requirements for patentability. The second is whether it is ethically appropriate to patent human genes. Third, does patenting of genetic information promote scientific research and contribute to human welfare? And fourth, does patenting human DNA and tissue demean human life and human dignity?

---

[1] *Universal Declaration on the Human Genome and Human Rights*, UNESCO 1997, UN General Assembly 1999.
(http://www.unesco.org/human_rights/hrbc.htm).
[2] See commentary in *Nature*, **409**, 745 (2001).

15.2   FULFILLING TECHNICAL REQUIREMENTS FOR PATENTABILITY

Historically, governments have sought to promote creativity, the dissemination of ideas, development of inventions, and scientific progress by providing time limited protection to creators and inventors in the form of intellectual property rights, such as patents and copyrights. There have been three primary criteria to qualify for patentability: novelty, utility, and non-obviousness (see also Chapter 10). To be 'novel' an invention must not have been known and available to the public at the time of the application. 'Utility' refers to usefulness. To qualify, a proposed patent must specify concrete function, service, or purpose. 'Non-obviousness' has a technical definition: an invention cannot obtain a patent if the differences between its subject matter and the prior art are such that 'The subject matter as a whole would have been obvious at the time the invention was made to a person having ordinary skill in the art to which said subject matter pertains.'[3] Legal doctrine also traditionally distinguished between objects someone designs from simpler materials by reason of a plan or principle and objects that an individual simply discovers or produces by applying a process or instrument to natural materials. The former was considered to be an invention and therefore patentable, but the latter was not.

Prior to 1980, some 200 years of legal doctrine conceptualized life forms as 'products of nature' and therefore unable to meet the criteria for patentability. However, in a landmark case in that year, *Diamond v. Chakrabarty,* the US Supreme Court ruled, in a narrow 5–4 decision, that genetically modified living organisms were not nature's handiwork but the result of human ingenuity and research and therefore patentable subject matter.[4] In making this decision, the justices apparently assumed that the refusal to accord patent rights in genetically engineered organisms would slow down the pace of research in this field and make US companies less competitive. In rendering its decision, the court explicitly refused to consider non-economic issues, holding that non-economic values, such as respect for life, were outside its purview in deciding cases related to patent law. The decision stated that only Congress could address matters of 'high policy' (Gold, 1996). However, the US Congress has preferred not to consider the appropriateness of granting life patents, and the US Patent and Trademark Office has been free to set policy without any meaningful ethical or political oversight by the courts or political representatives.

The lack of legislative guidelines in the US has meant that the US Patent and Trademark Office (PTO) has been free to determine policy on narrow technical grounds. After 1980, the PTO began to grant new kinds of biotechnology patent,

---

[3] 35 United States Code, Sec. 103.
[4] *Diamond v Chakrabarty,* 477 U.S. 303 (1980).

first on newly developed plant varieties and seeds (1985; see Chapter 10), then on non-naturally-occurring non-human multicellular living organisms, including animals (1987), and more recently to cover genes, gene fragments, and cell lines that are functionally equivalent to those occurring in nature. Accelerated international economic competition for markets has encouraged other patent offices in Europe and Japan to follow suit. In Europe, the European Patent Office granted the first patent for a microorganism in 1981, the first patent for a plant in 1989, even though the relevant legal provisions were not clear, and a patent for a genetically modified animal in 1992 (Salazar, 1999). In 2000, patents covering the methods of producing cloned non-human animals, human cell lines, and early human embryos, and also covering such animals, cell lines, and embryos as the products of cloning, were awarded in Britain despite their apparent violation of the European Directive on the Legal Protection of Biological Inventions. Patents on the same are pending in the US at the time of writing, early in 2001 (Poland, 2000). Currently, the major patent offices issue thousands of life patents every year.

To rationalize the granting of genetic patents, the PTO claims that a gene patent applies 'not to the DNA itself but to the process of altering it in some way using human knowledge to create something else' (Stone, 1995). In seeking to clarify eligibility for patenting, PTO officials distinguish between naturally occurring DNA sequences and sequences that are isolated and purified and claim that the latter meets the requirement of being distinguished from the natural state and are thereby patentable subject matter.

However, all that occurs in the process of identifying and cloning the coding region of a gene is the removal of the introns (the non-coding DNA sequences that interrupt the sequences that code for proteins) via the synthesis of complementary or cDNA (using messenger RNA as a template for this). Therefore this distinction can quite rightly be considered to be 'intellectually trivial.' Messenger RNA exists in nature and cDNA is just a copy of this sequence. One recent analysis likens claims that synthesizing cDNA is an alteration of nature, and as such a human invention, to saying that the same invention could be re-patented if translated into a different language (Bobrow and Thomas, 2001). Nevertheless the situation is that the PTO now issues patents on genes, gene fragments, proteins, and cell-lines that are functionally equivalent to their naturally occurring counterparts. Moreover, the patents issued cover the gene, gene sequence, or protein itself, not simply the process of reproducing, isolating, or purifying it.

Advocates of patenting also argue that it takes considerable ingenuity and investment to obtain genetic sequences and copy them and claim that this provides a sufficient warrant to support patenting. In the past, the process of isolating genes and determining sequences was cumbersome and time consuming, but new technologies have changed how discoveries are made. Currently, however, this work is done primarily by automatic gene sequencing machines that churn out sequences quite mechanically, incredibly quickly, and at a relatively low cost. DNA sequences identified through such 'high throughput' sequencing can perhaps best be characterized as new scientific information (Eisenberg, 2000) and scientific principles and knowledge are not patentable subject matter.

Criteria for patentability were developed to apply to mechanical inventions and from the above analysis it is questionable as to whether life forms, particularly genetic sequences, fulfill these conditions. Genetic patents would not seem to fulfill the first or third criteria, novelty and non-obviousness. Given the present state of genetic knowledge, few patent applications for genes or genetic fragments can specify the function of the genetic information in question or the likely applications. This obviously limits the utility of this information. The only clear value of much of this information is its potential utility through licensing access to the data and that benefit would accrue only to the patent holder, not to the general public (see also Chapter 10).

15.3 ETHICAL CONCERNS

That there are ethical implications of treating life forms as just another useful product was pointed out by critics at the time of the landmark *Chakrabarty* case and has been a source of ongoing contention. Ethical implications constitute grounds for excluding subject matter from patentability under some intellectual property regimes. The European Patent Convention Agreement and the European Directive on the Legal Protection of Biotechnological Inventions allow members to exclude subject matter from patenting 'to protect *ordre public* or morality, including to protect human, animal or plant life or health or to avoid serious prejudice to the environment'. This provision also appears in the World Trade Organization's Trade Related Aspects of Intellectual Property Agreement, generally known as TRIPS.[5] US patent law, however, does not condition patent recognition on ethical criteria. And even in Europe patent examiners have construed moral criteria so narrowly that few, if any, ethical considerations are likely to exclude patent applications. The European Patent Office, for example, only excludes patents whose exploitation would be 'abhorrent to the overwhelming majority of the public' or a contravention of the 'totality of accepted norms' (Drahos, 1999).

Three sets of communities have raised ethical issues about genetic patenting on an ongoing basis – indigenous peoples, the religious community, and public interest groups. A small but growing number of secular ethicists have also been critical of the ethical implications of patenting. The opposition of indigenous groups to genetic patenting reflects a belief that nature and all biological materials within nature are held as sacred trusts and therefore cannot constitute human property. According to one recent statement, 'nobody can own what

---

[5] Agreement on Trade-Related Aspects of Intellectual Property Rights (TRIPS Agreement) (1994), Section 5: Patents, Article 27 (2), published in a collection of documents compiled by the World Intellectual Property Organization, WIPO Publication No. 223 (E), Geneva, 1997.

exists in nature except nature herself. . . . Humankind is part of Mother Nature, we have created nothing and so we can in no way claim to be owners of what does not belong to us'.[6] The very concept of manipulating nature is also offensive to many traditional cultures. The perspective of one Native American is that 'DNA is not ours to manipulate, alter, own, or sell . . . It was passed on from our ancestors and should be passed on to our children and future generations with its full integrity.'[7] Because indigenous knowledge, culture, and resources are understood to belong in common to the community, members of these communities frequently object to the fundamental principle of patent law, vesting property rights in individuals, especially over living things. That there have been widely publicized examples of 'biopiracy' involving foreigners patenting plants cultivated by indigenous groups without consent or compensation, further alienates many groups. There is particular suspicion about human genetic patenting resulting from several high profile cases. In 1993–1994, for example, more than 30 organizations representing indigenous peoples objected to efforts by the (US) National Institutes of Health (NIH) to patent viral DNA taken from human subjects in Papua New Guinea and the Solomon Islands (Resnik, 1999). The outcry eventually prompted NIH to vacate the patent it received.

Although the religious community has generally been supportive of genetic research and applications, many churches and religious thinkers have been uncomfortable with genetic patenting, especially when the patents are composition of matter (or structure) rather than process patents. Shortly after the landmark 1980 Supreme Court decision allowing the patenting of life forms, the General Secretaries of the National Council of Churches, the United States Catholic Conference, and the Synagogue Council of America sent a letter to President Jimmy Carter warning that control of life forms by any individual or group poses a threat to all humanity:

> We know from experience that it would be naïve and unfair to ask private corporations to suddenly abandon the profit motive when it comes to genetic engineering. Private corporations develop and sell new products to make money, whether these products are automobiles or new forms of life. Yet when the products are new life forms, with all the risks entailed, shouldn't there be broader criteria than profit for determining their use and distribution?[8]

Echoing these concerns, a 1989 World Council of Churches publication, *Biotechnology: Its Challenges to the Churches and the World,* opposed the patenting of life-forms: 'The patenting of life encodes into law a reductionist conception of

---

[6] 'Indigenous Peoples' Statement on the Trade-Related Aspects of Intellectual Property Rights (TRIPS) of the WTO Agreement', signed by indigenous peoples' organizations and networks at the United Nations, Geneva, Switzerland, on 25 July 1999 and reprinted in *GeneWatch*, **12** (October 1999): 10–11.

[7] Debra Harry, the director of the Indigenous Peoples Council on Biocolonialism, was so quoted in Steve Olson (2001).

[8] 'A letter to the President of the United States', reprinted in Panel on Bioethical Concerns of the National Council of the Churches of Christ/USA (1984) *Genetic Engineering: Social and Ethical Consequences*. Pilgrim, New York, USA, Appendix A.

life which seeks to remove any distinction between living and non-living things
... This mechanistic view directly contradicts the sacramental, interrelated view
of life intrinsic to a theology of the integrity of creation' (World Council of
Churches, 1989).

The best known US initiative by the religious community on patenting was the
1995 'Joint Appeal Against Human and Animal Patenting' signed by the titular
leaders of some 80 religious faiths and denominations, including prominent
clergy from Protestant denominations and Catholic and Orthodox churches as
well as Jewish, Muslim, Buddhist, and Hindu leaders. The brief text calls for a
moratorium on the patenting of life:

> We, the undersigned religious leaders, oppose the patenting of human and animal
> life forms. We are disturbed by the U.S. Patent Office's recent decision to patent
> body parts and several genetically engineered animals. We believe that humans and
> animals are creations of God, not humans, and as such should not be patented as
> human inventions.[9]

According to this statement, patents on genes or organisms represent the
usurping of the ownership rights of the sovereign Creator. However, not all
religious thinkers or communities would agree with the claim that patents
amount to a symbolic demeaning or infringement of God's role as Creator.
Theologian Ronald Cole-Turner's view, for example, is that God's ownership
right does not entail exclusion or domination, but instead the right to define the
purpose and value of each creature, as well as to define the moral relationship
among creatures. Stressing that God's ownership is fundamentally different
from ours, Cole-Turner explains that 'God owns all things, not in an exclusive
sense, but precisely in the opposite way, that is, to give all the goodness of
creation as gifts to be shared by all creatures. . . . God's ownership does not
exclude but relativizes and qualifies human ownership' (Cole-Turner, 1999). He
goes on to conclude that there should be no theological objection to biological
patents as long as individuals and corporations exercise their intellectual property rights in a way that is consistent with the purposes that God defines.

Churches in Europe, as well as in the US, have taken strong positions against
patenting biological material. The European Ecumenical Commission for
Church and Society, a multi-disciplinary group of experts from a number of
Protestant, Anglican, and Orthodox national churches and councils of churches,
made several critical submissions to the European Commission and the Parliament at various stages of the drafting of a European Community Directive on
Biotechnology Patenting. Their concerns related both to the process, particularly the lack of adequate consultation with the public, and to substance. Their
critique included the failure to provide a proper basis for the key ethical
judgements implicit in the directive, beyond merely commercial arguments,
which were not deemed adequate. The Commission for Church and Society also

---

[9] 'Joint Appeal Against Human and Animal Patenting', text of the press conference announcement
made available by the Board of Church and Society of the United Methodist Church, Washington,
DC, 17 May 1995.

pointed out that the draft did not mandate the establishment of a body able to assess the ethical aspects of patents.[10]

Another source of opposition to gene patenting on moral grounds is the intuition that it is not appropriate to grant intellectual property rights over humanity's common heritage. In a 1991 letter to the journal *Science*, Hubert Currien, then the French Minister for Research and Technology, argued that 'It would be prejudicial for scientists to adopt a generalized system of patenting knowledge about the human genome. A patent should not be granted for something that is part of our universal heritage' (Currien, 1991). Philosopher Ned Hettinger uses a similar line of reasoning to oppose gene patents. Hettinger (1995) claims that proper appreciation for the three and a half billion year story of the development of life on this planet and respect for the processes of evolution and speciation preclude gene patenting. He goes on to observe that

> Just as it is presumptuous to patent laws of nature, so too it is presumptuous to patent genes, which are equally fundamental to nature. Ideally, gene-types should be treated as a common heritage to be used by all beings who may benefit from them. As previously existing, nonexclusive objects that may be used beneficially by everyone at once, no one should possess the right to monopolize gene-types with patents or to 'lock up' genes through any other property arrangements.

The Council for Responsible Genetics, a US based non-governmental organization, has actively campaigned against genetic patenting since 1995. Its statement 'No patents on life' argues that no individual, institution, or corporation should be able to claim ownership over species or varieties of living organisms or hold patents on organs, cells, genes, or proteins, whether naturally occurring, genetically altered, or otherwise modified (Wilson, 1999). Similarly, its 'Genetic Bill of Rights', intended to protect human rights, privacy, and dignity, states that 'all people have the right to a world in which living organisms cannot be patented, including human beings, animals, plants, and all of their parts' (Council for Responsible Genetics, 2000). Like the Council for Responsible Genetics, many Third World activists have participated in campaigns against patenting designed to affirm the integrity of nature and to protect their resources from exploitation by foreign corporations.

It should be noted that proponents of life patenting rarely contest the substance of the ethical and religious concerns noted here. Their approach generally has been to argue that patents are ethical because the patent system motivates corporations to make investments in expensive medical research that bring wide benefits. Another line of reasoning popular among patent supporters is that patenting should not be equated with ownership but instead merely constitutes the grant of an exclusive right to prevent others from making, using, or selling the protected invention without express permission for a period of 20 years (see

---

[10] 'A Submission to the European Parliament on the Common Position of the draft Directive on the Patenting of Biotechnological Inventions from the European Ecumenical Commission for Church and Society', Strasbourg, 28 March 1998, posted on the website of the Society, Religion and Technology Project, Church of Scotland, http://www.srtp.org.uk/eecept45.shtml

also Chapter 10).[11] But whether patents are a form of ownership depends on one's understanding of property. Here it is important to note that property is not an all or nothing concept; property rights can admit of degrees, and by this line of reasoning patents are a form of ownership (again, see also Chapter 10).

15.4   GENETIC PATENTS AND SCIENTIFIC RESEARCH

Traditionally, the central rationale for intellectual property protection is that incentives and rewards to inventors and creators stimulate economic and social development and thereby benefit human welfare, but an approach that works for artistic works and technology does not necessarily have the same benefits when applied to genetic knowledge. At the least, patenting complicates access and increases the cost to the consumer. The current patent system is also likely to inhibit new product applications of genetic knowledge.

The business community, as well as the patent offices, maintain that innovation in genetic and biomedical research and technology requires or is enhanced by patent protection. Industry representatives typically argue that strong patent protection is the only safeguard that can provide incentives to investors to take the long-term risks required for the development of new biotechnology products. They typically claim that it costs from $300 to $500 million in the US for a single product to move from inception through testing and FDA approval.[12] Writing in *Science* (Doll, 1998), the director of the US Patent Office's biotechnology division offered the following justification:

> Without the incentive of patents, there would be less investment in DNA research and scientists might not disclose their new DNA products to the public. Issuance of patents to such products not only results in the dissemination of technological information to the scientific community for use as a basis for further research but also stimulates investment in the research, development, and commercialization of new biologics. It is only with the patenting of DNA technology that some companies, particularly small ones, can raise sufficient venture capital to bring beneficial products to the market place or fund further research. A strong U.S. patent system is critical for the continued development and dissemination to the public of information on DNA sequence elements.

Many scientists though have questioned the appropriateness and scientific benefits of granting genetic patents, particularly on raw genomic data. The international Human Genome Organisation, often known as HUGO, has taken a very strong position against the patenting of partial and uncharacterized genetic sequences and argued that doing so would impede the development of diagnostics and therapeutics, which is clearly not in the public interest (The Human Genome Organisation, 1995). Both HUGO and the US National Genome Institute have been dedicated to the early release and public availability

---

[11] See, for example, the several articles by corporate representatives in Chapman (1999).
[12] This figure has been contested by patent critics, some of whom believe it is closer to $30 million. The issue is difficult to resolve because it depends on what items are included in the calculations.

of human genetic sequences.

Many individual scientists and organizations view patenting as contrary to the tradition of shared knowledge in scientific discovery. A number of studies have shown that life sciences faculties, particularly those with industry support, are far less likely to share research results with colleagues. Often this information is kept confidential to protect its proprietary value beyond the time required to file a patent (see, e.g., Blumenthal et al, 1996). This has become increasingly problematic as more and more academic scientists and institutions develop relationships with corporations.

Others fear that patenting will impede research and therapeutic applications of the knowledge generated by the Human Genome Project. During the Human Genome Project several major corporations, including Merck, supported efforts to keep genetic sequences in the public domain on a public database that would not be subject to patenting.

But again, there are differences of perspective in the scientific community and corporate world. Some genetic scientists have established or joined biotechnology companies and become part of the rush to patent. Craig Venter, a former staff scientist of the National Institutes of Health, who is now the chief of Celera Genomics, is a prime example. Corporations, such as SmithKline-Beecham and Celera Genomics, have sought patent protection for their sequences, and even corporations that oppose the patenting of genetic sequences, such as Merck, fully endorse patent protection for products developed from that knowledge.

Recently, there have been articles published in both *Science* and *Nature* that argue that current patent policy is likely to jeopardize rather than to facilitate medical progress. The authors share the concern that the proliferation of gene patents is resulting in a welter of claims and counter-claims to the same sequences. According to these analysts, the resulting fragmentation of property rights among too many owners will result in a situation where development of new products will require a complex and difficult bundling of agreements, something researchers may find difficult to achieve. Because each patent holder can potentially block the others and thereby deter applications of the knowledge, conflicting claims will undoubtedly produce legal challenges. Litigation is generally slow and expensive to wage, and the legal process therefore is likely to drain resources and block applications of genetic knowledge. As a consequence, intellectual property rights may lead to fewer useful products for improving human health (Heller and Eisenberg, 1998; Bobrow and Thomas, 2001).

15.5   PATENTING AND RESPECT FOR HUMAN DIGNITY

The concept of the inherent dignity of the human person is the grounding for internationally recognized human rights. It is also a central concept in most Western legal systems. A number of critics have raised concerns about whether the patenting of human genes infringes respect for human dignity. Clearly some potential applications of genetic patenting would be problematic. The 13th Amendment to the US Constitution prohibits the owning and selling of human beings. For this reason an application to patent an entire set of genes of a specific person would violate US law.

Some of those who object to patents on the grounds that intellectual property rights impair human dignity do so because of a commitment to preserve human genetic integrity. They anticipate that DNA sequences, once patented, might be altered either to eliminate flaws or to enhance human potential (see Chapter 14). Here the opposition to eugenics intersects with the patent debate. To protect the dignity and integrity of the person, the European Directive on the Legal Protection of Biotechnological Inventions, excludes the following from patentability: (i) processes for cloning human beings; (ii) processes for modifying the germ line genetic identity of human beings (i.e. processes for producing inheritable genetic modifications); and (iii) uses of human embryos for industrial or commercial purposes.[13]

The most immediate issue is whether patenting knowledge of a portion of DNA constitutes a direct threat to human dignity. Some opponents of patenting argue that it does and therefore advocate that protections should extend to human tissue, body parts, and genetic information (see, e.g., Mitchell, 1999). Most philosophers and theologians (e.g., Peters, 1999) disagree because they believe that the concept of human dignity applies to the whole person and not to component parts. Another distinction that some analysts propose is the difference between the status of human material in the body and outside of it. Thus, even if they recognize a moral basis for excluding patenting of human material, they argue that these protections do not extend to patenting *ex vivo* DNA sequences.

A related issue raised by religious thinkers and also by some secular ethicists is the concern that patenting demeans and commodifies life. Commodification is the process by which something previously valued in a non-economic manner comes to be understood as a commodity, that is, the appropriate subject of free market transactions. The legal theorist Margaret Jane Radin (1996) distinguishes between literal or narrow and broad or metaphorical senses of commodification. It is the latter, a worldview that conceives of human attributes as owned objects even where no money literally changes hands, that some critics fear will undermine the Kantian conception of the person as an end rather than as a means. As one religious critic pointed out, 'The patenting of genes, the building blocks of life, tends to reduce it to its economic worth.'[14] For some the 'conflict is between reverence for life and exploitation of life, life valued for its marketability and life valued as an intrinsic gift.'[15]

I concur with the view that there are categories of things, life forms in

---

[13] 'Directive 98/44/EC of the European Parliament', paras. 40, 41.
[14] Quoted by Ted Peters (1996).
[15] The statement by Bishop Kenneth Carder was made available to this author by the United Methodist Board of Church and Society.

particular, that by their very nature should not be treated as economic commodities. To do so is to adopt an instrumental or utilitarian perspective, which is to treat life as a means rather than as an end. Here I think it is useful to refer to Michael Walzer's concept of 'blocked exchanges' (Walzer, 1983). He notes that there are entities about which society has determined distribution should not be on an economic basis. His list of blocked exchanges, or things that cannot be bought and sold, includes human beings, criminal justice, freedom of speech, press, religion, and assembly, exemptions from military service or jury duty, political offices, and love and friendship. He does not, however, specifically mention genes, human tissue, or body parts, possibly because his book *Spheres of Justice* was published in 1983. Taking Walzer's insight, I believe that patenting is incompatible with reverence for life and human dignity.

15.6   CONCLUSION

Human genetic patenting as currently applied is counterproductive and raises serious ethical issues. To date patent regulations in the US and Europe have evolved through dialogue within a limited circle in which commercial interests have been favored and those representing the broader public interest virtually excluded. The result has been a tendency for the patent system to increasingly extend its sway to encompass life forms by lowering the requirements for novelty, inventiveness, and utility (Bobrow and Thomas, 2001).

What is needed now is nothing short of a major reform of the patent system, preferably through an international policy forum with broad representation of views. The nature of the patent system is too important to leave to the patent examiners to determine on narrow technical and economic grounds. Far from being an obscure technical issue, these standards will have major impacts on the lives and well-being of current and future generations. In undertaking such a review, there needs to be a reexamination of the costs and benefits, both economic and ethical, of intellectual property regimes, particularly how the current system applies to raw genomic information. The international character of genetic research underscores the need to develop common intellectual property standards. Otherwise individuals and corporations may seek patents in the venue with the lowest standards with other patent offices then pressured to follow suit. It would be important for the reform of the patent system to reflect the principles in The Universal Declaration on the Human Genome and Human Rights,[16] particularly the following two articles:

> Article 14: States should take appropriate measures to foster the intellectual and material conditions favourable to freedom in the conduct of research on the human genome and to consider the ethical, legal, social and economic implications of such research. . . .

[16] *Universal Declaration on the Human Genome and Human Rights*, 1997, 1999.

Article 18: States should make every effort, with due and appropriate regard for the principles set out in this Declaration, to continue fostering the international dissemination of scientific knowledge concerning the human genome, human diversity and genetic research and, in that regard, to foster scientific and cultural co-operation, particularly between industrialized and developing countries.

Certainly there are many measures that could be undertaken even within the current systems that would impose more stringent criteria based on ethical considerations. The principle of broad public access to and benefit from genetic information must be protected. As the UNESCO drafted Universal Declaration on the Human Genome and Human Rights recognizes, the human genome underlies the fundamental unity of all members of the human family, as well as the recognition of their inherent dignity and diversity. 'In a symbolic sense, it is the heritage of humanity'.[17] Implementing the ethical principles in the European directive and in TRIPS in a meaningful way could be a starting point. Doubtless that would require the establishment of mechanisms that would be competent to make ethical evaluations as part of the patent process. None of this will happen, however, without awareness of the problems in the current system and considerable public pressure.

REFERENCES

Blumenthal, D., Campbell, E.G., Causino, N. and Louis, K.S. (1996) Participation of life science faculty in research relationships with industry. *New England Journal of Medicine*, **335**, 1734–1739.

Bobrow, M. and Thomas, S. (2001) Patents in a genetic age: the present patent system risks becoming a barrier to medical progress. *Nature*, **409**, 763–764.

Chapman, A.R. (ed.) (1999) *Perspectives on Genetic Patenting: Religion, Science and Industry in Dialogue*. American Association for the Advancement of Science, Washington, DC, USA.

Cole-Turner, R. (1999) Theological perspectives on the status of DNA: a contribution to the debate on genetic patenting. In *Perspectives on Genetic Patenting: Religion, Science and Industry in Dialogue*, Chapman, A.R. (ed), American Association for the Advancement of Science, Washington, DC, USA, pp. 149–166 (see especially p. 152).

Council for Responsible Genetics (2000) The Genetic Bill of Rights. *GeneWatch*, **13**, 2–3.

Currien, H. (1991) The human genome project and patents. *Science*, **254**, 710.

Doll, J.J. (1998) The patenting of DNA. *Science*, **280**, 689–690.

Drahos, P. (1999) Biotechnology patents, markets and morality. *European Intellectual Property Review*, **21**, 441–449 (see especially p. 444).

Eisenberg, R.S. (2000) Re-examining the role of patents in appropriating the value of DNA sequences. *Emory Law Journal*, **49**, 783–800 (see especially p. 785).

Gold, E.R. (1996) *Body Parts: Property Rights and the Ownership of Human Biological Materials*. Georgetown University Press, Washington, DC, USA (see pp. 81–83).

Heller, M.A. and Eisenberg, R.S. (1998) Can patents deter innovation? The anticommons in biomedical research. *Science*, **280**, 698–700.

Hettinger, N. (1995) Patenting life: biotechnology, intellectual property, and environmental ethics. *Environmental Affairs*, **22**, 267–305.

[17] *Universal Declaration on the Human Genome and Human Rights.* 1997, 1999.

The Human Genome Organisation (1995) *HUGO Statement on Patenting of Genomic Sequences.* HUGO, Bethesda, MD, USA.

Marshall, E. (1997) Companies rush to patent DNA. *Science*, 275, 780–781.

Mitchell, C.B. (1999) A Southern Baptist looks at patenting life. In *Perspectives on Genetic Patenting: Religion, Science and Industry in Dialogue*, Chapman, A.R. (ed), American Association for the Advancement of Science, Washington, DC, USA, pp. 167–188.

Olson, S. (2001) The genetic archaeology of race. *The Atlantic Monthly*, **287**, 69–79.

Peters, T. (1996) Should we patent God's creation? *Dialogue*, **35**, 117–132.

Peters, T. (1999) DNA and dignity: a response to Baruch Brody. In *Perspectives on Genetic Patenting: Religion, Science and Industry in Dialogue*, Chapman, A.R. (ed), American Association for the Advancement of Science, Washington, DC, USA, pp. 127–136.

Poland, S.C. (2000) Genes, patents and bioethics: will history repeat itself? *Kennedy Institute of Ethics Journal*, **10**, 265–281.

Radin, M.J. (1996) *Contested Commodities: the Trouble with Trade in Sex, Children, Body Parts and Other Things.* Harvard University Press, Cambridge, MA, USA (see pp. 12–13).

Resnik, D. (1999) The human genome project: ethical problems and solutions. *Politics and the Life Sciences*, **18**, 15–23.

Salazar, S. (1999) Intellectual property and the right to health. In *Intellectual Property and Human Rights.* World Intellectual Property Rights Organisation and Office of the United Nations High Commissioner for Human Rights, Geneva, Switzerland, pp. 65–92 (the relevant passage is on p. 76).

Stone, R. (1995) Genetic engineering: religious leaders oppose patenting of genes and animals. *Science*, **268**, 1126.

Walzer, M. (1983) *Spheres of Justice: a Defence of Pluralism and Equality.* Basic Books, New York, USA (see pp. 100–103).

Wilson, K. (1999) CRG says: no patents on life! *GeneWatch*, **12**, 11.

World Council of Churches (1989) *Biotechnology: its Challenges to the Churches and to the World.* World Council of Churches Sub-Unit on Church and Society, Geneva, Switzerland.

16   Cloning of Animals and Humans

Harry Griffin

16.1   INTRODUCTION: A STAR IS BORN . . .

New reproductive technologies have attracted great interest ever since the birth of the first test tube baby in 1978. The current intense media and public fascination with cloning stems, however, from the birth not of a child, but of a sheep. Dolly was the first mammal cloned from an adult cell and she was born at the Roslin Institute near Edinburgh, Scotland on 5 July 1996 (Figure 16.1).

Dolly was derived from cells that had been taken from the mammary gland of a six year old Finn Dorset ewe and cultured in the laboratory. Individual cells were then fused with unfertilised eggs from which the maternal nucleus had been removed. Two hundred and seventy-seven of the 'reconstructed embryos' – each now with a diploid nucleus from the adult animal – were cultured for 6 or 7 days within the uterus of 'temporary recipient' ewes and then recovered by surgery. Twenty-nine of the embryos that appeared to have developed normally were implanted into 13 surrogate Scottish Blackface ewes. One became pregnant and gave birth to a live lamb, Dolly, some 142 days later (Wilmut et al, 1997).

When Dolly was announced to the world in February 1997, Roslin was besieged by the media (Wilmut and Griffin, 1997). Journalists and TV crews flew in from around the world and Dolly quickly became the most photographed sheep of all time. The Pope condemned cloning outright and President Clinton called on his recently established National Bioethics Advisory Committee to report on the ethical and legislative implications within 90 days. Dolly Parton, the actress and Country and Western singer, said that she was 'honoured' to have a sheep named after her and that there was no such thing as 'baaaed' publicity.

The immediate assumption was that a cloned human child could not be far behind and this triggered an explosion (at least in the media) of fears about the future. Clones would be created, it was said, for example as sources of 'spare parts', by dictators seeking immortality or to 're-create' dead children. However, the small size of the Roslin Institute's post bag suggested that the general public understood that most of the scenarios that the press were imagining (Figure

Bioethics for Scientists. Edited by John Bryant, Linda Baggott la Velle and John Searle.
© 2002 by John Wiley & Sons Ltd.

Figure 16.1.   Dolly as a lamb, with her Scottish Blackface mother

16.2) were unlikely to be realised and could, as a consequence, be enjoyed in safety – just like the *X Files*. A large part of the attraction for the media was that the story could be written from so many angles: articles appeared on the news, science, environment, health and business pages. Even the racing correspondents had their chance when cloning was proposed as a solution to the infertility of US champion racehorse Cigar.

The media's fascination with cloning has continued ever since, encouraged by a steady diet of newsworthy stories. With Dolly being a 'clone alone', some scientists speculated that she was not repeatable or that she was a 'mistake' or even a 'fake'. Publication of additional DNA evidence and news of the cloning of the first adult mice killed that particular angle. Stories followed about attempts to clone mammoths or to resurrect extinct species of birds and, slightly more credibly, to rescue endangered species. In October 1998, the publication of a paper describing the isolation of the first human embryonic stem cells (Thomson et al,1998) initiated a vigorous new debate about the possibility of cloning human embryos for stem cell therapy. Several groups claimed to be preparing for cloning a child, including elderly Chicago physicist Dr Richard Seed (see Cohen, 1998), a company, Clonaid, with ties to the previously unknown Raelian sect (www.clonaid.com) and, more recently, Italian IVF expert Professor Severino Antinori (see Abbott, 2001).

Figure 16.2. A few of the thousands of headlines on cloning

16.2 ETHICAL DEBATE IN THE MEDIA SPOTLIGHT

The origin of news stories in the media makes fascinating study. Investigative journalism is rare and almost all coverage originates from the hundred or so press releases sent to each journalist each week. The steady succession of cloning stories over the past four years represents the interplay of a wide range of vested interests: journalists wanting to make an impact with their editors, science journals competing with each other for subscriptions, new biotechology companies wanting to attract investors, pro-life and patients' groups seeking to influence public opinion, authors promoting their latest books, embryologists seeking 'immortality' and scientists and ethicists seeing a new opportunity to advance their careers.

Although much of the early coverage was more to do with providing entertainment than information, it is clear that cloning raised genuine disquiet among the vast majority of the population. The possibility of creating a child that was the genetic copy of an existing individual was seen as very different from assisted reproduction that, for all its sophistication, still represents the union of an egg and a sperm. Creation of a sense of urgency is almost obligatory for news reporters, but there is little doubt that many also saw cloning as yet another example of science progressing too far, too fast.

As current attitudes in Europe to genetically modified (GM) crops demonstrate (see Chapters 1, 2, 8 and 9) it is very much in the interest of scientists to engage in public discussion about the impact of new technologies earlier rather than later. Discussion directly with the public can engage only a tiny fraction of the population: if scientists are to improve the quality of debate about new technologies, then we have little alternative but to learn how to make our views heard through the media. This in turn means accepting its limitations: a focus on 'news', 30 second soundbites and an adversarial approach to debate. We also need to be prepared to comment on a much broader range of issues than our own narrow speciality and to address the social impact of our work in terms the public understands.

Like most scientists, I have had no formal training in ethics but over the past four years I have participated in a fair number of ethical debates around the world. One observation – perhaps now obvious to readers of this book – is that professional ethicists are most effective when they provide the rest of us with insight into how we might come to our own conclusions. The individual that had the most lasting impact on my own views was Professor Albert Jonsen of the University of Washington and formerly chair of the National Advisory Board on Ethics and Reproduction in the US. In a talk at the Institute of Bioethics in Madrid in 1988, he proposed three steps to ensure a disciplined approach to exploring the ethical implications of cloning. His advice seems equally applicable to any new technological advance.

• Firstly, to determine exactly what the scientists involved in the research are doing and what they believe they can do: in essence to separate sensible speculation from fantasy and develop an honest appreciation of the range and limits of the science.
• Secondly, to consider a limited number of applications in depth. The applications chosen should be realistic scenarios that have a significant impact on society rather than simply curiosities.
• Thirdly, to examine the ethical implications of each application by imaginative use of analogies from our own experience. In the case of the question of confused genetic identity of clones, for example, to refer to our experience of identical twins or to foster children.

The structure of the rest of this chapter attempts to follow these themes. It first describes the current state of the art of cloning and what might be possible with the technology in the foreseeable future. It then moves on to consider three possible applications – the cloning of farm animals, the cloning of children for infertile couples and the cloning of embryos in stem cell research.

16.3   THE CURRENT STATE OF THE ART

16.3.1   A BRIEF HISTORY OF NUCLEAR TRANSFER

Cloning simply provides genetic copies. Gardeners clone when they take cuttings, 'Nature' provides clones in the form of identical twins, molecular biologists clone genes and cattle breeders have been producing cloned calves by 'embryo splitting' for over 20 years.

Cloning by nuclear transfer is not novel. The technique was first reported in frogs in 1952 and has been used widely since in amphibians to study early development (see McKinnell, 1985, for a very readable review). These early studies showed that the first few cell divisions after fertilisation produce cells that are totipotent (i.e. they can develop into all of the cell types that make up the whole animal). As the embryo develops, the cells lose this property and the success of nuclear transfer rapidly declines. Some nuclear transfer experiments using cells from adult frogs produced viable embryos, but these never developed beyond the tadpole stage.

Nuclear transfer in mammals proved to be more difficult, in part because mammalian oocytes ('eggs') are so much smaller than those of amphibians. The cloning of mice using nuclei from very early embryos was *claimed* in 1977, but this work was not repeatable and interest among developmental biologists waned. Research on nuclear transfer in cattle continued, stimulated by the prospect of large commercial benefits of copying the very best performing animals. By the middle of the 1980s several research groups from around the world had produced cloned sheep and cattle by transferring nuclei directly from cells taken from very early embryos. Steen Willesden, working for Granada Genetics in the US, had produced live calves by nuclear transfer from embryos that had progressed to the 64- and 128-cell stage and this was the first suggestion that nuclear transfer in mammals was possible from at least partially differentiated cells. Many of the calves were larger than normal and had to be delivered by Caesarian section (see Kolata, 1997, Wilmut et al, 2000, for a full history of nuclear transfer up to the birth of Dolly).

In 1995, Keith Campbell, Ian Wilmut and colleagues produced live lambs – *Megan* and *Morag* – by nuclear transfer from cells from early embryos that had been cultured for several months in the laboratory. In 1996, they produced four lambs from embryo cells, three from foetal cells and, in collaboration with PPL Therapeutics, one, subsequently named Dolly, from an adult cell (see Figure 16.3).

16.3.2 THE SIGNIFICANCE OF CLONING FROM AN ADULT ANIMAL

For developmental biologists, the ability to clone from adult animals overturned one of the fundamental tenets of developmental biology. Most scientists had believed that in most, if not all vertebrates, differentiation – the gradual process of specialisation that allows the fertilised egg to develop into the 200 or so cell types that make up the whole animal – was irreversible. After all, even over a 90 year lifespan a liver remains a liver, a nerve cell a nerve cell. The production of a live lamb from a cell taken from the udder of a 6 year old ewe demonstrated that differentiated animal cells are not after all immutable.

16.3.3 SUBSEQUENT PROGRESS

At first Dolly was a 'clone alone' but in August 1998, a group in Hawaii published a report of the cloning of over 50 mice by nuclear transfer (Wakayama et al, 1998). Since then, research groups around the world have reported the

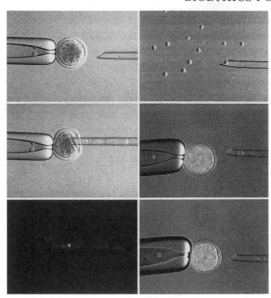

Figure 16.3. Cloning by nuclear transfer. The six images were taken through the lens of a microscope connected to a micromanipulator rig and show the key steps in nuclear transfer: *Top left*: an unfertilised sheep oocyte is held by gentle suction on the end of a broad-tipped pipette. *Middle left*: a fine pipette is inserted into the egg and the maternal nuclear DNA removed by suction with some of the cytoplasm. *Bottom left*: under UV light, the maternal DNA is shown in the pipette rather than in the oocyte. *Top right*: diploid mammary gland cells in culture. *Middle right*: one diploid cell has been picked up in the pipette and inserted into the enucleated oocyte. *Bottom right*: enucleated oocyte containing intact diploid cell. A short electric pulse is then used to fuse the membranes of the two cells together and to stimulate the single-celled 'reconstructed embryo' to begin to divide and multiply.

Simple? The creation of Dolly required the collection of 430 oocytes from over 40 super-ovulated ewes and these were used to create 277 'reconstructed embryos'. These were then surgically transferred to the uteri of temporary recipients and recovered six to seven days later, again by surgery. Only 29 embryos appeared to have developed normally to the blastocyst stage, and when these were implanted in 13 surrogate mothers only one became pregnant

successful cloning of adult cattle, sheep, mice and goats. Equally competent groups have had no success in cloning rabbits, rats or monkeys.

There are differences in early development between species that might influence success rate. In sheep and humans, the embryo divides to between the eight- and 16-cell stage before nuclear genes take control of development, but in mice this transition occurs at the two-cell stage. In 1998, a Korean group claimed that they had cloned a human embryo by nuclear transfer but their experiment was terminated at the four-cell stage (as seen in a British TV documentary in the BBC's *Panorama* series broadcast in February 1999) and more recently there has been a report of the cloning of several early-stage human embryos in a commercially-funded lab in the USA (Cibelli et al, 2001).

Success rates remain low in all species, with between 1 and 4% of 'reconstructed embryos' leading to live births as reported in the published data (see Pennisi and Vogel, 2000, for a detailed review). With unsuccessful attempts at cloning unlikely to be published, the actual success rate is certainly substantially lower. Many cloned offspring die late in pregnancy or soon after birth, often through respiratory or cardiovascular dysfunction. Abnormal development of the placenta is common and this is probably the major cause of foetal loss earlier in pregnancy. Many of the cloned cattle and sheep that are born are much larger than normal and apparently normal clones may have some unrecognised abnormalities.

The high incidence of abnormalities is not surprising. Normal development of an embryo is dependent on the methylation state of the DNA (i.e. the pattern of active and potentially active genes) contributed by the sperm and egg and on the appropriate reconfiguration of the chromatin structure after fertilisation. Somatic cells have very different chromatin structure to sperm and 'reprogramming' of the transferred nuclei must occur within a few hours of activation of reconstructed embryos. Incomplete or inappropriate reprogramming will lead to dysfunctional regulation of gene expression and failure of the embryo or foetus to develop normally or to non-fatal developmental abnormalities in those that survive.

Improving success rates is not going to be easy. At present, the only way to assess the 'quality' of embryos is to look at them under the microscope and it is clear that the large majority of embryos that are classified as 'normal' do not develop properly after they have been implanted. A substantial effort is now being made to identify systematic ways of improving reprogramming. One focus is on known mechanisms involved in early development, and in particular on the 'imprinting' of genes. Another is to use technological advances in genomics to screen the expression patterns of tens of thousands of genes to identify differences between the development of 'reconstructed embryos' and those produced by *in vivo* or *in vitro* fertilisation.

It is important to recognise the limitations of nuclear transfer. Plans to clone extinct species have attracted a lot of publicity: an Australian project aims to resurrect the 'Tasmanian tiger' by cloning from a specimen that has been preserved in a bottle of alcohol for 153 years and another research group announced plans to clone a mammoth from 20 000 year old tissue found in the Siberian permafrost. However, the DNA in such samples is hopelessly fragmented and there is no chance of reconstructing a complete genome. In any case, nuclear transfer requires an intact nucleus, with functioning chromosomes. DNA on its own is not enough: many forget that the Michael Crichton novel *Jurassic Park* was a work of fiction.

Other obvious requirements for cloning are an appropriate supply of oocytes and surrogate mothers to carry the cloned embryos to term. Cloning of endangered breeds will be possible by using eggs and surrogates from more common breeds of the *same* species. It may be possible to clone using a closely related species but the chance of successfully carrying a pregnancy to term would be increasingly unlikely if eggs and surrogate mothers are from more distantly related species. Proposals to 'save' the Panda by cloning, for example, would seem to have little or no chance of success because it has no close relatives to supply eggs or to carry the cloned embryos.

16.4   APPLICATIONS OF CLONING

Cloning by nuclear transfer has a wide range of possible applications. It is already being used as a research tool to understand the mechanisms involved in differentiation and de-differentiation of cells. It could, in principle, be used to produce unlimited numbers of genetically identical animals to improve, for example, the sensitivity of testing regimes for new pharmaceuticals or vaccination regimes (see Chapter 18).

Our own motivation for cloning was to develop better ways of genetically modifying farm animals. Until recently the only way to produce transgenic cattle, sheep or pigs was by injecting a suitable DNA construct directly into one of the pro-nuclei in a recently fertilised egg (see also Chapter 14). The injection tends to damage the genomic DNA and during the repair processes one or more copies of the injected construct can be incorporated into the animal's genome. When the injected embryos are implanted perhaps 2–4% of them give rise to transgenic offspring. Pro-nuclear injection provides no control of the number of copies of the construct or where they are incorporated and as a consequence many of the transgenic animals fail to express the gene at high enough levels. Importantly, pro-nuclear injection only allows genes to be added. Nevertheless, pro-nuclear injection has been used to create transgenic sheep and cattle that produce human proteins in their milk and transgenic pigs for possible use in xenotransplantation (i.e. as a possible source of organs for transplantation into humans).

More sophisticated genetic modifications are possible in mice using embryonic stem (ES) cells but, despite much effort, no ES cells have yet been isolated from farm animal species. Nuclear transfer allows cultured cells to be converted into live animals: if the cells are genetically modified first, then the cloned animals produced will also be genetically modified. The first transgenic sheep produced by nuclear transfer were born in 1997 and carried a gene coding for human blood clotting factor IX. Nuclear transfer has since been used to insert genes at a defined locus in the genome and such 'targeted insertion' is seen as a more reliable way of ensuring high levels of transgene expression (see Chapter 14).

All of these applications represent yet another use of animals in research and in the UK would be regulated under the Animals (Scientific Procedures) Act 1986. The ethical issues surrounding animal experimentation are covered in Chapters 6, 7 and 18; here we concentrate on the ethical implications raised specifically by cloning in cattle breeding, in human reproduction and in stem cell research.

16.5   CLONING IN CATTLE PRODUCTION

16.5.1   RATIONALE

Beef and dairy cattle account for 40% of farm income in the UK and cattle are the most important farm animal species worldwide. Much of the genetic progress in the breeding of cattle is made in relatively small elite herds, with the genetic merit of individuals being assessed on a combination of the animal's own performance and on the performance of its near relatives. Progress that is made within elite herds is then passed on to commercial farmers by use of the elite semen in artificial insemination. Semen provides only half the genes to the next generation and, as a consequence, the performance of the average dairy cow in the UK is estimated to be some 10 years behind the very best. With cloning, it would be possible to remove this difference within one generation. Farmers who could afford it would buy embryos that would be clones of the cows with greatest genetic merit. Frozen cloned embryos would be delivered to the farm much in the same way as semen is today, perhaps from breeders overseas.

In 1997, the UK's Ministry of Agriculture, Fisheries and Food commissioned the Farm Animal Welfare Council (FAWC) to report on the ethical implications of cloning in farm animal production. At the time, this request seemed premature. However, a large number of calves have since been cloned by research groups in Japan, New Zealand, France, Germany and the US and two cloned cattle were recently sold at auction in Wisconsin for over $40 000 each. Such prices represent the animals' novelty rather than their true economic value and costs would need to be brought down to about $100 for each elite embryo before cloning would be economically attractive for the average farmer.

16.5.2   ETHICAL CONCERNS

The FAWC's report (FAWC, 1998) listed a number of requirements that it considered would need to be fulfilled before cloning should be used routinely in livestock production. These included the elimination of the 'large calf syndrome', the avoidance of 'temporary recipients', as in the protocol that produced Dolly, and the use of non-surgical techniques for recovering oocytes and implanting the cloned embryos. The report also raised concerns about the dangers of inbreeding.

The conditions that FAWC sought to impose to protect the welfare of animals are exactly the same as those that would need to be met for cloning to be economically viable and are already part way to being met. *In vitro* maturation of oocytes is a well developed art in cattle (though not in sheep) and ovaries from the slaughterhouse provide an unlimited supply of eggs for nuclear transfer. Protocols for *in vitro* culture of embryos to the blastocyst stage are similarly well established in cattle and non-surgical implantation of embryos is already used routinely. The relatively low success rate and high incidence of developmental abnormalities remains a major challenge and can probably only be consistently addressed when much more is known about the mechanisms involved in reprogramming and the adverse effects of *in vitro* embryo culture.

If this research is successful, in 20 years time a majority of cattle in developed countries might be clones. The risks of inbreeding are well known to livestock breeders and there would need to be clones of many different elite animals to ensure that whole herds are not going to be vulnerable to diseases or nutritional inadequacies. Mistakes can occur. The widespread use of semen from a single Holstein Friesian bull from the US led to up to 25% of some herds suffering from bovine lymphocyte adhesion deficiency (BLAD) syndrome and the allele responsible had to be eliminated from affected herds by marker-assisted selection (Shuster et al, 1992). Cloning could increase the risk of such undetected defects and the proposal by FAWC that specific rules should be introduced to protect genetic diversity is a sensible precaution.

A more fundamental objection to cloning in livestock production was voiced by some who saw it as yet another step towards treating animals merely as machines, with the Church of Scotland viewing widespread cloning of livestock as a 'step too far' (Bruce and Bruce, 1998). Another concern was that cloning would provide yet another way for farmers to push animals even closer to physiological limits in their drive for productivity. This later criticism is of breeding goals rather than of the methods used to achieve them: cloning could be used, for example, to more rapidly disseminate genetic improvement in disease resistance, to the benefit of the animals as well as to farmers.

If cloning is to become practical in the future, then the techniques used in cloning would be little different in terms of their impact on animal welfare from methods already in widespread use in livestock production such as artificial insemination, multiple-ovulation embryo transfer (MOET) and cloning by embryo splitting (see MAFF, 1995). However, attitudes to farming practices are changing in the affluent part of the world and practices that were acceptable in the past may not necessarily be so in the future.

16.6 CLONING A CHILD

When Dolly was announced to the world in February 1997, the prospect of using the same technology to clone a child was met with almost universal condemnation. UNESCO declared that the cloning of a child would be contrary to 'human dignity' and the prospect was variously described as 'grotesque', 'revolting', 'appalling' and 'a nightmare scenario'. Opinion polls indicated the large majority of the public agreed.

The overwhelming evidence from the cloning of animals is that any attempt now at reproductive cloning in humans would be unlikely to succeed, and if a pregnancy were established that there would be a high risk of death of the child late in pregnancy or soon after birth. Dr Zavos, one of the consortium led by Professor Antinori, has implied that IVF clinics are able to screen human embryos for abnormalities (Stern, 2001). Methods are available for detecting major chromosomal abnormalities in early embryos but there are none that could detect the overall epigenetic state of the genome that appears to be critical for normal development of reconstructed embryos. In such circumstances, any group that intends to attempt to clone a child is being reckless. Some have correctly pointed out that fears on safety were used in the 1970s to try to prevent IVF. However, few experiments were carried out in animals before IVF was tried in humans and, while there was little evidence that IVF was safe, there was no evidence that it was unsafe either.

Research on cloning of animals is expected to improve success rates and reduce the incidence of developmental abnormalities and an inevitable question that will arise in the perhaps not too distant future is: is this technology now safe for use in humans? Those opposing new technologies often develop Olympian standards of safety to prevent developments (see, for example, the discussion about GM crops in Chapters 8 and 9), but an obvious threshold would be when abnormalities in cloned animals are reduced to the level of those in animals produced by natural mating. Even this may be too high a standard: no-one suggests, for example, that older women should be prevented from having children even though it is well known that the risk of Down's syndrome increases markedly with age.

With improvements in success rate likely, societies need to have developed clear and defendable public policies to address the inevitable pressure from prospective parents or IVF clinics to allow cloning. In Europe it seems safe to predict that public sentiment will remain opposed to cloning and that few (if any) parliamentarians will be prepared to campaign for a change in the law on behalf of the handful of parents that might benefit. By contrast, the emphasis in the USA on the right of the individual and the general antipathy to regulation by the state is likely to put the burden of proof on those wanting to restrict cloning.

One of the difficulties in translating the immediate visceral reaction (gut reaction or 'yuk' factor) against reproductive cloning into a more logical view of the rights and wrongs is the wide variety of scenarios envisaged for its use. In his book *Remaking Eden* (1997) for example, Professor Lee Silver saw cloning being used in the future to create 'designer babies' with improved intelligence and physique and the gradual emergence of two types of human: the privileged 'genrich' and the unimproved rest. Such speculation almost certainly misrepresents what science will be able to do. While we know that intelligence is an inherited trait, it is controlled by several if not many genes and proving that any one specific gene has a significant effect would depend on our willingness to carry out long term, controlled experiments on hundreds if not thousands of children.

While this might make a good story line in an episode of *X Files*, it is unlikely ever to be acceptable in real life. The sort of future imagined in Professor Silver's book also misrepresents human motivation. Cloning by nuclear transfer is not going to be any more efficient than IVF: the idea that many (or any) women would be prepared to go through several rounds of an IVF-like protocol on the off-chance that her child would have two to three more IQ points seems a very *male* fantasy. More importantly, such speculation into the distant future distracts from real ethical issues that we should be facing up to now.

Some have suggested that cloning would allow couples to 'replace' a child that had tragically died or to allow lesbian couples to have children without involving a man. Interestingly, a study by the Wellcome Trust (Wellcome Trust, 1998) found no evidence that either lesbians or women who had lost children through miscarriage showed any more enthusiasm for cloning than the rest of the population. In considering such cases, it may well be difficult to separate our views on cloning from our attitudes to the decline in 'family values' or the misguided logic of the parents who believe they could recreate their dead child.

A simpler and therefore perhaps more useful case would be the married couple in their early 30s. The woman is fertile but the man's testes failed to develop properly in childhood so that he cannot produce sperm. They both desperately want a child but have rejected anonymous sperm donation. They view the cloning of a son using a donor cell from the 'father' as an acceptable option, since it would provide them with a child that both partners had contributed to creating. For the purpose of this study let us look forward to the year 2020 by which time the success rate of cloning in animals has improved dramatically. The IVF clinic involved is well respected and is deliberately avoiding publicity.

Arguments used against cloning in such circumstances tend to focus on the interests of the child. UNESCO, for example, has proposed that cloning violates a basic human right to a 'unique genetic identity' and to 'an open future'. Critics of this line of argument rightly point out that this 'right' is not one that Nature accords to identical twins, who in any case can and do grow up with personalities and behaviours all of their own (Harris, 1998; Pence, 2000). The concept of an 'open future' is also going to be increasingly called into question as we understand more about the genes that influence our health and well being. Nevertheless, the factors that influence our sense of identity are complex and it may well be that cloned children will believe that their uniqueness has been compromised.

Some have argued that a cloned child would suffer psychological harm because of confused and ambiguous relationships with other members of the family. Parents might have unfilled expectations of the child based on their experience of the partner from which he or she was cloned. The mother of a cloned son might have difficulties in reconciling her relationship between the clone she married and the clone she carried. Counselling of prospective parents might reduce the risk but others have suggested that the confused heritage of clones would be little different from children who have been fostered or adopted (Green, 1999). However, fostering and adoption are devices to accommodate unforeseen and unplanned circumstances – they are attempts to make the best of an otherwise intolerable situation (O'Neill, 2000; see also Chapter 12). By contrast, the confused heritage created for the clone child is the result of a deliberate act and one that the child might resent.

Some have asserted that the most important thing for a child is the unconditional love of its parents, rather than the manner or circumstances of its birth. Experience with anonymous sperm donation suggests that those of us who have never had the need to question our genetic inheritance can underestimate its importance. Two thousand children have been conceived in this way in the UK but only a very small minority of their parents have found it possible to tell their child of his or her origins and children who do find out inadvertently or later in life find the knowledge deeply disturbing. This experience suggests that cloned children will have, on average, rather more to cope with in their adolescence than the normally conceived child and, although some parents might be well equipped to handle the additional challenges involved, most will find it very difficult.

Although considerations of safety and the interests of the child are important, they in no way account for the strength of feeling expressed by so many against reproductive cloning. Leon Kass attempted to address this discrepancy by asserting that 'repugnance is the emotional expression of deep wisdom, beyond reason's power to articulate' (Kass, 1997) and Timothy Renwick invited comparison of cloning with other human activities that provoke or have provoked bitter condemnation, including incest, bestiality and inter-racial marriage (Renwick, 1998). Society's condemnation of incest has some similarities to cloning in that it results from knowledge of its genetic consequences and the disruption of family relationships and is supported by a combination of social mores, religious taboos and legislation in all societies. By contrast, inter-racial marriages, which have provoked extreme reactions in the past, for example, in the Southern USA, are now rightly accepted as a part of the richness of modern life.

Is our opposition to cloning one that will stand the test of time or will it be like our attitudes to homosexuality or IVF, where the majority do not wish to become involved themselves but will accept – perhaps reluctantly – the rights of others to do so?

16.7 CLONING OF HUMAN EMBRYOS

16.7.1 STEM CELL THERAPY

Many common degenerative diseases are a consequence of the failure of just one of the 200 or so types of cell that make up our bodies. *Type 1 diabetes*, for example, is caused by the failure of the islet cells in the pancreas to produce enough insulin. *Parkinson's disease* is largely a consequence of the inability of certain neurones in the brain to produce enough neurotransmitter. The longer term effects of *strokes* and *heart attacks* are the result of the death of a relatively small number of brain or heart cells downstream of the blocked blood vessels.

Attempts to treat such diseases using drugs have had limited success and research groups around the world have begun to develop novel treatments based on transplants of healthy cells. Injection of neuronal stem cells into the brains of rats with experimentally induced stroke has been shown to cause significant improvement in performance. The transplanted cells seem to be recruited into the existing repair mechanisms, migrating small distances within the damaged tissue and converting to the most appropriate cell type through the action of locally produced growth factors. In studies in mice, cardiomyocytes injected into the damaged areas of the heart after experimentally induced ischaemia have been shown to integrate with neighbouring healthy tissue and function normally. Results such as these are encouraging the search for cell therapies for a much wider range of degenerative diseases including spinal cord injury, congestive heart failure, osteoarthritis and osteoporosis, hepatitis and muscular dystrophy.

The cells being used in these early experiments come from a variety of sources. Parkinson's disease patients, for example, have been treated with cells from human or pig foetuses. The so-called 'Edmonton protocol' involves infusion of diabetic patients with a suspension of pancreatic islet cells obtained from cadavers. These sources are clearly not practical for routine treatment of the hundreds or thousands of patients that could ultimately benefit from cell therapy and the search is now on for alternatives which will be both safe and cost effective. Much current attention is focused on stem cells, and in particular on human embryonic stem (or ES) cells.

16.7.2   EMBRYO STEM CELLS

All mammals start life as a fertilised egg or zygote. After five to six days and seven or eight cell divisions, the developing embryo consists of a small ball of cells about a tenth of a millimetre in diameter (i.e. smaller than the full stop at the end of this sentence). The outer cells of this 'blastocyst' are destined to become part of the placenta, whereas the embryo stem cells (ES) of the inner cell mass will go on to form the foetus proper. The ES cells are therefore the progenitors of all of the 200 or so terminally differentiated cell types in the complete animal.

Embryo stem cells were first isolated from mice in the early 1980s. They can divide and multiply indefinitely in the laboratory and, when grown in appropriate cocktails of growth factors, can be directed to differentiate into a wide variety of different cell types. Putative ES cells were isolated from human embryos in 1998 and have similar characteristics. Human ES cells are of particular interest for cell therapy because they can be obtained free of contamination from other cell types, will multiply indefinitely in culture and can potentially be converted to any cell type needed by patients (Pederson, 1999).

A major problem to be addressed is that of immune rejection. Cells injected into the brain are partially protected from immune rejection by the blood–brain barrier. By contrast, cells transplanted to other tissues would be recognised as foreign and immediately destroyed by the complement-mediated hyper-acute response. In theory, human stem cells could be genetically engineered to reduce their immunogenicity and provide 'universal donor' cells that could be used in any patient. A more practical approach would be to create a bank of human ES cell lines representative of the major tissue types in the population as a whole and then withdraw cells of the most appropriate cell line as and when required. Both strategies would require patients to remain on immuno-suppressive drugs for the rest of their lives, adding to both the cost of treatment and the risk of infection or carcinoma.

16.7.3 THE ROLE OF NUCLEAR TRANSFER IN STEM CELL THERAPY

The ability to clone animals from adult cells suggests a radical new approach to the problem of tissue incompatibility. Perhaps 10–15 years into the future when cells are needed for transplant, skin cells could be obtained from patients simply by scraping the inside of the mouth. These skin fibroblasts would then be multiplied many times in the laboratory before being converted to the specific cell type needed for the disease being treated. When these cells were returned to the patient, they would not be rejected: they are the patient's own cells.

At present the only way to achieve such a transformation would be to use part of the process used to clone whole animals. The skin cell from the patient would be introduced into a human egg from which the nucleus had been removed and the resulting human embryo incubated for six to seven days before recovery of ES cells. Incubation of these cells with appropriate growth factors would then be used to obtain the desired cell type.

Such 'therapeutic cloning' is an aid to research rather than a routine way of creating stem cells for every patient. Very few human eggs are surplus to requirements from IVF clinics in the UK whereas the numbers of potential patients for cell therapy runs into hundreds of thousands. On practical grounds alone it will therefore be essential to find ways of avoiding the use of human eggs if stem cell therapy is to achieve its true potential. It may be possible to use other types of cell – perhaps embryo stem cells – to reprogramme the donor nucleus. More radically, if we understood the mechanisms by which the cytoplasm of the oocyte is able to facilitate the reprogramming of differentiated cells, we may be able to recreate appropriate conditions for reprogramming in the test tube. Such an ability would escape the chronic limitations on supply of human eggs and avoid the routine creation and destruction of human embryos that many people would find ethically unacceptable.

16.7.4 AN ETHICAL DEBATE BROUGHT TO A CONCLUSION

Research on human embryos is already allowed in the UK under the Human

Fertilisation and Embryology Act 1990. Under the 1990 Act human embryos are given 'special status' but not that of a human being. This designation was consistent with UK attitudes to abortion and the 'morning-after' contraceptive pill. Research on human embryos was allowed only up to 14 days of development and only under licence from the Human Fertilisation and Embryology Authority. Research was also only allowed for only five specific purposes connected with fertility, contraception and genetic disorders.

In December 1998, the Human Fertilisation and Embryology Authority and the Human Genetics Advisory Commission published a joint report recommending that two new purposes should be added to the 1990 Act: to progress new therapies for 'diseased or damaged tissues or organs' and mitochondrial diseases (HGAC, 1998). The government responded by asking Professor Liam Donaldson, the UK's Chief Medical Officer, to convene an 'expert panel' to review the case in detail. The *Donaldson Report* (CEGC, 2000) was published on 16 August 2000 and fully endorsed the proposed changes in the HFE Act. The government immediately accepted all the recommendations of the report and, as with the original 1990 Act, the government offered a 'free vote' in both Houses of Parliament.

The Catholic Church and 'pro-life' groups were most active in opposing the proposed amendment to the 1990 Act, arguing that human life began at conception and therefore that any research on human embryos was wrong. Others were content that embryos that were surplus to requirements from IVF clinics could be used for research (with appropriate consent) rather than simply discarded, but some who accepted this position were uneasy about the deliberate creation of human embryos for research purposes. The creation of human embryos for research purposes is allowed under the original 1990 Act but only 108 embryos have been created in this way since the HFEA started to grant licences. Some argued that the cloning of embryos would be the start of a slippery slope to reproductive cloning and others actively promoted an overly optimistic view that stem cells for therapy might be obtained from adult tissue or umbilical cord blood rather than from embryos.

The end result of four parliamentary debates at the end of 2000 and beginning of 2001 was a decisive vote in favour of the amendment in both the House of Commons and House of Lords. Researchers in the UK can now apply to the HFEA for a licence to carry out research on human embryos to progress stem cell therapy. Applications have first to be passed by an ethics committee created by the research organisations involved before being submitted to the HFEA. It then sends the application to external referees and only when their opinion is available is the application considered by the licensing sub-committee of the HFEA.

Other countries including Sweden, Denmark, the Netherlands and France already allow research on human embryos (EC, 2000) and some are considering similar amendments to their national legislation as introduced in the UK. In marked contrast, the legal position in the USA is very confused, with research on human embryos being allowed in the private sector but not with federal funds. In 2000, the Head of the National Institutes of Health, Dr Harold Varmus, fearing researchers in the USA might be left behind, proposed a 'compromise' that would allow federal-funded researchers to carry out research on human embryo stem cells that had been isolated in the private sector. At the time of writing (early 2001), it seems highly unlikely that this device will survive the new Bush administration's sympathy for the pro-life lobby.

16.8 COMMENT

The existence of bodies such as the Human Fertilisation and Embryology Authority and the Farm Animal Welfare Council has greatly helped to provide a focus in the UK for ethical debate on cloning. Such government-sponsored vehicles, however, have their limitations. The almost universal negative reaction to human reproductive cloning when Dolly was announced to the world meant that there was little or no incentive for anyone within the political or scientific establishment to formally question whether this response was justified. As a consequence, the reaction to more recent announcements of proposals to clone children has simply been to repeat expressions of outrage and the opportunity to deploy more thoughtful and arguably more persuasive arguments for not proceeding along this particular path is being missed through lack of appropriate preparation.

REFERENCES

Abbott, A. (2001) Trepidation greets plans for cloning humans. *Nature*, **410**, 293.

Bruce, D. and Bruce, A. (1998) Animal ethics and human benefit. In *Engineering Genesis: the Ethics of Genetic Engineering in Non-Human Species.* Bruce, D. and Bruce, A. (eds), Earthscan, London, and Church of Scotland, Edinburgh, pp. 127–15.

CEGC (2000). Stem cells: medical progress with responsibility. *Report of the Chief Medical Officer's Expert Group on Therapeutic Cloning* on www.dti.gov.uk/cegc.

Cibelli, J.B., Kiessling, A.A., Cuniff, K., Richards, C., Lanza, R.P. and West, M.D. (2001) Somatic cell nuclear transfer in humans: pronuclear and early embryonic development. *The Journal of Regenerative Medicine*, **2**, 25–31.

Cohen, P. (1998) Crossing the line. *New Scientist*, 17 January, 4–5.

EC (2000) Ethical aspects of human stem cell research. *Opinion of the European Group on Ethics and Science and New Technologies to the European Commission*, No. 15.

FAWC (1998) *Report on the implications of cloning for the welfare of farmed livestock*, Farm Animal Welfare Council, Surbiton, Surrey, UK.

Green, R. (1999) I, clone. *Scientific American Presents*, **10**, 80–82.

Harris, J. (1998) Rights and reproductive choice. In *The Future of Human Reproduction: Ethics, Choice and Regulation.* Harris, J. and Holm, S. (eds), Clarendon, Oxford, UK, pp. 5–37.

HGAC (1998) *Cloning Issues in Reproduction, Science and Medicine.* Joint report by the Human Genetics Advisory Committee and the Human Fertilisation and Embryology

Authority on www.dti.gov.uk/hgac

Kass, L. (1997) The wisdom of repugnance. *New Republic*, **2**, 5–12.

Kolata, G. (1997) *Clone. The Road to Dolly and the Path Ahead.* Penguin, London.

MAFF (1995) *Report of the Committee to Consider the Ethical Implications of Emerging Technologies in the Breeding of Farm Animals* (Banner Committee Report). HMSO, London, UK.

McKinnell, R.G., (1985) *Cloning: of frogs, mice and other animals.* University of Minnestoa Press, Minneapolis, MN, USA.

O'Neill, O. (2000) The 'good enough parent' in the age of new reproductive technologies. In *Ethics of Genetics of Human Procreation*, Haker, H. and Beyleveld, D. (eds), Ashgate, London, pp 33–48.

Pederson, R. (1999) Embryonic stem cells for medicine. *Scientific American*, April, 45–49.

Pence, G.E. (2000) *Re-Creating Medicine: Ethical Issues at the Frontiers of Medicine*, Rowman and Littlefield, Lanham, MD, USA (see especially 'Recreating our genes: cloning humans', pp.119–135).

Pennisi, L. and Vogel, G. (2000) Clones: a hard act to follow. *Science*, **288**, 1722–1727.

Renwick, T.M. (1998) A cabbit (cat/rabbit) in sheep's clothing: exploring the sources of our moral disquiet about cloning. *Annual of the Society of Christian Ethics*, **18**, 259–274.

Shuster, D.E., Kehrli, M.E., Ackermann, M.R. and Gilbert, R.O. (1992) Identification and prevalence of a genetic-defect that causes leukocyte adhesion deficiency in Holstein cattle. *Proceedings of the National Academy of Sciences*, **89**, 9225–9229.

Silver, L. (1997). *Remaking Eden: Cloning and Beyond in a Brave New World.* Avon, New York, USA.

Stern, A. (2001) *The Boston Globe*, 26 January.

Thomson, J.A., Itskovitz-Eldor, J., Shapiro, S.S., Waknitz, M.A., Swiergiel, J.J., Marshall, V.S. and Jones, J.M. (1998) Embryonic stem cell lines derived from human blastocysts. *Science*, **282**, 1145–1147.

Wakayama, T., Perry, A.C.F., Zuccoti, M., Johnson, K.L., and Yanagimachi, R. (1998) Full term development of mice from enucleated oocytes injected with cumulus cell nuclei. *Nature*, **394**, 369–374.

Wellcome Trust (1998). *Public Perspectives on Human Cloning. A Social Research Study.* Wellcome Trust, London, UK.

Wilmut, I., Campbell, K. and Tudge, C. (2000) *The Second Creation: Dolly and the Age of Biological Control.* Headline, London, UK.

Wilmut, I. and Griffin, H. (1997) Seven days that shook the world. *New Scientist*, 22 March, 22.

Wilmut, I., Schnieke, A.E., McWhir, J., Kind, A.J. and Campbell, K.H.S. (1997) Viable offspring derived from fetal and adult mammalian cells. *Nature*, **385**, 810–813.

17 Dealing with Death: Euthanasia and Related Issues

John Searle

17.1 INTRODUCTION

There are only two absolute certainties about life; it begins and it ends. We are born and we die. Until the early part of the 20th century, death was an accepted part of life. Many diseases were incurable. People died at home rather than in institutions. War took its toll of young men – most notably the First World War of 1914–18. Life expectancy was much shorter compared with today. Infant death rates were high. By the end of the 20th century, advances in therapeutics, surgery and medical technology had conquered many diseases and enabled those with chronic disease to live for a long time, often with a good quality of life. With these advances came the popular view that death could be postponed almost indefinitely. The current search is for an understanding of the genetic, nurturing and environmental components of the ageing process with a view to slowing it down and prolonging life even further. As a result of these changes there have been two parallel but different developments.

First, the terminally ill have often been neglected. In the 1960s and 70s hospital staff commonly hurried past the beds of the dying because they represented a failure of modern medicine which was difficult to face. The distressing symptoms, emotional turmoil and social disturbance of dying were frequently left unattended. Two responses followed this neglect. First, the Hospice Movement, pioneered by a doctor, Cicely Saunders, took the care of the dying seriously. Distressing and painful symptoms were relieved; the emotional and social upheaval alleviated and spiritual comfort provided. Those close to the dying person were well supported. The hospice emphasis is that although the person is dying, they can live fully to the end of their lives within the limitations of their illness.

Others however, have taken the view that the only certain way to control the pain and distress of dying is to end the person's life. This argument has been bolstered by the prevailing ethical principle of Western society which is that of individual autonomy. This principle says that my life is my own to do with it

Bioethics for Scientists. Edited by John Bryant, Linda Baggott la Velle and John Searle.
© 2002 by John Wiley & Sons Ltd.

whatever seems best for me. Despite the remarkable effectiveness of hospices this view is increasingly promoted.

Secondly, one outcome of advances in medicine is that life can be prolonged, but in some cases with a very poor quality of life. In intensive care units, people who would have died within a few days can be kept alive by drugs and machines, only to die a few weeks later. Victims of dementia and severe strokes may survive long beyond the time when they have lost the ability to know about or respond to the world around them. People with advanced cancer may be subjected to surgery, radiotherapy and chemotherapy with little prospect of a reasonable quality of life being secured.

Again, two views have emerged in response to these developments. There are those who say that life must be preserved at all costs. Others believe that 'nature should be allowed to take its course'. There then follows a further question in the debate: if in some people, treatment is either not to be instituted or withdrawn and the result of that is that they die, what is the difference between this and euthanasia – that is actually killing them? After all the outcome is the same. They die.

The dilemma that confronts society at the beginning of the 21st century is how to find a way through this complex web of ethics and medicine. This chapter will explore ways of doing so by looking first at the euthanasia debate and then discussing the grey area of when continuing medical treatment is prolonging dying rather than promoting life. It will conclude with some case studies to illustrate these dilemmas.

17.2   THE EUTHANASIA DEBATE

17.2.1   WHAT IS EUTHANASIA?

The English word *euthanasia* is derived from two Greek words that mean 'a quiet and easy death'. That is something for which most people hope as they do not want their dying to be prolonged, painful or violent. However, in the present debate, 'euthanasia' means something different. Voluntary euthanasia is defined as 'the deliberate ending of a person's life, at their request, because they find their illness and/or disability intolerable'. The assumption is that as doctors have both the knowledge and the means for doing this, they will normally administer it. Such an act by a doctor is, in English law, unlawful. However, the British Courts have increasingly taken a lenient view when euthanasia has been administered and the doctor prosecuted. Two cases illustrate this point.

In 1991, Dr Nigel Cox, a consultant physician in Winchester, gave Mrs Lilian Boyes a lethal injection of potassium chloride. She was in severe pain from rheumatoid arthritis. The pain made her 'howl and scream like a dog'. Dr Cox believed that the only way to end this appalling suffering was to terminate her life. He was convicted of murder but given a suspended prison sentence. However, the regulatory body of the medical profession, The General Medical Council, found him guilty of serious professional misconduct and required him to undergo training in the control of severe pain and palliative care before allowing him to return to work.

On Boxing Day (26 December) in the same year, Paul Brady killed his brother James, by giving him five times the normal dose of the sleeping drug, temazepam, washed down with alcohol. He finally smothered James with a pillow. James was suffering from Huntington's disease and had begged his family to help him to die. At the High Court, in Glasgow, Paul Brady was initially charged with murder. This was reduced to culpable homicide, thereby allowing the court to admonish him and not to impose a custodial sentence on him. The judge accepted that there were powerful mitigating factors. He said 'you brought your brother's life to an end at his own earnest and plainly heartfelt request' (Christie, 1996).

However, the debate is in fact wider than a handful of specific cases. Indeed, the issue has been brought the attention of wide television audiences. In the UK for example, the television 'soap operas' *Brookside* and *EastEnders* have, during the period 1995–2000, both dealt with the issue of helping somebody to die because their suffering in a terminal illness was severe.

The momentum in support of voluntary euthanasia in the developed world has increased considerably. In 1996, euthanasia became lawful in the Northern Territories of Australia. Seven people took advantage of this change before it was overturned by the Federal Australian Parliament (Kissare et al, 1998). In the United States of America the debate was brought powerfully into the public arena in April 1999. Dr Jack Kevorkian, a pathologist, is a passionate advocate of the right to die. He had assisted over 100 patients in bringing about their own death. In Michigan, he was found guilty of second-degree murder and sentenced to 10 to 25 years in prison. In the State of Oregon, on the other hand, physician-assisted suicide is lawful for patients judged to have less than six months to live.

In The Netherlands, voluntary euthanasia has been practised by doctors for two decades and therefore provides the largest single source of data about the practice. Until the end of 2000 euthanasia was actually unlawful although the Dutch Courts recognised it as acceptable medical practice. However, the practice has now been formally legalised.

So far, there has been no serious call for involuntary euthanasia. This is defined as 'the deliberate ending of a person's life, without their request, because some other party considers their life intolerable or its quality not worth having.' One of the key issues in the debate is whether the legalisation of voluntary euthanasia would lead to the practice of involuntary euthanasia.

17.2.2   THE CASE FOR VOLUNTARY EUTHANASIA

There are three main arguments for making voluntary euthanasia lawful:

• autonomy

- necessity
- openness.

Autonomy

Autonomy is the key ethical argument in the debate. The basic question is whether or not human beings have the right to decide themselves how they will end their lives. This has been vividly expressed by Brian Clarke in his play *Whose Life is it Anyway?*, which played in the West End of London in the 1970s. The central character, Ken Harrison, is paralysed from the neck downwards and has no use in either his arms or his legs. He wants to die and says, 'I have coolly and calmly thought it out and I have decided that I would rather not go on. Each must make his own decision' (Clarke, 1978). The argument is that what is intolerable for me is for me to decide and if I wish to be delivered from that by dying, that is my decision.

There is growing support for this view. As long ago as 1985, 85% of people questioned in a survey in the United States of America believed that they had the right to decide for themselves how they died (The New York State Task Force on Life and the Law, 1987). This is frequently cited in feature articles and correspondence columns of newspapers and magazines.

Necessity

If autonomy is the key philosophical argument in favour of voluntary euthanasia, necessity is the most emotionally powerful. Case histories are cited of people dying in great pain and distress. In 1995, the BBC television series *Panorama* screened a programme about euthanasia in The Netherlands. A man had motor neurone disease, which causes progressive muscular paralysis but leaves sensation and mental function intact. His general practitioner told him that not only would he become increasingly disabled, but also his breathing would become more difficult as his lungs became waterlogged and that he would eventually choke to death. Faced with this the patient asked to have his life ended, which the doctor did, injecting him with a lethal combination of anaesthetic drugs. The protagonists of voluntary euthanasia almost always quote cases where the suffering has been terrible and pain poorly controlled. The motivation is compassion and a desire to see people die without pain and with dignity.

Openness

There is some evidence that doctors in the UK do practice euthanasia. It is clearly difficult to obtain evidence about an unlawful practice but in one study (not peer reviewed) carried out through anonymised questionnaires, of 273 doctors questioned, 163 had at times been asked by patients to end their lives. Of these, 124 had taken steps to hasten death and 38 had actually complied with the request (Coulson, 1995). In another study, 50% of doctors questioned wanted the law changed. The argument is that if some doctors are covertly practising euthanasia, the law should be brought into line with such compassionate medical practice, thereby removing the threat of prosecution from doctors.

These arguments were dramatically illustrated by the death of King George V in 1936.

On the night of 20 January 1936, the royal physician, Lord Dawson of Penn, was summoned to Sandringham by Queen Mary. King George V was dying from heart failure. The following morning, *The Times* carried the first news that the King was dead. The assumption was that he had died naturally in his sleep. What actually happened was only made public 48 years later with the publication of Dawson's diaries. Of the evening of 20 January 1936 he had written

> at about eleven o'clock it was evident that the last stage might endure for many hours, unknown to the patient, but little comporting with the dignity and the serenity which he (the King) so richly merited and which demanded a brief final scene. Hours of waiting just for the mechanical end when all that is really life has departed only exhausts the onlookers and keeps them so strained that they cannot avail themselves of the solace of thought, communion and prayer. I therefore decided to determine the end.

Dawson then gave the King intravenously a lethal dose of morphine and cocaine and he died. A timely public announcement in the appropriate newspaper was assured and Dawson was able to return to London by the morning to his private practice (Ramsay, 1994). Dawson's brief entry encapsulates all the main arguments in favour of euthanasia:

- the patient was terminally ill
- his dignity was assured
- his suffering was relieved
- the relatives were spared a long bedside vigil.

17.2.3 THE CASE AGAINST EUTHANASIA

The triad of autonomy, necessity and openness is a formidable one. But those who oppose the legalisation of voluntary euthanasia raise two important questions.

- Is there a downside to the exercise of autonomy?
- Are there effective ways of controlling the pain and distress of terminal illness apart from killing the person? Is voluntary euthanasia necessary?

The downside to autonomy

That each person should be able to make their own decisions about their own lives is an important general principle in the conduct of civilised human affairs.

However, men and women are not only individuals; they are part of society. Human responsibilities therefore extend beyond self to others, to the community. Thus a second important principle in the conduct of civilised human affairs is that individuals may only exercise their rights in so far as they do not infringe the rights of others. So does the right to die by voluntary euthanasia infringe the rights of others to live?

In 1989 the Dutch Government appointed a commission under the chair of the Attorney General to report on 'the extent and nature of medical euthanasia practice'. The report was published in English in 1992 and its findings were as follows:

Total number of deaths in The Netherlands in 1990	12 900
Deaths not terminated or assisted by the doctor	9200
Termination of life at patient request	2300
Assisted suicide	400
Termination of life without patient request	1000

28% of all deaths in The Netherlands in 1990 involved the active participation of a doctor, but 7.7% of all deaths were actually procured by a doctor without the request of the patient (Van Der Maas et al, 1996). In four out of five cases the doctor said that there was no alternative way of relieving symptoms.

This practice continues in The Netherlands and has been extended to other conditions (Hendin et al, 1997). In 1994 a Select Committee of the British House of Lords, while recognising the right of every competent person to refuse medical treatment, rejected any proposal to make voluntary euthanasia lawful because such a law would threaten the weak, the vulnerable and the mentally incompetent.[1] What is voluntary for some would be assumed to be in the best interest of others.

Is voluntary euthanasia necessary?

If the pain and other distressing symptoms of terminal illness and chronic disease cannot be effectively controlled, the case for voluntary euthanasia on compassionate grounds is strong. The House of Lords Select Committee took a great deal of evidence on this point. It concluded that 'through the outstanding achievements of those who work in the field of palliative care, the pain and distress of terminal illness can be adequately controlled in the vast majority of cases' (House of Lords, 1994). However, there remain about 4% of people in whom pain control is very difficult. This has always been acknowledged by palliative staff (Gilbert, 1996), but they are equally clear that therapeutic strategies can be used to give these people a reasonable quality of life and that certainly nobody need die in pain (Saunders, 1999). The Exeter Hospice (which is typical of hospices in the UK) contributes to the care of 700 people with

---

[1] The correct term is 'lacking capacity'.

terminal illness every year, 300 of whom require in-patient care at some time. The staff do not encounter any persistent rational demand for euthanasia (Gilbert, 1996).

In the 1995 *Panorama* programme about euthanasia in The Netherlands the patient with motor neurone disease was told that he would become increasingly breathless as his lungs became waterlogged and that eventually he would choke to death (as mentioned earlier), but the fact is that in the world's largest series of deaths from this disease these end-stage complications were never encountered (O'Brien et al, 1992). It is difficult not to conclude that either the doctor was ignorant or he misled the patient.

A frequent criticism of those who work in palliative care is that, in order to relieve pain, they progressively increase the dose of powerful pain-relieving drugs and thereby shorten the person's life. They are charged with hypocrisy as it is claimed that they are administering euthanasia covertly. Two objections are put forward to counter this charge (Gilbert, 1996). Firstly, there is a wide individual variation in response to these drugs. Huge doses can be given to some individuals with seemingly little effect while others respond to much smaller doses. Part of the skill in palliative care is to titrate the dose of drug against the person's response. Secondly, pain acts as a physiological antagonist to the depressing effects of these drugs on respiration.

These arguments against voluntary euthanasia being necessary to ensure 'death with dignity' also cut across the argument about openness. Once the effectiveness of good palliative care is recognised, this argument evaporates and becomes one not about changing the law but about ensuring that doctors and other healthcare workers are properly trained in the care of the terminally ill, and that such care is regularly reviewed and audited.

Pressure continues to be put on many governments in the developed world to legalise voluntary euthanasia. Those who see such a move as a threat to the weak, the vulnerable and the terminally ill oppose it with equal vigour. The debate is set to go on for some time. However, the history of the last 50 years is that public policy is eventually determined by the ethos of perceived human rights and personal autonomy (but see the discussion on utilitarianism in Chapter 1).

17.3 PROLONGING DYING

17.3.1 INTRODUCTION

The arguments for and against voluntary euthanasia are fairly sharply defined. However, the area of withholding or withdrawing medical treatment from a person whose illness or disability has a very poor outlook and renders miserable their quality of life is much more complicated, although the arguments are no less vigorously pursued. The pro-life groups have taken a passionate stand against any act of omission that seems to them to result in the shortening of human life, however tenuous the hold on life may be. Healthcare workers and the English courts have taken the view of 'principled pragmatism', trying to balance a number of ethical principles, which are sometimes in conflict.

These principles, which underlie contemporary medical and healthcare practice, were set out by Beauchamp and Childress in 1979 and are widely accepted. There are four main principles.

- Beneficence: a doctor must always act in a person's best interest, that is in a way which confers benefit upon that person.
- Non-maleficence: a doctor must not harm a person. That can never be in a person's best interest.
- Autonomy: a doctor may not do anything to a person without their consent. Any person has an absolute right to refuse medical treatment.
- Justice: a doctor must act in such a way that resources are apportioned equitably within a society.

The application of these is not always straightforward. Medical and surgical treatment always involves risk. No drug is devoid of side effects and no operation can be undertaken with an absolute guarantee that there will not be any complications. Consent to treatment can only be given to persons who are competent; that is to say they understand what is being offered to them and they understand and can articulate the consequences of refusing it. Where a person lacks the capacity to do this because of age, mental dysfunction or unconsciousness, doctors have to consult with those who have an interest in that person's welfare; however, ultimately the law accords to the doctor the privilege of deciding what is beneficial to that person and what the balance is between benefit and harm. All healthworkers recognise that resources are limited and that healthcare provision cannot do all that it is possible to do. Nonetheless, when facing an individual person they want to do what they consider to be in that person's best interests, irrespective of cost, notwithstanding that doing one thing for this person may prevent something else being done for another person.

It is within this context that decisions have to be made about withholding or withdrawing treatment in some cases. There are five areas to consider:

- resuscitation
- brain stem death
- permanent vegetative state
- intensive care, cancer treatment and heroic surgery
- conjoined twins.

17.3.2   RESUSCITATION

Since 1960 the technique of external cardiac massage and expired air ventilation and electrical defibrillation has saved the lives of many people who would otherwise have died suddenly because their heart stopped beating from some reversible cause. Since the introduction of this technique the question 'who should be resuscitated?' has been asked. A consensus emerged that it should not be embarked on if

- the patient did not wish it
- it would be futile
- the costs were too great
- the subsequent quality of life would be unacceptable.

Despite this consensus there has been recent public concern about resuscitation. On the one hand people continue to be resuscitated in hospital when there is little or no hope of securing a reasonable outcome because the person has in fact reached the end stage of their disease. Their cardiac arrest is in fact death. Clearly to embark on resuscitation under these circumstances does not benefit the person and may cause them harm.

On the other hand in the late 1990s it became clear that doctors were making decisions not to resuscitate patients should they have a cardiac arrest, without any discussion with them beforehand. This clearly infringes the principle of consent and betrays the partnership of trust between doctor and patient.

17.3.3 BRAIN STEM DEATH

Death occurs when one of three organs cease to function – the brain stem, the heart and the lungs. If one of these stops working the other two cease to function rapidly thereafter and death occurs. By the early 1970s it became possible to interrupt this process where the organ initially failing was the brain either because of accident (a severe head injury) or disease (a severe stroke). By passing a tube into the person's trachea and attaching it to a ventilator ('life-support machine'), it is possible to inflate the lungs with oxygen and thereby maintain the function of the heart, in the hope that recovery of the brain would occur and all three organs be able to function near normally again. A key question then arose: was it possible to distinguish between those people in whom such recovery would occur and those in whom it would not? That is to say, is it possible to identify those in whom the brain damage is fatal and the inevitable process of death has simply been interrupted? When this happens the higher centres of the brain, which are the neurological basis of personhood, can neither receive nor transmit information.

In 1976 (Conference of Medical Royal Colleges and their Faculties, 1976) a series of reliable tests was developed, which enabled this distinction to be made by demonstrating whether or not the brain stem, which connects the higher centres of the brain to the rest of the body, had been destroyed. When brain-stem death is shown to occur, the ventilator is turned off and the heart and lungs rapidly cease to function. Such an act is not causing death but recognising that death has already taken place.

However, not everybody has accepted this concept, maintaining that death can only be said to have occurred when there is total cessation of brain function. When the stem of the brain has been destroyed, electrical activity of the higher levels of the brain may persist for some time. Their view is that death can only properly be said to have occurred when there is no electrical activity in the brain and thus to turn off a ventilator before this happens is tantamount to euthanasia.

The criteria for establishing death of the brain stem and the actions which follow have never been enshrined in statute. The legal position seems to be that a person is dead when a doctor says they are dead.

17.3.4   PERMANENT VEGETATIVE STATE (PVS)

In PVS the brain stem is intact so that the heart and lungs continue to function normally without artificial support. It is the higher centres of the brain that have been destroyed. People with PVS display sleep–wake patterns and respond reflexly to stimulation but show no evidence of cognitive function. They are unable to swallow and have to be fed through a tube placed in the stomach via the nose and the oesophagus. With expert nursing care they can be kept in this condition for years and indeed many have been in the past. However, by the early 1990s two key questions were being asked by those looking after them and their families.

- Is such a person alive?
- Is feeding through a tube when the person is unable to take food and hydration normally through the mouth an artificial means of support?

These questions were highlighted by the case of Tony Bland in 1993. He had been a victim of the Hillsborough Football Stadium (Sheffield, UK) disaster in 1989. He had been severely injured but resuscitated and treated. However, he never regained consciousness and had been in a PVS for three years. The doctors looking after him and the hospital in which he was resident took the case to the courts. The two key questions were addressed, firstly in the Family Division of the High Court, then in the Court of Appeal and finally in the House of Lords which confirmed the judgement of the lower court.

The Law Lords took the view that Tony Bland was not alive in any normal meaning of the word. They also accepted that feeding him via a tube was a form of medical treatment that was not only futile since Bland could never recover but was also being administered without his consent. They made it clear that in a person without capacity, doctors have a clear duty to do that which is beneficial to a person. However, feeding in this way could not benefit Bland. Indeed there was no benefit that could be conferred upon him. They therefore ruled that it was lawful to withdraw food and hydration (Airedale NHS Trust, 1993). This was done and Bland died some days later.

While many regarded this ruling as both wise and compassionate, others were profoundly disturbed by it (Fergusson, 1993; Hume, 1997) for three reasons.

They argued that this was in fact euthanasia, if not openly at least by the 'back door', because of the following factors.

- Hitherto death had been defined as irreversible cessation of the activity of the heart or the lungs or the brain. In this sense Bland was not dead.
- Food and hydration are basic human needs and while someone is alive there is a moral duty to provide them.
- The intention of stopping food and hydration was that Bland would die.

However, the counter-argument to this is that the Law Lords based their decision on the answers to two vital questions. First, was there any possibility of Bland recovering? Second, could he swallow food and fluid when it was offered to him on a spoon or in a cup? The answer to both questions was 'no'. If the answer to either question had been 'yes' then there was a duty to continue feeding him. But as the answer to both questions was 'no' withdrawing food and hydration was reasonable. It is principled pragmatism.

However, the decision is absolutely dependent on the accuracy and reliability of the diagnosis. The Royal College of Physicians issued strict diagnostic criteria for PVS (Recommendations and Standards, 1996). The British Medical Association recommended two safeguards in making decisions to withhold food and hydration.

- These cases should be subject to formal clinical review in each case by a senior doctor with experience of the condition from which the person is suffering. This doctor must not be part of the team treating the person.
- All cases should be reviewed regularly to ensure that appropriate procedures and guidelines are followed.

In England it remains a requirement that the permission of the courts is necessary before food and hydration can be withheld in cases of PVS. It is no longer necessary to do so in Scotland. The situation is anomalous, as in other conditions, such as end stage dementia and severe stroke, where the answers to the two key questions are 'no', fluid and hydration can lawfully be withdrawn without recourse to the courts.

17.3.5   WITHHOLDING AND WITHDRAWING TREATMENT

Finding the balance of what constitutes a person's best interests extends to other areas of medical practice. Examples include intensive care, the treatment of advanced cancer and the care of the new-born with severe handicap. One way of approaching this has been to quantify the risk of death and placing patients thereby into categories of the likelihood of death occurring irrespective of the treatment being proposed or given. This does not give an absolutely certain prediction about what is going to happen to each individual person but together with the progress of the condition over a longer or shorter period of time enable the best judgement to be made. This judgement will never be perfect but the nature of medicine is such that the most that can be achieved is a 'good enough' judgement. Of course at some date in the future that judgement may be different and a different decision made. A good example of this is the respiratory complications of AIDS. In the mid–1980s the onset of respiratory failure in patients with AIDS was inevitably fatal and therefore to embark on intensive care treatment only prolonged the time of death by a short time, while also imposing on the person a considerable burden of suffering. By the late 1990s the prognosis for patients with AIDS related respiratory disease had improved considerably and therefore such treatment became justifiable.

17.3.6   CONJOINED TWINS

A recent British example of the dilemmas at the end of life that modern medicine has generated is well illustrated by the case of the conjoined twins, Mary and Jodi. The issue at stake was 'should the twins be separated by a surgical operation?' The expert medical opinion in this case was that Jodi did have a chance of a reasonable quality of life if she were separated from her twin, even though that separation would cause the certain death of the other twin, Mary. Both twins would inevitably die without such separation since Mary was only oxygenated through the heart and lungs of Jodi and Jodi's heart would eventually fail irreversibly if such a burden were imposed on it indefinitely.

In his summary (as given to *The Times* newspaper, 2000), Lord Justice Ward said, 'In my judgement it is overwhelmingly in Jodi's interest that she be given the chance to live a normal life with a normal expectation of life. It is certainly not in her best interest to be left to die'. He went on, 'In my judgement it cannot be in Mary's best interest to undergo surgery which will terminate her life'. He described the dilemma thereby facing the court as follows: 'It is in the best interests of Jodi that separation takes place. It is in the best interests of Mary that it does not. . . . The only solution is to balance the welfare of each child against the other to find the least detrimental alternative. . . . One cannot escape from the fact that Mary has always been fated for early death: her capacity to live has been fatally compromised. . . . Nobody but the doctors can help Jodi. Mary sadly is beyond help. The best interests of the twins is to give the chance of life to the child whose actual bodily condition is capable of accepting the chance to her advantage, even if that has to be at the cost of the sacrifice of a life which is so unnaturally supported. . . . The least detrimental course is to permit separation to take place'. While many people welcomed this judgement, others were deeply disturbed by it. The court's view was determined by trying to solve the dilemma of benefit versus harm. If the operation did not take place Jodi and Mary would be harmed. Jodi would be harmed by the inevitable harm which was going to befall Mary. However, if the operation took place, Jodi would benefit but Mary would be irreparably harmed. The argument is a utilitarian one, 'what benefit is obtained by each course of action?'

This approach lacks deontological principles and raises many questions. Was

Mary of less value than Jodi? How can an operation be ethical if it results in the death of one of the twins? Mary was being used instrumentally in order that Jodi might benefit, thereby turning on its head the traditional view that the end can never justify the means. On what grounds did the court rule against the parents, who did not wish the operation to take place? Is death the worst fate that could happen to Jodi?

Questions also arose about the possibility of harming Jodi. Were the doctors being overly optimistic about her future quality of life? May she not suffer psychological harm in later life knowing that she lived only because her twin sister was killed?

The case illustrates the increasing gap in society between traditional deontological ethics and contemporary utilitarian ethics.

17.4   CONCLUSION

Modern medicine has brought huge benefits to human beings, especially in the developed world. For many people life expectancy has been extended with a good quality of life. But with this has come the ability to sustain life when there is little or no prospect of recovery or of anything approaching a reasonable quality of life. Dying is being prolonged. Ancient philosophers understood this well; 'For in much wisdom is much vexation and those who increase knowledge increase sorrow' (Ecclesiastes, Chapter 1, verse 18). Doctors and their health-worker colleagues have three duties in the face of this dilemma: first to maintain a person's right to life, second to preserve their right to die and third to distinguish between the two (Dunstan, 1985).

APPENDIX: SOME CASE HISTORIES

CASE 1

You are parents. Two days ago your 18 year old son had a motor cycle accident. He has severe head injuries and has been in an intensive care unit on a life support machine for the last 48 hours. The doctors tell you that his brain is dead and can never recover. Will you agree to the life support machine being turned off?

CASE 2

You are a general practitioner. You have a patient who is a man of 40 with a wife and two children aged 13 and 11. He has battled with cancer for three years and he is now dying. He has, you think, four to six weeks to live. One day when you are visiting him at home he says to you 'Doctor, my wife and I have talked about this and we realise that there is nothing more that can be done for me. Please would you give me an injection so that I can die peacefully in my sleep? We have a written request here which we have both signed and our next door neighbours have witnessed it.' Would you give the injection?

CASE 3

You are 50. Your widowed mother is 85 and you are her only child. She is blind. She had a stroke two years ago and is confined to a wheelchair. She is mentally alert and cheerful and well looked after in a local nursing home.

She has no financial difficulties. You will inherit a substantial capital sum after her death. Three days ago she had a major bowel operation for cancer which the doctors say was curative. She has developed complications and is confused. The doctors want to do another operation to repair a leak in her bowel but they are not sure whether or not she will survive the operation. She will certainly die if she does not have the operation. Should she have the operation?

REFERENCES

Airedale NHS Trust v. Bland (1993) Appeal Cases 835.

Beauchamp, T. and Childress, J. (1979) *Principles of Biomedical Ethics*. Oxford University Press, Oxford, UK.

Christie, B. (1996) Man walks free after Scottish euthanasia case. *British Medical Journal*, **313**, 961.

Clarke, B. (1978) *Whose Life is it Anyway?* Amber Lane, Ashover, UK.

Conference of Medical Royal Colleges and their Faculties (1976) Diagnosis of brain death. *Lancet*, **ii**, 1069–1070.

Coulson, J. (1995) Doctors oppose legal mercy killing for dying. *British Medical Association News Review*, March, 15.

Dunstan, G.R. (1985) Hard questions in intensive care. *Anaesthesia*, **40**, 479–482.

Fergusson, A. (1993) Should tube-feeding be withdrawn in PVS? *Journal of the Christian Medical Fellowship*, April, 4–8.

Gilbert, J. (1996) Palliative medicine: a new speciality changes an old debate. In *Euthanasia: Death, Dying and the Medical Duty*. Dunstan, G.R. and Lachmann, P.J. (eds), *British Medical Bulletin*, **52**, 296–307.

Hendin, H., Rutenfrans, C. and Zylicz, Z. (1997) Physician assisted suicide and euthanasia in the Netherlands. *Journal of the American Medical Association*, **277**, 1720–1722.

House of Lords (1994) Report of the Select Committee on Medical Ethics. HMSO, London, UK, para 241.

Hume, B. (1997) The death of trust. *The Times, London*, 27 November.

Kissare, D., Street, A. and Nitschke, P. (1998) Seven deaths in Darwin: case studies in the Rights of the Terminally Ill Act, Northern Territory, Australia. *Lancet*, **353**, 1097–1102.

The New York State Task Force on Life and the Law (1987) Life-sustaining treatment: making decisions and appointing a health care agent.

O'Brien, T., Kelly. M. and Saunders, C. (1992) Motor neurone disease: a hospice perspective. *British Medical Journal*, **304**, 471–473.

Ramsay, R. (1994) A king, a doctor, and a convenient death. *British Medical Journal*, **308,** 1445.
Recommendations and Standards (1996) The permanent vegetative state. *Journal of the Royal College of Physicians of London*, **30**, 119–121.
Saunders, C. (1999) *Euthanasia: the Heart of the Matter*. Hodder and Stoughton, London, UK.
The Times, London (2000) Siamese twins: the judgement. 23 November, 7.
Van der Maas, P.J., Van der Wal, G., Haverkate, I., de Graff, C.L.M., Kester, J.G.C., Bregje, D., Onwuteaka-Philpsen, D., Van der Heide, A., Bosma, J.M. and Willems, D.L. (1996) Euthanasia, physician-assisted suicide, and other medical practices involving the end of life in The Netherlands. *New England Journal of Medicine*, **335,** 1699–1705.

18 Animal Experimentation in Biomedical Research

Linda Baggott la Velle

18.1 ANIMAL RELATIONSHIPS: HUMAN AND NON-HUMAN

Bioethical issues arise when there are questions about what is right and wrong for people to do with other organisms. This chapter will discuss human relationships with other species of the vertebrate class to which we ourselves belong – the mammals. Few bioethical issues arouse the passion of some people so strongly as does the use that others make of mammalian species in sport, fashion, agriculture and scientific research. We need look no further than the current UK parliamentary debates about foxhunting, the use of animals in testing cosmetics, the mass culls of livestock in the attempt to eradicate foot-and-mouth disease and the vociferous demonstrations against the work of the pharmaceutical company, Huntingdon Life Sciences. These issues are presented to us daily in the media. To do justice to each of these would need more than just one chapter of this book. After an overview of the historical and philosophical thinking that has informed our present relationship with other mammals, this chapter will look in more detail at human use of animals in biomedical research, and will touch briefly on the role of animals in education.

All ethical decisions are formed from a range of choices. These choices can best be made from within an informed ethical framework. But how is this framework constructed? An important part of the history of the developed world has been the transition of human activity from a series of strategies for survival towards a civilisation with human welfare at its centre (see Table 18.1). But this benefit has a cost. Scientific (biomedical) advances aimed at alleviating human suffering and premature death are brought about as a result of the public demand for them. Implicit in this is the acceptance of the processes of science by which these advances are brought about. (See, for example, Chapter 2.)

There are three important elements in the practice of science. These are the following.

Bioethics for Scientists. Edited by John Bryant, Linda Baggott la Velle and John Searle.
© 2002 by John Wiley & Sons Ltd.

Table 18.1 Levels of organisation and science

Level of organisation	Studied by
International organisations	
Nations	Politics
Societies	
Individuals	Social sciences
Organs	
Tissues	
Cells	Biomedical sciences
Subcellular fractions	
Molecules	
Atoms	Chemistry
Subatomic particles	Atomic physics

1. The acquisition of new knowledge and understanding.
2. The application of that knowledge and understanding (note that some would call this 'technology').
3. The ethical aspect.

In the context of the use of animals in the first and second elements of science, this chapter will examine the third by discussing the values that inform opinion about what is right or wrong.

In order to begin a discussion about whether other (non-human) forms of life should be exploited by humans we might begin by considering some aspects of the characteristics of life. What distinguishes the living from the non-living? Pupils in school biology classes will readily answer, 'Mrs Gren!' (movement, respiration, sensitivity, growth, reproduction, excretion and nutrition), and of course these are the basic qualities that humans share with most other organisms. However, the quality of our relationship with those other organisms is not based on these common features. As David de Pomerai has argued in Chapter 6, we feel closer to other mammals than we do, say, to molluscs or plants. Although in the 19th century Darwin said 'the difference in mind between man and the higher animals, great as it is, certainly is one of degree and not kind', surely the complexity of this relationship has its foundations in more than phylogenetic taxonomy?

So what, precisely, distinguishes humans from other members of the animal kingdom? The main characteristics are probably those of rationality and its vehicle, language. Many philosophers have claimed that humans alone have reason – this is all-or-nothing. That we are endowed with the ability to reason enables us consciously not merely to make decisions pertinent to our survival, because other animals are manifestly able to do this. We are also able to decide what is right or wrong. This heightened ability to moralise has, probably from the beginning of the history of *Homo sapiens,* shaped the development of human community and ultimately civilisation.

It is clear even to the casual observer that animals can communicate with one another, but it also seems obvious that their level of communication has a limited vocabulary. There is evidence that our closest relatives among the primates can understand human language. For example, Washoe the chimpanzee can understand 150 meaningful signs and words, and Koko the lowland gorilla around 500 signs using language (Dolins, 1999). There are many other reported instances of communication in other animals. So while we may not be alone in being able to generate, learn and use language, there seems little doubt that the level of sophistication reached by even the simplest of human languages far exceeds any communication system developed in any other species.

The development of language has facilitated, enhanced and augmented the ability to moralise, an idea that resonated with early religious thinking. Speech and reason are linked in the notion of *logos* – an ancient Greek word simultaneously meaning 'reason', 'word', 'discourse', and 'saying'. For example, the first version of the New Testament contained the word *logos* in the opening words of the gospel according to St John:

> In the beginning was the word, and the word was with God and the word was God (*Holy Bible*, Revised Standard Version).

Philosophical and religious thinkers, from the earliest days of recorded language, have wrestled with the question of what it means to be human. Christians, for example, believe that alone in the animal kingdom humans are spiritual beings. The question of the nature of the spirit and its relationship to mind and consciousness are well beyond the scope of this chapter, but remain among the most profound questions facing philosophers, scientists and theologians today. It is interesting to note that the rise of neuroscience has given recent impetus to the debate. Whether self-consciousness is a uniquely human characteristic is an almost unanswerable question because of the layers of complexity associated with the concept. One might consider the extent to which some form of consciousness might be applied to any organism that responds to stimuli, and even extend this question to a consideration of the status of robots and computers in this context.

It is difficult then, to escape from the conclusion that the ability to moralise mediated by self-consciousness and language is a uniquely human characteristic. As moral agents, therefore, having a perception (knowledge) of good and evil, justice and injustice and so on, how has our relationship with other animals developed through the course of history?

18.2   HUMAN–ANIMAL RELATIONSHIPS: HISTORICAL PERSPECTIVES

There has been a long tradition in Western thought that emphasises the differences between humans and animals. These differences, it is said, are significant enough to justify our using them for food, experiments, fun and fashion (but see Chapter 7). Across the world, within all cultures people use animals as pets, subjects in experiments, objects of study, objects of reverence, the basis for subsistence and objects of financial gain. This implies that humans have taken on the role of 'creator, protector, and/or steward' over the animal world. Some may even add the word exploiter. How has this perceived right to exercise control over animals' lives – and even over the existence or disappearance of an entire species – arisen? In order to start to answer this it is necessary to look at Judaeo-Christian tradition that has influenced much thinking in this area.

The author of the book of Genesis, the first book in the Jewish and Christian scriptures, says in Chapter 1 verse 28

> Be fruitful and multiply, and fill the Earth and subdue it; and have dominion over the fish of the sea and over the birds of the air and over every living thing that moves upon the Earth. (*Holy Bible*, Revised Standard Version).

The important word here is dominion, and the idea that as humans it is right that we should dominate all other species probably springs from this. Indeed, as discussed by Christopher Southgate in Chapter 3, some environmental ethicists blame this Judaeo-Christian concept of dominion for the present ecological crisis. However, for those who attempt to base their bioethical and environmental ethical thinking on biblical principles, it is worth noting that another theme of the creation narratives, and indeed throughout the biblical text, is that humans are to exercise stewardship rather than dominion over the creation.

Many belief systems about nature, Christianity included, are based at least loosely on, or have some similarity to the thinking of the ancient Greek philosopher Aristotle (384–322 BCE). He first proposed the idea that humans alone are rational, suggesting the natural hierarchy illustrated in Table 18.2.

Aristotle saw rationality as a divine virtue, so humans alone among organisms have a divine element within them. This divine element is seen in Christian thinking as the uniquely human spiritual dimension. This idea was reinforced in early Christian thinking by the influential St Augustine (CE 354–430), who taught that the commandment 'You shall not kill' (Exodus 20 v13, RSV) does not apply to animals, because animals are irrational, and therefore dissociated from us by lack of rationality (reason). Augustine thus believed that God

Table 18.2 Aristotle's natural hierarchy

Mind/God		Perfect reason
Humans		Rational
Animals		Sentient
Plants		Alive
Stone		Inanimate

subjected animals to humans for their use and that it was also right for people to keep them alive for their own uses.

The next major interpretation of the relationship between humans and other animals was by the French philosopher, René Descartes (1596–1650). The Cartesian theory was based on the ideas that there is no distinction between mind and spirit and that possession of a mind or spirit is an all-or-nothing matter, uniquely human. He thought that animals were akin to machines – operating without consciousness, and that it was not morally wrong to exploit them. Descartes was, however, isolated in his view. Most philosophers of the time agreed that animals could suffer and that inflicting suffering on them was wrong.

Emmanuel Kant (1724–1804) believed that humans had no *direct* duties to animals because they are not self-conscious, so they cannot judge. He further believed that animals are a means to an end and that end is man, so animals are man's instruments. However, he qualified this by stating that it *is* sometimes wrong to hurt animals. Kant believed that how we treat an animal affects or determines how we treat other humans. This has become known as the *indirect duty* view. Implicit in this is the idea that wanton inflicting of suffering harms the perpetrator. In other words, we should be kind to them not because of our duty to them directly, but because it is good practice for being kind to humans, those who *can* judge us. We have no duties directly towards animals, not even those of compassion or sympathy, but we do have a direct moral obligation to other humans for compassion, because this will improve our society.

By the end of the 19th century people were thinking more in terms of animal welfare. Early works, such as Lewis Gompertz's *Moral Inquiries on the Situation of Man and Of Brutes* (1824) and Henry S. Salt's *Animals' Rights* (1892), defended the rights of animals. However, it was not until the late 20th century that a more forceful defence of animals gained significant ground and the notion of animal welfare became an increasingly important issue in the public debate. In 1975 the contemporary Australian ethicist and philosopher Peter Singer put forward the view that animal and human interests are comparable in moral terms (Singer, 1975). He believes that the principle of equal consideration in interest cannot be limited to humans. Whether or not an animal is self-conscious enables the distinction of personhood to be drawn but in Singer's view this is irrelevant to animal welfare. Echoing the words of the English utilitarian philosopher Jeremy Bentham (1748–1832), 'The question is not, can they reason? Nor, can they talk? but, can they suffer?' Singer argues that the capacity for suffering is the vital characteristic that entitles a being to equal consideration (Singer, 1991). This, he contends, is because sentience is the only defensible boundary. If we define suffering as the susceptibility to pain, awareness of being in pain, or of about to be in pain, there is little doubt that most vertebrates can suffer. However, the extent to which they are aware is questionable, but there is good evidence to suggest that the great apes have a high degree of self-awareness. This raises the question of the extent to which sentience contributes to self-consciousness, and thus the extent to which self-consciousness is a purely human characteristic. Controversially, Singer goes on to argue that logically, the intellectually impaired, the disabled, infants, and embryos are akin to non-human animals as far as justifying experiments are concerned. A central tenet of his opinions is that if we say it is acceptable to use animals for our own ends in ways which cause them to suffer just because they belong to another species – because they are *only animals* – then we are showing a form of prejudice akin to racism or sexism (see also Chapter 7). Thus the word *speciesism* was coined, and to its adherents underlies all our uses of animals that cause them harm. However, we should note that Singer's position embodies an element of utilitarianism and does not in fact prohibit the use of animals by humans.

In his book *The Case for Animal Rights* (1985), the American philosopher Tom Regan offers a detailed analysis and critique of Peter Singer's philosophy, and then proposes an alternate route towards an understanding of humanity's moral obligations to animals. He develops the idea of animal rights, arguing that animals possess morally important characteristics, and those that we use for food, experiments, sport and fashion all have inherent value, equal to our own. Animals have an equal *right* to be treated with respect, not to be used as mere resources. Regan argues that this right is violated by our current practices and goes on to call for a total abolition of the use of animals in science, agriculture, and sport. Regan rejects the indirect duty view of Kant and Singer's utilitarianism because he believes that a good end does not justify evil means.

So far in this chapter, I have given an overview of the issues in our relationship with other animals, and the historical development of the range of attitudes held by people today. Many fundamental problems, such as what is it to be human, as well as bioethical questions, such as whether animals have equal rights, have been raised. Clearly there is not sufficient space here (but see Chapters 6 and 7) to explore all these avenues, so because of its far-reaching relevance to humankind I shall go on to discuss the issue of the role of animals in the advancement of biomedical science.

18.3   THE USE OF ANIMALS IN BIOMEDICAL SCIENCE

18.3.1   INTRODUCTION

As this brief overview of the development of human–animal relationships over the course of history has shown, the acceptance that human life is more intrinsically valuable than that of other animals is deeply rooted. An important part of the history of the developed world has been the transition of human activity from a series of strategies for survival towards a civilisation with human welfare at its centre. In this context, the purpose of biomedical research and testing is to increase our knowledge and understanding of the normal structure and functioning of the body and how this changes as a result of disease or injury, and to develop safe and effective treatments for the various malfunctions. Table 18.3 lists some of the major medical advances that would have been impossible without experiments performed on living animals.

Since the beginning of the 20th century, two-thirds of the Nobel Prizes awarded for medicine have been for discoveries and advances in which laboratory animals played a crucial role. So perhaps it is not surprising that those with knowledge of medical history and medical research agree that animal research is central to medical progress.

Table 18.3 Examples of medical milestones reached as a result of animal research

1500s	First use of tourniquets to staunch bleeding from wounds
1600s	Circulation of blood described
1700s	First vaccination
1800s	First use of anaesthetics
	First use of aseptic technique
1906	Corneal transplants
1907	Blood transfusion
1912	Kidney transplants
1914	Kidney dialysis
1922	Insulin isolated to treat diabetes
1929	Penicillin to treat infections
1937	Heart lung machine, open heart surgery, valve replacements, pacemakers
1937	Anticoagulants
1940	Whooping cough vaccine
1941	Diphtheria vaccine
1948	Drugs for high blood pressure
1950	Drugs to control transplant rejection
1956	Polio vaccine
1967	Heart transplants
1973	Treatments for leukaemia
1979	Drugs for asthma
1992	Hib meningitis vaccine
1995	Understanding of programmed cell death, with implications for treatment of e.g. Alzheimer's disease, rheumatoid arthritis, stroke
1998	First cloned mammal – 'Dolly' the sheep
2000	After natural mating and gestation, Dolly gives birth to lamb 'Polly'

18.3.2 WHAT ANIMALS ARE USED IN RESEARCH?

In Chapter 6, David de Pomerai discusses some of the invertebrate animals that
have been so valuable in the elucidation of basic biological processes. In the UK
the law affords protection only to vertebrates and a very limited number of
invertebrates in the context of scientific research. Parliament first passed an act
in 1876 to control the use of animals in medical research. This was strengthened
in 1986 when the Animals (Scientific Procedures) Act passed into law. This
requires that the Home Office must license all scientific procedures carried out on
any vertebrate (and certain invertebrates such as the octopus). A procedure is
defined as any intervention that causes pain or suffering, and is somewhat
broader than an 'experiment', including, for example, the use of animals to
produce natural products for research or treatment (about 10% of all procedures). A research project, the premises in which the animals are kept, and the
principal researcher all have to be licensed for the procedure, and all are
regularly inspected by Home Office Inspectors, who have medical or veterinary
as well as scientific qualifications. Table 18.4 shows the proportion of vertebrate
types involved in regulated procedures in Great Britain in 1999.

Scientific research, particularly that involving the use of animals, is very
expensive, and the funding available is limited. This means that research must be
worthwhile, and before granting funds for a project, the trustees of grant-giving
bodies, such as the Biotechnology and Biological Sciences Research Council
(BBSRC) in the UK, must satisfy themselves of the potential value of the
research. To do this, they undertake a cost–benefit analysis in which they
consider whether the benefits (positive consequences) outweigh the costs (negative consequences). This approach was first outlined by Bateson (1986), and
involves ranking the proposed research on three axes: quality of research,
certainty of medical benefit and animal suffering. In this model, a high quality
project with high possibility of medical benefit, might be permitted even if the
animal suffering score was high. However, if a project was only of average

Table 18.4 Animals used in scientific procedures

Vertebrate group	% total regulated procedures
Primates[1] (e.g. marmosets and macaques)	0.2
Dogs and cats[2]	0.4
Rodents[2] (vast majority are rats and mice)	86
Small mammals other than rodents, (mostly rabbits and ferrets)	1.7
Large domestic mammals (vast majority sheep, cows, pigs)	2.4
All other vertebrates: fish, amphibians, reptiles and birds[3]	9.0

Figures from the Research Defence Society (2001).
[1] The great apes, chimpanzees, orang-utans and gorillas have not been used for research in this country for over
20 years and their use is now banned.
[2] All specially bred laboratory species.
[3] Includes many fertilised hen's eggs after the half-way stage of incubation.

quality, with a medium expectation of useful outcome, it might not be permitted even if animal suffering was also rated low. The main problem with this algorithmic approach is that of deciding on the score values for each axis – how is the balance of animal suffering evaluated against the benefits? This is the crux of the problem: the issue around which the controversies of animal research centre.

18.3.3 AREAS OF RESEARCH AND TESTING IN WHICH LABORATORY ANIMALS ARE USED

In the Western world, by law, any chemical or medicine destined for human use must be rigorously tested before it can be made available to the market (see Figure 18.1). Developing countries, which do not have the resources to undertake this work, rely on a safety guarantee for drugs, toiletries, household cleaning materials, and cosmetics and so on from such countries as the UK, USA, France, and Germany.

Animals are used by the scientific community in basic biomedical research aimed at finding out about the mechanism of disease, and also in product safety testing. Using the statistics for 1995, it is possible to break down types of animal procedure as shown in Figure 18.2.

When a pharmaceutical company decides to fund the development of a new drug targeted at a specific disease, traditionally the first, or biochemical stage is to synthesise about a thousand different compounds. However, the information arising from the various genome projects is leading to a different approach known as rational drug design. This developing science, also called pharmacogenetics, will inevitably raise its own bioethical issues. In whatever way the candidate compounds are produced, they are tested on animals in which the disease condition is reproduced in a series of experiments designed to find out whether the compound under test has any beneficial effect on the condition. This brings the number of test compounds down to about a hundred. The next set of animal experiments elicits the maximum therapeutic dose for the minimum toxic

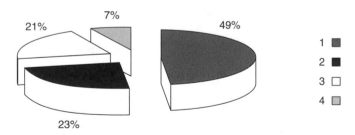

Figure 18.1. Sectors using scientific procedures involving animals in the UK (data from RDS, 2001)
1. Pharmaceutical industry
2. Fundamental scientific/medical research (non-university)
3. Fundamental scientific/medical research (university)
4. Safety testing centres

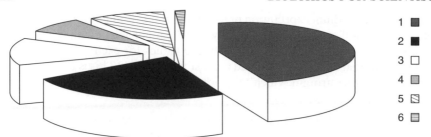

Figure 18.2. Areas of research and testing in which animals are used (data from RDS, 2001)

1. Developing new treatments for diseases, or ways of preventing diseases
2. Fundamental biological and medical research
3. Preparing natural products used in medical research and treatment
4. Safety testing (less than 0.2% on cosmetics and toiletries)
5. Animals with an inherited genetic defect bred for medical research
6. Developing new methods of diagnosis

effect. Before the potential drugs can go on to clinical trials, the company must be sure that through animal testing it has answered the basic questions: does it treat the disease? What is the optimum dose? What are the long-term effects? How does the body break down the compound? Failure to do this can have disastrous consequences as was seen in the early 1960s when about 10 000 children were born in the UK with phocomelia (deformed, flipper-like limbs) as a result of their mothers taking the drug thalidomide to ease the morning sickness of early pregnancy. In the UK rigorous testing of all medicines is now required under the Medicines Act of 1968. It has been estimated that it takes an average of 7–10 years to develop a new drug from the time of first identifying a novel compound through to its successful use in patients at a financial cost of up to US$250 million.

18.3.4   ARE THERE ALTERNATIVES TO RESEARCH INVOLVING ANIMALS?

Those opposed to the use of animals in this way sometimes claim that there are alternatives, frequently citing *in vitro* techniques, computer modelling and epidemiological methods. The scientific community remains unconvinced, for two major reasons. Firstly, the complexity of the body, in both health and disease, cannot be reproduced by any of these means; to see the effect in the living body, animals are used because to use people would be unethical (but see Chapter 7). The second main argument for using animals in scientific research is that these other techniques are extensively used as well, and are complementary with animal work, contributing vital evidence to the overall conclusions.

The phrase *in vitro* literally means 'in glass', and it has come to be used for many test-tube, or laboratory bench procedures. These techniques involve the study of isolated molecules, cells and tissues, obtained from human, animal, micro-organism or plant sources. Such research gives base line information about the interactions between molecules, within or between cells, or about organ function. Evidence from Home Office and Medical Research Council reports as well as from published biomedical literature all suggests that *in vitro* methods are extensively used (Paton, 1993). Over the last 20 years the annual number of animal experiments has almost halved, mainly because of the resources needed for the higher standards of animal welfare, scientific advances and stricter controls. It is interesting to speculate on the extent to which the decline in the actual number of procedures carried out on live animals may be related to the decreasing rate of drug innovation. In evaluating these numbers it should be noted that the total number of animals used in scientific procedures is very much less than that involved in human food supply: recall, for example, that in the UK 800 million chickens are raised (mostly under conditions *much* less 'humane' than are required for animals used in research or testing), slaughtered and eaten every year, let alone the millions of pigs, sheep and cattle. As is evident from Figure 18.3, about 2.3 million animal 'procedures' are carried out each year, of which 86% involve rodents and a further 9% involve non-mammalian vertebrates (as in Table 18.4). Even given that each procedure may require several animals (in order to provide replicates and controls in properly conducted experiments) these numbers are but a very small fraction of those relating to use of animals as food.

Computer modelling is also a widely used approach. It is possible to build up an extensive bank of information from previous studies, and then interrogate this database, for example to discover the fate of drugs in the body such as the accumulation in different tissues and the rate of breakdown and excretion. Analysis of experimental data is made easier by computers, and is often used in this way to predict whether an experiment involving animals is necessary.

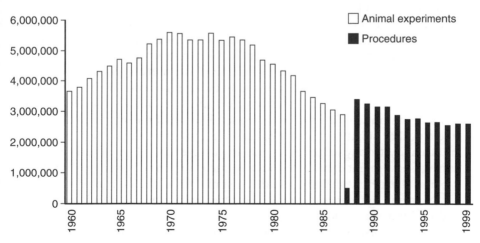

Figure 18.3. Numbers of 'animal experiments' and 'procedures' over the past 40 years (RDS, 2001)

Computers can also increase the efficiency of some research techniques, for example by high-speed computation, and for control of complex experiments. Software can mimic physiological processes such as respiration, and can be used to test theories about the action of new drugs. In none of these uses can the computer be said to be replacing a whole animal in an experiment, and in many of the uses described above computer methods serve merely as a prelude to an investigation using animals.

Epidemiological studies give information about the body in health and disease, and about the distribution of diseases in society in time and space (Smyth, 1978). It enabled for example the link between smoking and lung cancer to be made (Royal College of Physicians, 1962), and over the last decades has established causal relationships between many other diseases and environmental factors. Epidemiological methods are also able to provide evidence in the search for adverse reactions to new drugs, but only at the stage of clinical trials, i.e. when the animal work is already done. Although epidemiology can show useful information on cause, it is relatively powerless to provide any evidence for the mechanisms of these cause–effect relationships, or how this may be important in developing treatments for diseases. The way that smoking causes cancer was, controversially, determined by a series of experiments on tissues and animals to isolate the carcinogenic substances in tobacco.

18.3.5   ARE FINDINGS FROM RESEARCH ON ANIMALS RELEVANT TO HUMANS?

An argument often put forward by those opposed to animal experiments is that they produce little effect on human life expectancy or disease rate, because animals are different from people (Ryder, 1975). Most biologists now believe that the variety of mammals in the world today, humans included, have evolved over the last 120 or so million years from tree-shrew-like common ancestors. Apart from the obvious differences of body size, shape, and covering, we are anatomically and physiologically very similar to other mammals. Advances in molecular genetics show that the structure of the human genome is more like that of other animals than was previously thought possible. Whilst the similarities are of obvious use to the researcher wishing to mimic a human condition in say a rat, the differences can also give useful information about human diseases and how they might be treated. For example, important clues from a mouse with muscular dystrophy suffering less muscle wastage than a human patient with the same disease might contribute to a treatment for this devastating disorder.

Animal experiments have revealed a similarity in the functioning of certain vitamins and hormones in people. This had led to the successful treatment of several human conditions, such as diabetes (with insulin from the porcine or bovine pancreas), and thyroid disease (with thyrotropin from the bovine pituitary). We should note however that many human therapeutic proteins are now produced in genetically modified micro-organisms. These similarities mean that veterinary drugs, for example antibiotics and analgesics, are often identical to those used in human medicine. Disease is a universal mammalian problem, and maladies such as cancer, heart failure, asthma, rabies and malaria are suffered by many species. Most human diseases exist in at least one other mammalian species (Cornelius, 1969).

18.3.6 SOME ISSUES OF TOXICITY TESTING

Multinational agencies such as the European Commission and the OECD have produced a formidable battery of regulations and guidelines on the toxicity testing that must be carried out not only on new drugs, but also on agrochemicals, industrial chemicals, substances for household use, foodstuffs, and food additives. Similar codes of practice exist for the testing of pollutants. Toxicity testing on animals gives information on toxic action and its cause, and can be used to make predictions about the risks to other biological systems – including those in other species – in different circumstances. Tests are carried out at all levels of biological organisation, from the organelle to the ecosystem (see Table 18.5).

The test substance can be administered by the oral route – directly into the stomach, or in the diet/drinking water; by inhalation; ectopically – by administration directly to the skin or body surface – or parenterally – e.g. by injecting intraperitoneally or intravenously. Clearly, to be sure of the safety for human use of any substance the range of possible tests is very extensive. This is where ethical problems become apparent.

Table 18.5 Types of toxicity testing

Type of effect	Test on animals
Acute toxicity	The effects of limited dosage
Subacute/chronic toxicity	Exposure from a few days – most of life span
Carcinogenicity	Specific tests of the potential to produce tumours lasting for not less than 2 years in rodents
Reproductive toxicity	Action on gametogenesis, mating, foetal development, development to sexual maturity
Topical toxicity	Local effects at the site of application, e.g. skin, eye, lungs
Immunotoxicity	Effects on immuno-depression, autoimmunity, hypersensitivity
Genetic toxicity	Study of damage to DNA, inherited abnormalities, onset of cancer
Neurotoxicity	Study of the effects on the nervous system
Ecotoxicity	Study of effects on population, health of ecosystem
Toxicokinetics	Detailed analysis of uptake, distribution, metabolism, and excretion of the test substance by a cell, tissue, organism, or ecosystem

After Dayan (1993).

The main reason for using animals in these tests is the same as that for using animals in research: that the body is a complex system, which cannot accurately be reproduced. The other main argument in favour of animals in research, namely that information is directly transferable from animal to human, does not hold so good in this type of testing. This is because the differences in physiological pace and anatomical structure between individuals and species have a marked effect on the toxicity effect.

Consider these results of LD50* tests of dioxin on various animals:

Guinea pig – 1 μg kg^{-1}
Hamster – 5000 μg kg^{-1}
Female rat – 45 μg kg^{-1}
Male rat – 22 μg kg^{-1}.

This vast difference in toxicity among such closely related animals clearly shows the problems in extrapolating this sort of data to human beings. However, this test, once notorious amongst animal rights campaigners, is now all but abandoned in the UK. In 1999 the UK Home Office announced that licences for the LD50 acute oral toxicity test '. . . will no longer be granted if a suitable alternative is available'. There are however some tests, e.g. for botulinum, where the LD50 cannot be replaced, and licences can still be granted whilst many other countries still continue to use the LD50 routinely. The British Toxicological Society has developed the fixed dose procedure, a milder and more humane test, which uses fewer animals and is designed such that none receives a fatal dose of the test substance.

Another infamous procedure is the Draize eye irritancy test, often used to test cosmetics and household products. The test substance is placed in one eye of a rabbit and the other is left untreated for comparison. This has attracted high levels of public concern, and has been strongly criticised on the grounds of cruelty, and claims have also been made that it produces unreliable results that might bear little relation to human responses. Animal campaigners have focused public attention on such toxicity and irritancy tests with the result that several major cosmetic and household product companies began a serious search for non-animal methods to fulfil their scientific and corporate objectives. Alternatives to the Draize test, such as the chorio-allantoic membrane (CAM) test that uses 10 day old hen's eggs, and EYTEX, which uses a clear gel from jack bean protein to mimic the cornea of the eye, have been developed. Although this has meant that the Draize test is now less widely used, no alternative gives the quality of results obtainable from the Draize, so work continues to find a good replacement. The Draize test still continues to be used by companies because of the necessity to cover themselves against possible litigation, particularly in view of the increasingly litigious public in the USA and now in the UK.

The ethical issue here is that if we accept that we have the right to know

---

* The LD50 test (defined by OECD Guideline 401) determines the substance dose required to kill 50% of the test animals.

whether a substance will harm us, then we have little alternative in the present scientific climate to accept the results of toxicity testing. This raises the issue of what should or should not be tested. Where is the line to be drawn across a list that includes, *inter alia*, medicines (human and veterinary), pollutants, agrochemicals, household products, toiletries, building and household materials, cosmetics, and children's toys? If testing on non-essential substances – those below where we choose to draw our line – were banned, then we would have to continue to use those products already available, and could not enjoy the development of improved products, for example less allergenic cosmetics, more efficient kitchen floor cleaners, improved sun blocks, and so on. Alternatively, it could be argued that we do not need these new items, because the cost, i.e. the need to use animals in testing, is too high.

18.4   THE GUIDING PRINCIPLES IN ANIMAL RESEARCH

The vast majority of responsible scientists do not want to use animals unnecessarily or to cause them unnecessary suffering. Russell and Burch (1992) drew up the three guiding principles of *refinement*, *reduction* and *replacement* in animal research, which are used today.

Refinement aims to reduce animal suffering to a minimum. In order for a UK Home Office licence to be granted any research involving animals must be designed to minimise distress or suffering. For example, anaesthetic or analgesic is normally given if any painful procedure is to be carried out. When levels of substances, such as hormones in the blood, need to be regularly measured, the animal can be fitted under anaesthetic with an in-dwelling catheter, so that it does not have to undergo the repeated stress of being caught, held and having blood drawn. Animals such as the oncomouse, specially bred to inherit fatal diseases such as some forms of cancer, can be humanely killed earlier rather than later in an experiment. The fixed dose procedure, described above, is another example of refinement. This notion extends to the husbandry of the animals in research holdings. The vast majority of laboratory animals spend most of their lives simply living in their cages and not being used in an experiment, and efforts to enhance their environment are often made in an attempt to improve their quality of life.

Reduction is the principle in which the number of animals used is minimised. As already discussed, *in vitro* and computer-based investigations often precede animal work, and the information gained from these sources informs the design of the proposed experiment, so the optimum number of animals for statistical validity is found. Variation of individuals within a species is a basic problem in biological investigation, but this can be overcome by using genetically identical animals, and laboratory bred animals that are free from infections or illnesses that might skew the experimental results. This can considerably reduce the numbers of animals needed for meaningful results.

Replacement is the principle of replacing animal procedures with non-animal techniques wherever possible. As discussed, the complexity of the whole organism is such that efforts to find effective non-animal techniques have been disappointing. Notable successes include the LAL test for pyrogens, which can be performed *in vitro*, and testing the purity of insulin by chromatography. Both these tests were previously carried out on animals. As science progresses, new non-animal techniques will continue to develop, but in some cases the new knowledge of gene manipulation may lead to the production of genetically tailored animals for specific research. This may lead to an increase in the number of some animal procedures, almost exclusively involving rodents, particularly mice. For instance it is now possible to breed mice with cystic fibrosis, having the same symptoms as children with cystic fibrosis. Gene therapy, first applied to the mice, may offer a medical breakthrough for this disease (Boyce, 1999).

The principle of the 3Rs, refinement, reduction and replacement, is the basis of the ethical consideration given to the use of animals in biomedical research and testing. Research into refinement and reduction of procedures continues, but so does the development of new products, which may involve more new tests. It seems unlikely that replacement will ever be total. Scientists have a strong ethical responsibility to predict and warn of the harmful effect of toxins, but must balance this with the need to minimise the scale of their experiments, and any suffering to animals. Nevertheless, for those ideologically opposed to the use of animals in this way, any such use is simply too much.

18.5   EDUCATION

Although this chapter focuses mainly on the relationship of humans and animals in scientific research and testing, the role of animals in education is pertinent. One powerful argument in favour of the use of animals in education is that new medical practitioners and scientists cannot be trained without practising on animals. In higher education, a considerable number of practical sessions involving freshly killed whole animals and tissues are routine. Students may opt out of these classes in certain programmes, but this is not possible in some biomedical courses. In school science in the UK, live vertebrates are never used in any practical work. No school can be licensed to undertake a regulated procedure under the 1986 Act, but traditionally biology lessons have often included dissection – the cutting up of a dead animal or animal organ. There has been a reduction in the use of animals in school science in many countries, because of a shift in the perspectives of both pupils and teachers. Fewer live animals are kept in schools and colleges, because of ever more prescriptive health and safety regulations, and ever tightening curricular constraints (Lock and Millet, 1992). In the UK, examination boards at school level do not now require dissection.

Some teachers rue this move away from direct contact with biological materials, arguing that it sanitises the study of biology. Traditionally, the arguments in favour of using dissection as a teaching and learning strategy are that it uniquely provides knowledge and understanding of internal structures and of the quality of tissues and an appreciation of the whole organism. Because of its hands-on approach, the learning is through active involvement in investigation. Some teachers argue that dissection provides an opportunity to overcome revulsion, and by so doing engenders a greater understanding and respect for life. Those opposed to dissection claim that it involves the taking of life, because animals supplied to educational institutions are specially reared and killed for the purpose, and this is morally wrong. They claim that the students can become desensitised, because animal life is devalued and treated as expendable, so less respect for life is created as a result of dissection classes. They believe that dissection might offend the sensitivities and conscience of some students, thus alienating them from the study of life sciences, as a result of which some able students may choose careers in other fields. Arguing that dissection rarely involves more than observation and memorisation, they suggest that students are not challenged with higher-order intellectual activity, such as forming hypotheses or analysing and interpreting data.

18.6 CONCLUSIONS

This chapter has explored some of the ethical dilemmas that arise in the quest for scientific understanding through research upon mammals. In recent years the ethical debate has gathered momentum, sometimes becoming acrimonious and even giving rise to acts of terrorism as evidenced by recent (early 2001) events at a UK testing laboratory. It is virtually impossible to set out a completely neutral position, but as well as supplying some basic information about animal-based science, I have attempted to put forward the main ethical arguments. However, in the end, the moral position one chooses to adopt is a personal decision. It behoves everyone who is willing to accept the safety of virtually every consumable product to engage with these difficult matters. The debate is certain to become increasingly important as we learn more about the lives of all animals.

REFERENCES

Bateson, P. (1986) When to experiment on animals. *New Scientist*, 20 February, 30–32.

Boyce, N. (1999) The gene healer. *New Scientist*, **163**, 43–45.

Cornelius, C.E. (1969) Animal models – a neglected medical resource. *New England Journal of Medicine*, **281**, 934–945.

Dayan, A.D. (1993) Safety testing: problems of perceptions, definitions and understanding. In *Ethical Issues in Biomedical Sciences: Animals in Research and Education.* Anderson, D., Reiss, M. and Campbell, P. (eds), Institute of Biology, London, UK.

Dolins, F.L. (1999) *Attitudes to Animals: Views in Animal Welfare.* Cambridge University Press, Cambridge, UK.

Lock, R. and Millett, K. (1992) Using animals in education and research; student experience knowledge and implications for teaching the National Science Curriculum. *School Science Review*, **74**, 115–123

Paton, P. (1993) *Man and Mouse: Animals in Medical Research.* Oxford University Press, Oxford, UK.

Regan, T. (1985) *The Case for Animal Rights.* University of California Press, Berkeley, CA, USA.

Research Defence Society (2001) http://www.rds-online.org.uk/home.html

Royal College of Physicians (1962) *Smoking and Health: a Report of the Royal College of Physicians of London on Smoking in Relation to Cancer of the Lung and Other Diseases.* Pitman, London, UK.

Russell, W.M.S. and Burch, R.L. (1992) *The Principles of Humane Experimental Technique.* Universities Federation for Animal Welfare, Wheathampstead, Hertford, UK.

Ryder, R.D. (1975) *Victims of Science.* Davis-Poynter, London, UK.

Singer, P. (1975). *Animal Liberation: a New Ethics for Our Treatment of Animals.* Avon, New York (see especially pp. 27–91).

Singer, P. (1991) *Animal Liberation.* Avon–Hearst, New York, USA.

Smyth, D.H. (1978) *Alternatives to Animal Experiments.* Scholar, London, UK, in association with the Research Defence Society, London, UK.

FURTHER READING

In addition to the above references the following are recommended for further reading.

Anderson, D., Reiss, M. and Campbell, P. (eds) (1993) *Ethical Issues in Biomedical Sciences: Animals in Research and Education.* Institute of Biology, London, UK.

Carruthers, P. (1992) *The Animals Issue: Moral Theory in Practice.* Cambridge University Press, Cambridge, UK.

Glossary

Allele: one of any pair of alternative hereditary characters; many genes can exist in two (or sometimes more) forms at a locus (i.e. position on a chromosome), each of which is an allele. See also *polymorphism*.

Allergen: a substance that induces allergy.

Amniocentesis: procedure in which a sample of amniotic fluid is withdrawn from a pregnant woman's womb. Cells in the fluid are then tested for chromosomal or genetic abnormalities.

Amniocytes: cells in the amniotic fluid.

Androgynous: having sex organs of both female and male.

Andrology: the study of male reproduction.

Annelids: invertebrate phylum of segmented worms.

Antihelminthic (drugs): drugs toxic to parasitic flat worms.

Apomixis: the formation, without fertilisation, of seeds that contain viable embryos.

Arthropods: phylum of jointed legged invertebrates.

Asthenospermia: abnormally formed spermatozoa.

Autosome: typical, i.e. non-sex, chromosome.

Azoospermia: total lack of spermatozoa in the seminal fluid.

Back-cross: to mate a hybrid to one of the parental stocks; a hybrid resulting from such a mating.

Biodiversity: the range of living organisms in a particular habitat, community or biosphere.

Bioethics: the ethics related to biology and medicine and to medical and biological research, or the informing of ethics by biological knowledge.

Bioinformatics: study of biological systems using the tools of information technology.

Blastocyst: a mammalian embryo at around the time of implantation when it forms a hollow ball of cells.

Blastomere: undifferentiated cell of an embryo during the early cleavage stage.

Bovine spongiform encephalopathy (BSE, 'mad cow disease'): a degenerative disease of the central nervous system of cattle, which is thought to be transmissible to humans; related to Creutzfeldt–Jakob Disease.

Carbon dioxide sink: an organism, community or ecosystem that takes up carbon dioxide from the atmosphere.

Carcinogen: substance that induces the formation of malignant tumours.

Cardiomyocytes: heart muscle cells.

cDNA: complementary DNA, a DNA copy (made in the laboratory) of the natural sequence of messenger RNA.

Centric fusion: fusion of the centromeres of chromosomes.

Centromeres: the constricted regions of chromosomes at which pairs of new ('daughter') chromosomes are held together in the early stages of cell division.

Chorionic gonadotrophin: hormone of pregnancy.

Chorionic villus: fold of the chorion, a membrane external to and enclosing the amnion.

Chromosome: one of a number of structures in the nucleus of a cell containing genetic material, the number of which is characteristic for the species. Each chromosome is, in effect, a subset of the total number of genes possessed by the organism.

Clone: organisms identical in genetic make-up, produced asexually from one stock or ancestor. May also be used as the verb, to clone. Both noun and verb are also used in the context of DNA, as in 'molecular cloning', the multiplication of particular DNA sequences by 'growing' them in genetically modified bacterial cells.

Coelenterates: phylum of hollow bodied invertebrates having body walls of two cell layers, for example jelly fish.

Consequentialist: line of argument that follows directly as a result of foregoing events or factors. Ethical system based on evaluation of the consequences of an action.

Creutzfeldt–Jakob Disease: a degenerative, incurable disease of the human central nervous system, related to BSE (see above).

Cryopreservation: preservation of cells or tissues at very low temperatures, e.g. in liquid nitrogen.

Cultivar: a plant variety used in agriculture or horticulture.

Deontology: the study of duty and/or obligation as an ethical concept.

DNA vectors: carriers of DNA sequences into the genomes of other organisms, e.g. bacterial plasmids.

Ecosystem: a biological community of interacting organisms and their physical environment.

Embryo: the stage of life between the first cell division after fertilisation until the completion of organogenesis. In the human this is during the first eight weeks of pregnancy.

Embryologist: scientist who studies early development; scientist who manipulates human gametes and embryos under license in fertility treatment.

Enabling technology/technologies: the basic scientific techniques that can be applied for use e.g. in medicine or agriculture.

Endometriosis: a condition in which tissue identical to the endometrium is found outside the uterus, e.g. in the ovaries or Fallopian tubes.

Endometrium: the lining tissue of the uterus.

Epididymis: part of the male mammalian reproductive tract between the testis and sperm tube in which the spermatozoa mature and acquire the ability to fertilise an egg.

Epistemology: the theory of knowledge, especially with regard to its methods and validation.

Ethics: study of sets of moral principles; systematisation of the principles involved in moral decision making.

Eugenics: the process of improving the human population by genetic selection for desirable inherited characteristics.

Excurrent duct: tubes of the reproductive tract in which there is an out-going flow.

Exogenous genes: literally, genes from outside; in general this refers to genes that are transferred from one species to another.

F1: the first filial generation; hybrids arising from a first cross. Successive generations are denoted F2, F3 etc. P1 denotes the parents of the F1 generation, P2 the grandparents etc.

Fetus/foetus: the stage of development between organogenesis and birth.

Gamete: a reproductive cell (spermatozoon or ovum) containing half the number of chromosomes of a somatic cell and able to unite with one from the opposite sex to form a new individual.

Ganglion: a mass of nerve cell bodies giving origin to nerve fibres.

Gene: an individual hereditary unit in a chromosome consisting of a characteristic sequence of DNA.

Gene flow: the spreading of genes resulting from out-crossing and from subsequent crossing within a population.

Gene silencing: the switching off of genes (not usually used in the context of the 'normal' on–off control mechanisms).

Genome: the total genetic content of an organism.

Genomics: the study of the genomes of organisms, particularly in relation to information content and sequence organisation.

Germ cell: a cell belonging to the specialised cell lineage that gives rise to gametes (sperm and eggs) in a multicellular animal or plant.

Germinal vesicle: structure within an ovum, containing the genetic material, before the extrusion of the polar bodies (see below).

Graafian follicle: ovarian structure in which an ovum is brought to maturity.

Heterozygous: having dissimilar, alternative forms of a gene for a given characteristic.

Homozygous: having identical genes for a given characteristic.

Huntington's disease: hereditary, incurable and degenerative disease of the human central nervous system.

Hybridisation: process of interbreeding organisms to produce heterozygotes.

Implantation: process by which an embryo becomes embedded in the lining of the womb.

Inbreeding: breeding through a succession of parents belonging to the same stock/lineage, or within very closely related lineages.

Intra-cytoplasmic: within the cytoplasm of a cell; inside the cell surface membrane.

Introgression: the introduction, by plant breeding, into elite crop lines/varieties of genes/traits from related varieties (or even from closely related wild species). Has also been used recently to describe (the possibility of) transfer of transgenes (see below) from GM crops into wild populations.

Intron: region of a gene containing a sequence of DNA which does not code for a protein but is transcribed as part of a 'pre-mRNA' molecule and then excised by RNA splicing to produce mRNA; introns are so called because they intervene between exons, or sequences of DNA that do code for proteins.

Karyotype: group of individuals with the same chromosome number and similar linear arrangement of genes in homologous chromosomes; the chromosome complement of such a group.

Laparoscopy: examination of the interior organs by a fibre optic instrument (laparoscope) inserted into the abdominal cavity.

Mendelian: the normal pattern of inheritance of single genes, as originally described by Gregor Mendel.

Monogenic diseases: diseases resulting from the action of a single (defective) gene, e.g. sickle-cell anaemia.

Morals: the study of goodness or badness of human character or behaviour, or of the distinction between right and wrong; the outworkings in action of systems of ethics.

Mutagen: a substance capable of inducing a heritable change in the nucleotide sequence of DNA.

Mutant: an organism that has undergone a permanent, heritable change in structure and/or function based on a change – mutation – in DNA.

Nematodes: invertebrate phylum of round worms.

Neoplasm: a mass of cells, the growth of which is abnormal, excessive, persistent and un-coordinated. Benign neoplasms grow only at the site of origin; malignant neoplasms spread to other sites in the body.

Neurone: a nerve cell.

Oligospermia: low sperm count.

Oncogene: gene that makes a cell cancerous; typically a mutant form of a normal gene (proto-oncogene) involved in the control of cell growth or division.

Oncomouse: a genetically modified strain of mouse programmed to form malignant tumours.

Oocyte: female gamete; egg, ovum.

Organelle: an organised structure with a specific function or functions within a cell, e.g., a mitochondrion.

Out-crossing: breeding with new stock.

Penetrance: the frequency, measured as a percentage, with which a gene shows any effect.

Phylogenetic: concerned with the history of development of a species or race.

Plasmid: small circular DNA molecule that replicates independently of the genome. Used extensively as a vector for transfer and cloning of DNA.

Pluripotent: capable of forming many different types of cell (see also *totipotent*).

Polymorphism: occurrence within a population of an organism of different varieties based on different alleles of a given gene.

Promoter: the tract of DNA adjacent to a gene that contains the gene's 'on–off' switch.

Pronucleus: egg or sperm nucleus in the zygote prior to nuclear fusion.

Pyrogen: substance capable of producing heat in the body.

Recombinant DNA: novel combination of DNA sequences made by recombining different DNA sequences in the laboratory, e.g. insertion of a mammalian gene into a bacterial plasmid.

Restriction endonuclease: enzyme that recognises a specific short sequence of nucleotides in DNA and cleaves the DNA wherever this sequence occurs; extensively used in recombinant DNA technology.

Restriction fragment length polymorphism: a change in the nucleotide sequence of DNA such that a recognition site for a particular restriction endonuclease (see above) is either lost or gained. This will cause the generation of different fragment lengths around that site when the DNA is cleaved with the endonuclease.

Semen: seminal fluid containing spermatozoa.

Somatic: concerned with the body cells (as opposed to the germinal cells).

Spermatogenesis: sperm formation in the testes, from spermatogonia, through primary and secondary spermatocytes and spermatids to spermatozoa.

Surrogacy: process by which a woman bears a child on behalf of another woman, from an egg fertilised by the other woman's partner.

Syngamy: fusion of gametes during sexual reproduction; point at which the male and female genetic contributions fuse to become the zygotic genome.

Teratogenic: term applied to a substance that induces gross abnormalities during development.

Totipotent: genetically capable of forming a whole organism.

Transgenes: genes that have been transferred via a vector from one organism to another.

Utilitarianism: the doctrine that actions are right if they are useful or for the benefit of the majority; guiding principle of the greatest happiness for the greatest number.

Vasectomy: surgical procedure in which the *vas deferens* is severed and tied to stop sperm cells entering the seminal fluid. Carried out for contraceptive purposes.

X chromosome: one of the two sex chromosomes, singly represented in the heterogametic sex (male in humans) and paired in the homogametic sex (female in humans).

Xenotransplantation: transplantation of cells, tissues or organs from one species into another.

Y chromosome: one of the two sex chromosomes, in mammals exclusive to males.

Zona pellucida: protein coat surrounding an oocyte and early embryo, shed immediately prior to implantation.

Zygote: cell formed by the union of spermatozoon and oocyte; a fertilised egg.

Index